A World-Systems Reader

A World-Systems Reader

New Perspectives on Gender, Urbanism, Cultures, Indigenous Peoples, and Ecology

edited by
Thomas D. Hall

ROWMAN & LITTLEFIELD PUBLISHERS, INC.
Lanham • Boulder • New York • Oxford

ROWMAN & LITTLEFIELD PUBLISHERS, INC.

Published in the United States of America
by Rowman & Littlefield Publishers, Inc.
4720 Boston Way, Lanham, Maryland 20706
http://www.rowmanlittlefield.com

12 Hid's Copse Road, Cumnor Hill, Oxford OX2 9JJ, England

British Library Cataloguing in Publication Information Available

Library of Congress Cataloging-in-Publication Data
A world-systems reader : new perspectives on gender, urbanism, cultures, indigenous peoples, and ecology / edited by Thomas D. Hall.
    p.   cm.
    Includes bibliographical references and index.
    ISBN 0-8476-9183-7 (cloth : alk. paper) — ISBN 0-8476-9184-5 (paper : alk. paper)
      1. Social systems.   2. System theory.   3. Sex role.   4. Indigenous peoples.
   5. Urbanization.   6. Social ecology.   I. Hall, Thomas D., 1946–

   HM701 .W67   2000
   301—dc21
                                     99-048269

Printed in the United States of America

♾™ The paper used in this publication meets the minimum requirements of American National Standard for Information Sciences—Permanence of Paper for Printed Library Materials, ANSI/NISO Z39.48-1992.

Contents

Preface

For those of us who teach courses, or parts of courses, that tap into world-systems analysis, there has been a dearth of broad readings that sample the very wide range of world-systems-based research and theorizing. In part, this is because world-systems analysis and political economy in general is a "bookish" field. Often articles are spin-offs or extracts from longer works. Many teachers resort to course packs or lots of reserve readings tailored to their specific course. Others assign a few books and lecture about others. None of these are satisfactory solutions for introducing world-systems analysis to students, especially undergraduates who are loathe to wade through lots of readings that do not hang together well. Thomas Shannon's *Introduction to the World-Systems Perspective* (2d ed., 1996, Westview) has helped fill this gap in admirable ways, but it is a textbook and summarizes arguments, research, and critiques. It does not *do* world-systems analysis.

A second difficulty is the disciplinary range of world-systems analysis—which is found in all the social sciences—here broadly conceived to include history, which many would place in the humanities. One of the strong features, to my view and that of many other world-systems writers, is this inter- and multidisciplinary quality. Indeed, it derives directly from Immanuel Wallerstein's views on social science disciplines (see, e.g., *Open the Social Sciences: Report of the Gulbenkian Commission on the Restructuring of the Social Sciences* [Stanford University Press, 1996]). But this, too, can create problems for students who have yet to master the ins and outs of any one discipline, not to mention leaping blithely over disciplinary fences. While this quality makes world-systems research ideal for a liberal education, it does require some aids for students entering it for the first time.

My goal in organizing this reader is to help fill these two lacunae simultaneously. First, I present brief examples of world-systems research that can take students into the nuts and bolts of the research. I sought examples that were different, yet interesting and topical. Second, I provide students (and sometimes their teachers) brief introductions to how world-systems analysis is used outside sociology—the "home discipline" of world-systems analysis. These chapters describe a bit about how those other disciplines work and use world-systems analysis, introducing basic readings and a road map to help students to use those disciplines in their own research papers. Third, I intentionally sampled widely, rather than in depth, and intentionally sought the unusual rather than concentrating exclusively on the mainstream—if there really is such in world-systems research. Basically, I wanted to demonstrate the invalidity of statements such as "World-systems analysis is about . . . ," where any brief description is tossed into the final phrase. Fourth, I wanted to showcase some of the exciting new work being done within a world-systems framework. Finally, and far from least important, I wanted the collection—unlike many course packs—to be held together by more than the binding.

How well this collection meets those goals must be left to others to decide. I would say that it barely sketches the proverbial "tip of the iceberg." Thus, many relevant, interesting, and important areas are barely mentioned. The rather long bibliography that accompanies chapter 1 is a more reasonable sample, but it is just that—a sample. In trying to collect materials for that list, I constantly discovered new ones. World-systems research is a very active and rich area with new materials appearing daily. There are several on-line resources listed in note 4 in chapter 1 that have links to many other resources. Two appendices list the compilations that grow out of the Political Economy of the World-System annual meeting and the books selected as outstanding contributions by the Political Economy of the World-System section of the American Sociological Association every year. All these resources indicate the breadth and richness of this approach to social sciences.

In short, this collection is an introduction, a tease, and, most important, an invitation. A careful perusal of this collection and its resources will negate most statements of the type "But world-systems analysis does not address _____." In those few cases where it does not, consider that an invitation to roll up your sleeves and go to work on it.

Finally, let me be clear that I am *not* claiming that world-systems analysis replaces, supplants, or usurps all other social science schools. Rather, I am making a different argument. It is a perspective, a way of examining social issues that is a vital component to our understanding of them. In the buzzwords of logic, it is a *necessary* but far from *sufficient* set of tools in the rather large armamentarium of social scientists.

All this explains *why* this collection was assembled but does not explain how. It began as the result of a suggestion from Joane Nagel who had just begun her term as coeditor of *Sociological Inquiry,* the official journal of Alpha Kappa Delta (AKD), the International Sociology Honor Society. To do something new with the journal, she invited several scholars to edit special sections of each issue on various areas of sociological research. She asked me to edit one on world-systems theory. It was published in fall of 1996 (66:4:440–519) and included four articles, plus an introduction, and reviews of four recent books. The goals for that special collection were much the same as those outlined here, but more modestly approached.

The process of assembling that collection, and subsequent reactions to it, led me to consider expanding it into a book. Discussions with Dean Birkenkamp, longtime editor and friend, suggested that this was a project worth pursuing. As is always the case with edited collections, there were many stops, starts, and changes of direction along the way. Originally, we had intended to republish all four articles from *Sociological Inquiry*. When this approach proved impractical, some of the contributors were gracious enough to write new articles (Fred Shelley and Colin Flint, chap. 4; Wilma A. Dunaway, chap. 11) on very short notice. My introduction (chap. 1) is a considerably expanded and updated version of my introduction to the special section. The other chapters, thirteen in all, were commissioned for this collection.

As anyone who has ever commissioned and edited a collection knows, everything is not always completed in as timely a manner as one would wish. Editors typically blame contributors, but I blame a hard drive crash and myself. The authors all worked hard to meet the various goals of the collection, which sometimes entailed extensive revisions. They all did so with good grace. How well each chapter meets the goals varies. Some are directed more at one than another goal. They all reflect their authors' individual styles, the styles of their home disciplines, and their schools within those disciplines. Many of the chapters report and summarize a large body of the authors' own work. Others are early reports or self-contained parts of larger works. The number of references to in press, forthcoming, and unpublished manuscripts indicate how much of this is continuing work. As is inevitably the case in such situations, some of the "forthcoming" works have since appeared, occasionally changing titles or venues along the way; others remain in one or another pipeline.

This point suggests a word or two about references. One of my chores as editor was to go through the entire collection and transform all the references into a single format. I have tried, to the extent possible, to give references as completely as possible. This means that where journals have a volume, a number, an issue (typically month or season), page numbers, and full author names, I have included all this information. While sometimes redundant, this often makes the task of finding these materials easier for researchers and for librarians—especially those who must resort to interlibrary loans. Again, while redundant, I have included full URLs, each with a reference to an electronic resource. These, of course, change. As of publication, they are all correct.

I must also pay some debts and acknowledge the contributions of others to this collection. Obviously, all the authors helped immensely by producing chapters that fit the collection. Many others were consulted along the way. At DePauw University Darrell La Lone, Eric Silverman, Thomas Ewing, and Paul Watt contributed comments and suggestions. The collection also reflects many long conversations with Al Bergesen, Chris Chase-Dunn, Wilma A. Dunaway, James Fenelon, Wally Goldfrank, Patricio Korzeniewicz, Terry Kandal, Les Laczko, John Markoff, Joya Misra, Peter Peregrine, Carol Ward, and David Wilson. They, of course, are not accountable for any of my failures to heed their sound advice. Finally, I thank Dean Birkenkamp and Rebecca Hoogs for their help and patience with this project. Any author would be hard put to find better editors.

I would also like to thank the Faculty Development Committee and the John and Janice Fisher Fund for Faculty Development at DePauw University for support in producing this collection and for support in attending many social science meetings where I benefited from innumerable discussions and papers.

Part I

Overview and Introduction

1

World-Systems Analysis: A Small Sample from a Large Universe

Thomas D. Hall

Many writers have pronounced Marxist analysis dead with the end of cold war. Albert Bergesen (1992c) has noted "now that the [Berlin] wall is down," left analyses—including world-systems analysis—are perceived to have lost any relevancy they might once have had. Others, such as Fukuyama (1992), have gone so far as to proclaim the end of ideology and the end of history. These claims assert that now that capitalism is triumphant, there will be no more social change. Thus, history and ideology will end because everything will remain the same, just more so. This sound bite version is clearly nonsense, but even in the more sophisticated versions it amounts to an overly easy dismissal of all social thinking that traces any roots to Karl Marx. I firmly disagree with this dismissal. But I do agree with Al Bergesen that those of us who trace such intellectual roots have an obligation to explain their continuing relevance. This collection shows that the burial of world-system analysis, like that of Mark Twain, is premature. World-system analysis frequently suffers from another egregious inaccurate assessment: that everything it has to say was said by Immanuel Wallerstein twenty-five years ago (1974a, 1974b). Nothing could be further from the truth. Indeed, to have read only these two publications would barely scratch the surface of Wallerstein's own work, not to mention that of scores of other world-system analysts.

By presenting important new research that is accessible to students and useful to their teachers, I hope to illustrate the myriad relevancies of world-systems analysis. An equally important goal, which reflects my own predilections, is to sample the range of world-systems research. In doing so I pursued a "variance-maximizing" strategy. Rather than attempting to capture the "central tendencies" of world-systems literature, I solicited essays that map the wide ranges of world-systems analysis. In short, I want to map the wide range of world-systems scholarship, rather than attempt a definitive summary. I also want to reflect the disciplinary diversity of world-

3

systems analysis and conversely provide convenient maps of relevant research and approaches in other social sciences besides sociology.

Because world-systems research is a "book" rather than article field (see Clemens et al. 1995), I requested articles that synthesized the work of many scholars. I also asked the authors to indicate readings they thought were especially accessible and useful to the research, analyses, and debates each discussed. Several have done this in a special section, a few through elaborate citations or notes. This collection is thus an invitation, not a definitive summary.

WHAT IS WORLD-SYSTEMS ANALYSIS?

This chapter is a "quick and dirty" overview of world-systems theory for those not already familiar with it. The second chapter by Peter Grimes elaborates several of the concepts sketched here. More detailed introductions may be found in the works of Immanuel Wallerstein, Alvin So (1990), Thomas R. Shannon (1996), Christopher Chase-Dunn (1998), and other works cited later. World-systems theory is contentious: articles and books dispute virtually every statement in this summary.

Immanuel Wallerstein defines a world-system as an intersocietal system marked by a self-contained division of labor. That is, all the roles and functions necessary to maintain this "world" are contained within it—though, to be sure, they are spread very unevenly across it. Thus, it is a "world," with some degree of internal coherence and forming a complete unit. Wallerstein says, "my 'world-system' is not a system 'in the world' or 'of the world.' It is a system 'that is a world'" (Wallerstein 1993b, 294). Hence the hyphen in the term. This hyphenation has itself become a politicized issue (Thompson 1983a, 1983b; Wallerstein 1983). Only in the late twentieth century has this "world" become truly global.

According to Wallerstein the world-system is the fundamental unit of analysis within which all other social processes and structures should be analyzed. Wallerstein argues there are three fundamental types of world-systems: world-economies, world-empires, and minisystems. A world-system typically is composed of many different states. These states trade and war with one another. This is called a *world-economy*. If one state succeeds in conquering the others, thus uniting it politically, Wallerstein calls it a *world-empire*. Typically, one state in a world-economy eventually conquers, or at least dominates, the others, transforming it into a world-empire. Often, world-empires break apart. This can be due to any number of factors; a crisis of succession for the ruler or ruling family is typical. A group of interacting, nonstate groups constitutes a *minisystem*. Wallerstein does not analyze them in detail. Wallerstein claims that the modern world-system is unique because it is based on capitalism, has not become a world-empire, and has become truly global.

Wallerstein developed world-system theory to explain the origins and interrelations of what were typically called the first, second, and third worlds and their roles in the rise to dominance of capitalism and industrialization (Wallerstein 1974a,

1974b, 1979, 1980, 1984, 1989, 1991, 1992, 1993a, 1993b). He also sought to overcome the artificial compartmentalization of the social sciences, arguing that they are one unified field (Wallerstein 1991, 1997, 1998). He drew inspiration from the *Annales* school of French historiography and from dependency theory.

In his account the "modern world-system" originated in western Europe between 1450 and 1640 C.E., often called "the long sixteenth century." Early capitalists and merchants needed labor, raw materials, and markets. These needs fueled the expansion of trade networks and often led to colonization of many areas of the world. Expansion was not steady but cyclical, a fundamental quality of the world-system. This "world"-system finally became truly global in the twentieth century.

The division of labor has three components:

1. a core that employs advanced industrial production and distribution and has strong states, a strong bourgeoisie, and a large working class;
2. a periphery that specializes in raw materials production and has weak states, a small bourgeoisie, and many peasants; and
3. a semiperiphery that is intermediate between core and periphery, in its economic, social, and political roles and its own internal social structure.

Core capitalists typically use some form of coercion to force peripheral producers to accept lower prices and lower wages. This allows the core to accumulate capital at the expense of the periphery. Such uneven terms of trade are known as *unequal exchange*. Thus, unequal exchange simultaneously promotes core development and peripheral impoverishment.

The claim that unequal exchange promotes impoverishment in the periphery is also known as the "development of underdevelopment," a phrase coined by Andre Gunder Frank (1966). One set of vociferous debates about world-systems analysis focuses on the necessity of core exploitation of the periphery for either core or system development. Daniel Chirot (1977, 1986) adds an interesting wrinkle to this debate. He argues that exploitation was not fundamentally necessary for the development of the core or the system. However, because most core leaders, especially in the late nineteenth and early twentieth centuries, believed it was, they acted on that belief and scrambled to colonize the world, especially Africa. Chirot claims that the wealth that developed the core and fueled the Industrial Revolution primarily came from intracore trade. At best, wealth drawn from the periphery merely assisted what was a process internal to the core. Most world-systems analysts disagree strongly with this view.

The debate continues to rage. What is clear, however, is that resources extracted by the core from the periphery via power and coercion did occur and did benefit at least some classes in the core. Beyond this, conclusions must be drawn very carefully based on precise concepts and measurements. While in principle such distinctions are clear and so are the derived hypotheses, the data to test them precisely have not been readily available, nor have available data been unambiguously clear. (The following chapter has more to say on unequal exchange.)

A major premise of this mode of analysis is that the world-system must be studied as a whole. Therefore, the study of social, political, economic, or cultural change in any component of the system must begin by understanding that component's role within the system, whether it be a nation, state, region, ethnic group, class, gender role, or nonstate society.[1] World-systems analysis has a dual research agenda: (1) How do the processes of the system affect the internal dynamics and social structures of its components? and (2) How do changes within its components affect the entire system?

WORLD-SYSTEMS DYNAMICS

The world-system is not a static structure that explains everything. The dialectic between local and global implicit in the dual research agenda is the heart of world-systems analysis. The system itself exhibits several trends with embedded continuing cycles.[2] The net effect produces a spiral (somewhat akin to the wire that holds spiral notebooks together). The trends are as follows:

- Commodification
- Proletarianization
- State formation
- Increasing size of enterprises
- Capital intensification
- Globalization

Commodification is the familiar process of more and more exchanges being based on money, rather than moral or personal obligations. *Proletarianization* is the process in which more and more people become wage workers and fewer are slaves or peasants. Citizenship—that is, membership in a state—has become almost universal. Multinational corporations (MNCs) and transnational corporations (TNCs) have grown continuously larger. Some are so large that only the largest states have economies that are greater than those of the largest corporations. Human productivity increasingly depends on capital in the form of equipment but also human capital in the form of learning and skills. Increasing use of capital in the form of machinery causes increasing drain on environmental resources, especially energy sources. An important insight from world-systems theory is that globalization is not just a "current thing" but has been going on for centuries or, in some views, millennia. In the late twentieth century all these processes became increasingly planet-wide and readily visible to everyone.

At least two types of cyclical process characterize the world-system. The best known and studied are the Kondratieff long wave or K-wave, named for the Russian economist who first documented its existence early in the twentieth century.[3] K-waves are approximately fifty-year cycles in prices. The upswing is called the A-phase; the downswing, the B-phase. Apparently, the world-system is moving from the end of one B-phase to the start of another A-phase. I say "apparently" because K-waves are notoriously difficult to pin down precisely because they must be measured indirectly via trade volume and prices (see chap. 2 in this volume; Boswell and Misra 1995; Goldstein 1988; Barr 1979). The basic dynamic seems to be that as a new technology is discovered and implemented, the economy expands. As the market saturates and competition increases, the expansion slows until another cycle, based on new or renewed technologies, begins.

A related but different cycle is the hegemonic sequence. *Hegemony* is a condition in which one state in the core dominates the world-system typically without overt coercion but through its sheer economic and political power. Once its power peaks, hegemony is lost, or at least lower, and the core is marked by much more intense interstate rivalry and competition. Hegemons usually come to power as the result of a global war. Here, "global" must be read in the sense of "world" in a world-system. The United States is in a phase of hegemonic decline. The decline is relative. The United States is in no danger of falling into the periphery, or even the semi-periphery. Rather, it is far less powerful than it was in the decades immediately after World War II. This can be illustrated by comparing the degree to which the United States consulted allies when it invaded Vietnam with the level of consultation in the Gulf War. Hegemonic cycles appear to be about twice as long as K-waves, approximately one hundred years, and are related to them in complex and controversial ways.

These two dynamic cycles give rise to other cycles of war, state formation, colonization, decolonization, and social movements. They are a major context within which all other social processes occur. These cycles do not determine or cause these other processes. Rather, they create conditions that are more or less conducive to their occurrence and sometimes success. Research on the various cycles and their correlates is relatively recent (for summaries, see Boswell 1989b; Boswell and Sweat 1991; Boswell and Chase-Dunn 1999, chap. 2).

Several questions permeate world-system literature:

1. Is poverty or underdevelopment in the periphery *necessary* for core development, and if so, to what degree?
2. Are exogenous factors (primarily markets) or endogenous factors (e.g., class) the main agents of change?
3. Is socialism possible for a region within a capitalist world-system, or must the entire system become socialist simultaneously?
4. Is world-system theory a useful extension or crude distortion of Marxist theory?

The most enduring and valuable contribution of world-system theory to the social sciences is that it compels social theorists to attend simultaneously to historical processes and global interconnections in constructing theories of all social processes. World-system theory is a major contributor to the dismantling of the parochialism of the 1950s and 1960s when American sociology was largely the sociology of one case: post–World War II America. (For more on this criticism, see chap. 10 in this volume; Bergesen 1995c; Blaut 1993.)

Recently postmodernism and deconstructionism have engaged in a similar "project," although it has not been limited to sociology but attacks all social science. Their attacks, however, are often too extreme, declaring all theorizing, especially evolutionary theorizing, suspect. Wallerstein (1998) says about postmodernism, "I am sympathetic to many of their critiques (most of which, however, we have been saying more clearly, and indeed earlier). However, I find them on the whole neither sufficiently 'post'-modern nor sufficiently reconstructive" (110).

Ironically, there is a rather simple world-system explanation for these movements. Albert Bergesen argues in chapter 10 (and 1995c, 1996) that in periods of hegemonic decline, such as the current one, dominant paradigms for organizing information break down. This gives rise to a proliferation of viewpoints and modes of analysis, with no criteria for discriminating among them. This is the current situation in literary studies in which "anything goes" and "all truth is relative"—at least that is how my colleagues in literature describe it to me.

This is not to say that some of the criticisms, especially those directed at exposing ethnocentric, or core-centric, biases in theorizing, choice of research topics, and modes of analysis are without merit. In that vein, world-systems theory also has been criticized for being overly economistic and for being Eurocentric (or core-centric), state-centrist, and for paying too little attention to states, culture, and gender. However, in the last decade or so world-system analysts have addressed these, and many other issues.

NEW DIRECTIONS IN WORLD-SYSTEMS RESEARCH

There is so much new work that Wallerstein's writings, including *Modern World-System, I, II, III*, barely sample world-systems research (1974b, 1980, 1989). Christopher Chase-Dunn's *Global Formation* (1998) summarizes much of world-systems research. Giovanni Arrighi's *The Long Twentieth Century* (1994) explores the oscillations between production capital and finance capital within the modern world-system. William Martin (1994) and Chase-Dunn and Grimes (1995) review recent world-systems work. Philip McMichael (1996) also summarizes much relevant work.[4]

There are so many variations in world-system analysis that it is not appropriate to refer to it as a "theory." It is better called a perspective or, in Thomas Kuhn's sense (1970, 1977), a paradigm. A paradigm is more general than a theory. It is a set of assumptions that guide questions and development of theories. The confusion between theory and paradigm or perspective is the source of the misperception that Wallerstein's early works (1974a, 1974b) encompass the whole of world-system analysis. Because there are now so many versions of the approach, many of which argue that there have been more than one world-system, the plural is used, except when referring explicitly to Wallerstein's original formulation, in which case the singular is retained. Indeed, the world-system perspective can no longer be associated exclusively with his work alone. In the chapters that follow the terms *world-system theory*, *world-system analysis*, and *world-system perspective* are used synonymously in most cases. If readers keep in mind the meaning discussed here for all three, the confusion should be minimal.

Researchers within the world-systems perspective address many subjects. Some of the new topics are: cyclical processes in world-systems (Dassbach et al. 1995; Dassbach 1993; Suter 1992; Goldstein 1988); decolonization (Bergesen and Schoenberg 1980; Boswell 1989b; Martin and Kandal 1989); cycles of revolution (Boswell 1989a; Boswell and Dixon 1993; Foran 1993; Kowalewski 1991; Robinson 1996); cycles of war (Chase-Dunn and O'Reilly 1989; Chase-Dunn and Podobnik 1995); hegemonic cycles (Boswell and Sweat 1991; Goldfrank 1987, 1994, 1999; Goldfrank and Maki 1995; Robinson 1996; Silver 1995; Taylor 1996, 1997); inequality and democracy (Markoff 1998; Korzeniewicz 1996; Korzeniewicz and Moran 1997; Korzeniewicz and Awbrey 1992); interactions of class and trade (Aston and Philpin 1985; Denemark and Thomas 1988); the roles of women, households, and gender in the world-economy (Ward 1984, 1990, 1993; Smith et al. 1988; Kandal 1990–91); race and ethnicity (Denemark 1992, 1995; Hall 1989a, 1998; Kandal 1990–91); commodity chains (Gereffi and Korzeniewicz 1994); socialism and the consequences of the collapse of the Soviet Union (Bergesen 1992a; Smith and Böröcz 1995; Chase-Dunn 1992a; Goldfrank 1987); cities in world-systems (Timberlake 1985; Kasaba 1991; Bosworth 1995; Smith 1996; Sokolovsky 1985); the role of culture in the world-economy (Kiser and Drass 1987; Bergesen 1990, 1991, 1992b, 1995a, 1995b, 1995c; Boli and Thomas 1997); the environment (Bergesen 1995a, 1995b; Bartley and Bergesen 1997; Chew 1992, 1995a, 1995b; Goldfrank et al. 1998); and subsistence (Bradley et al. 1990). Many case studies offer fine-grained analyses of the complex functioning of the world-system with respect to slavery (Morrissey 1989; Tomich 1990), agrarian capitalism (Goldfrank and Gomez 1991; Goldfrank et al. 1995; McMichael 1984; So 1986), peasants (Bunker 1987; Troillot 1988), changes in east Asia (So and Chiu 1995; Deyo 1987), changes in Latin America (Korzeniewicz and Smith 1996), technological change (Hugill 1993), and relations with nonstate or aboriginal peoples to the world-economy (Baugh 1991; Dunaway 1994, 1996a, 1996b, 1996c, 1997; Hall 1986, 1987, 1989a, 1989b, 1991; Peregrine and Feinman 1996; Harris 1990; Kardulias 1990; Mathien and McGuire 1986; Meyer 1990, 1991, 1994; Peregrine 1992, 1995; Faiman-Silva 1997; Himmel 1999).

World-systems theory has even inspired at least one novel. W. Warren Wagar's *A Short History of the Future* (1999) depicts a sequence of one dystopia followed by two utopias. His novel is loosely based on world-systems analysis. An entire section of *Journal of World-Systems Research* discussed the novel from various perspectives (1996). Boswell and Chase-Dunn's (1999) *The Spiral of Capitalism and Socialism* discusses alternative futures. David Waller (1998) discusses China's future from a world-systems perspective.

In political science and international relations George Modelski, William Thompson, and Karen Rasler have developed a number of world-systems theories (Thompson 1983a, 1988; Rasler and Thompson 1994; Modelski 1987; Modelski and Thompson 1988, 1996). These writers emphasize geopolitics rather than economics and focus on war and international relations per se more than other world-systems analysts. Modelski has deeper interest in long-term evolution than does Wallerstein (1995a, 1995b). Thompson and Modelski contribute to the analysis of precapitalist world-systems discussed later (see chap. 5).

Eric Wolf has suggested that world-systems theory is one way of cumulating anthropological knowledge and building explanations for cultural phenomena (Wolf 1990, 594; see Blanton et al. 1997). I argue that much work in anthropology that does take cognizance of external connections is world-systemic but often does not employ an explicitly world-systems perspective (Hall 1997, 1998, 1999).

Archaeologists, in particular, have found considerable potential in world-systems analysis but have been dissatisfied with the results (Hall and Chase-Dunn 1993). All have recognized, to some degree, that world-system theory cannot be applied wholesale to precapitalist settings.[5] Pailes and Whitecotton (1975, 1979) were the first to modify world-system theory for use in precapitalist settings. Jane Schneider (1977) wrote one of the most insightful critiques of early world-system theory, questioning Wallerstein's emphasis on bulk to the neglect of luxury goods. Blanton and Feinman (1984), Kohl (1987), and Santley and Alexander (1992) have also made important critical statements. Schortman and Urban (1994a, 1994b) developed important critical insights in core-periphery relations in their study of southeast Mesoamerica. Kardulias (1999) edited a volume on the uses and problems of using world-systems analysis to explain archaeological processes. Peter Peregrine discusses these contributions in more detail in chapter 3.

The concern with precapitalist settings is a major new area in world-systems analysis (see the special issue of *Review* edited by Chase-Dunn [1992b]; and Hall and Chase-Dunn 1993, 1994). Janet Abu-Lughod (1989) argues for twelfth-, thirteenth-, and fourteenth-century roots for the modern world-system, deeper than Wallerstein's long sixteenth century. She also argues that the entire conception of the "rise of the West" is mistaken and suggests that the "East fell," or at least withdrew (Abu-Lughod 1989, 1993; Fitzpatrick 1992).

Andre Gunder Frank and Barry K. Gills (1992, 1993) argue that there has been one continuous "five thousand–year world system" since the first appearance of states in Mesopotamia. Thus, they use "world system" without a hyphen because they see it as one growing system that becomes the world. They also see capitalism as struggling to dominate this system and finally achieving domination in what Wallerstein calls the "modern world-system."

Frank goes beyond Abu-Lughod's fall of the East claim and argues in *ReOrient* (1998) that China has been the center of an Afroeurasian world-system for millennia. Indeed, he argues China was ahead of Europe in development well into the eighteenth century. Thus, he suggests that the late twentieth-century economic growth in Asia (the downturn in the late 1990s notwithstanding) marks not the rise of a new core but the return of a former core.

Christopher Chase-Dunn and I (Chase-Dunn and Hall 1991, 1993, 1994, 1997a) have argued that the extension of world-systems analysis to precapitalist settings requires that many of the assumptions of the theory of the modern world-system must be transformed into empirical questions. We argue that there have been many types of world-systems, which we classify into four broad types: kin ordered, tributary, capitalist, and a potential socialist. Each type has considerable range of variation in subtypes (see Chase-Dunn 1996). Each type is characterized by a dominant mode of accumulation, a term that emphasizes how wealth or capital is amassed, as opposed to "mode of production," which refers to how capital is produced.

Kin-ordered world-systems are composed of small, stateless groups of sedentary foragers, have very little differentiation or hierarchy (for a detailed example, see Chase-Dunn and Mann 1998). Then some seven to five thousand years ago chiefdom-based world-systems began to appear that were marked by sharper hierarchy throughout the system and some degree of differentiation into core and periphery. Out of the tension and dynamics of these systems the first states emerged some five thousand years ago and tributary world-systems developed. These lasted until the seventeenth century when the Dutch developed the first state dominated by a capitalist elite. This marked the emergence of the capitalist world-system. Finally, we discuss the possibility of a fourth type of world-system based on global democracy and collective rationality, which we call, for want of a better term, a "socialist" world-system. Terry Boswell and Christopher Chase-Dunn analyze these possibilities in more detail in their chapter.

Among the many findings in our work a few are worth noting. First, modes of accumulation typically encompass more than one type of "mode of production." Tributary world-systems typically include some kin-ordered subsections—most often in peripheral areas, but not exclusively. They also contain pockets of capitalist modes. Still, amassing wealth through tribute paid to a central ruler is the dominant mode of accumulation.

Second, trade in bulk goods is not the only form of exchange that constitutes a world-system. Rather, we see three other types of system exchange, and consequently, boundaries: political/military exchanges; luxury, preciosity, or prestige goods exchanges; and information exchanges. Each of these three boundaries is successively larger, and all are larger than the bulk goods exchange network. How they relate to each other is a major issue for further research. (For examples, see Chase-Dunn and Hall 1997 and that work's chap. 3, especially fig. 3.1.)

Third, we found that all world-systems "pulsate"—that is, expand then contract, or expand rapidly then more slowly. This pulsation is the engine of world-system expansion, which is sporadic and cyclical rather than linear. The finding is interesting in that *all* types of world-systems pulsate. Hence, the explanation for these pulsations cannot reside in the mode of production or, as we prefer, mode of accumulation.

Fourth, we find that all of Afroeurasia (in conventional terms Asia, Europe, northern Africa) has been linked, at least at the information and luxury goods exchange levels, for over two millennia. Hence, one *cannot* explain events and processes in Europe by only examining European processes (for more details and maps, see Chase-Dunn and Hall 1997a, chap. 8). Among the puzzles this approach uncovered was that empire size and city size distributions at the western and eastern ends of Eurasia have been linked for two millennia. This, of course, is not news. Frederick Teggart (1918, 1925, 1939) observed this early in the twentieth century. What is puzzling is that there appears to be no linkage to processes in South Asia (see Chase-Dunn and Hall 1995, 1997a, chap. 10; Chase-Dunn, Manning, and Hall in press).

As the foregoing suggests, our analysis differs significantly from Wallerstein's, Frank's, Gills's, Modelski's, and Thompson's in several ways. First, we argue that there have been many world-systems, and several broad types, each with many variations. We do see these as world-systems. Hence, we retain the hyphen and the plural. Second, we argue there is a distinct, if complex, evolutionary process driving the changes from one type of world-system to another.[6] Third, we argue that the semiperiphery—which we see as underanalyzed and undertheorized—is a major locus of change. In particular, the capitalist world-system was the attempt of several semiperipheral areas in western Europe to get into the lucrative Asian trade.

Fourth, as noted earlier, this occurred in the seventeenth century when the capitalist class took control of the Dutch state. Thus, we disagree fundamentally with Frank and Gills that capitalism has always been a characteristic of the system, and with Wallerstein that the capitalist world-system originated in the long sixteenth century. While the difference is sharp, it is not unbridgeable. We recognize that there have been pockets of capitalism embedded within tributary systems for millennia. Our argument, however, is that they did not dominate the system but were subordinate pockets within it. We also recognize with Wallerstein, Abu-Lughod, Gills, and Frank that the modern, capitalist world-system has deep roots.

Indeed, a fifth difference is that we see the roots as far deeper than any of the other analysts. Much of Frank and Gills's analysis is readily acceptable to us. But in our view, they start *after* one of the most important and thorniest changes in human history had occurred: the invention of states. This, of course, is a major problem in archaeology and world history. Archaeologists have found some use in this approach (Kohl 1987; Kardulias 1999).

With the extension of world-systems analysis into precapitalist settings, world historians and civilizationists have developed a wary interest in world-systems analysis, first in a special issue of *Comparative Civilizations Review* (Hall 1994; Melko 1994; Sanderson 1994), then in an edited collection (Sanderson 1995). Chapters by Matthew Melko (1995), William McNeill (1995), and David Wilkinson (1995) highlight the similarities and differences between world-systems and civilizations. Articles by Frank and Bergesen (Frank 1995; Bergesen 1995d) outline key issues and do much to combat the Eurocentrism that infects much social theory.

The dialogue with world historians and civilizationists tends to reopen issues of nomathetic versus ideographic approaches to history, but not entirely. World historians, in general, are amenable to some broad theoretical principles. However, they tend to prefer rich description and primary documents more than social scientists. Yet they are aware that world history requires considerable synthesis if it has any hope of being concise.

An excellent sampling of world historical approaches to the past is available in the American Historical Association pamphlet series "Essays on Global and Comparative History," under the general editorship of Michael Adas. Two works in this series are particularly useful. Michael Adas's "'High' Imperialism and the 'New' History" (1993) summarizes the many controversies that concern historians about the rise and spread of the modern world-system. He is moderately critical of social science approaches. Jerry Bentley's "Shapes of World History in Twentieth-Century Scholarship" (1996) surveys the many approaches, including many from social sciences. It is an excellent road map to the approaches, debates, and divides in world history. For the social scientist first encountering world history, it is an excellent introduction. His *Old World Encounters* (1993) is a good example of how world historians approach the study of interactions. In 1998 he extended his argument for looking at "Hemispheric Integration." Bentley also edits *Journal of World History,* which publishes many useful papers and often features articles by social scientists.[7]

The division of labor within world-systems (modern or ancient) into core, periphery, and semiperiphery suggests that world-systemic processes have spatial dimensions. Few sociologists have attended to this. Not surprisingly, however, many geographers have (Flint and Shelley 1996; Agnew 1982, 1987; Hugill 1993; Knox and Agnew 1994; Knox and Taylor 1995; Taylor 1993, 1996; Terlouw, 1992). They have made major contributions to world-systems theorizing, many of them of value to sociologists. Fred Shelley and Colin Flint discuss those contributions in detail in chapter 4.

A FEW EXAMPLES

This collection consists of the two introductory chapters and four additional parts. The second essay in the introductory part, by Peter Grimes, reviews much recent world-systems research and elaborates on several concepts mentioned in this essay. Part II surveys the various social science approaches to world-systems analysis and related topics. Part III presents summaries and overviews of world-systems analysis from a variety of angles. Part IV presents several "case studies" or specific areas of research. Part V presents two visions of the future of the world-system and of world-systems analysis. For readers who prefer starting with concrete examples then building to theoretical analyses, Part IV is the place to begin. For those whose interest is to "reconnoiter the lay of the land" in world-systems analysis, Part III is the place to start. For those who want to know how social scientists who are not sociologists use and discuss world-systems analysis, the current order is best.

Part II, "From Many Disciplines," begins with Peter Peregrine's argument that sociologists can benefit from perusing archaeological research that uses or critiques world-systems analysis. His article introduces many of the fundamental issues that archaeologists discuss and provides helpful guidance to that literature. He suggests that some topics, such as the ramifications of types of goods exchanged for the functioning of world-systems, are best explored with archaeological rather than sociological data.

Fred Shelley and Colin Flint present a similar discussion of the contributions of geographers to world-systems analysis. As noted earlier, the tripartite division of world-systems into core, periphery, and semiperiphery and creation of frontiers by incorporation of new areas suggest the need for spatializing world-systems analysis. They discuss how world-systems interactions are manifested in spatial patterns and how local and global processes shape each other. This chapter along with their 1996 article is a thorough overview of geographical approaches to world-systems analysis.

William Thompson summarizes his work, often with George Modelski or Karen Rasler on global wars. In doing so he shows how world-systems analysis (with or without the hyphen, which Thompson does not use) relates to the field of international relations in political science. He analyzes the relations of K-waves to cycles of war and hegemony and global leadership. He presents a very interesting camera angle on world-systemic processes.

Joya Misra takes on the challenge of explaining why world-systems theory and feminism have not been able to work together as much as one might hope. She not only reviews world-systems work on gender but summarizes how feminists have examined many of the same issues. She then shows the convergences and divergences between feminist and world-systems, indicating many areas ripe for fruitful development. This is an excellent summary of a rapidly expanding area. Two new contributions to it appear in Part IV (Dunaway and Ward, Stander, and Solomon).

Part III, "World-Systems Overviews," begins with Leslie Laczko's analysis of the role of Canada in the modern world-system. He provides a useful summary of the distribution of the types of social organizations in the world and how they have changed, as background to explaining how and why Canada is unusual. In short, it is one of the modern core states with the highest level of ethnic pluralism and is an immediate neighbor of the United States. Both are significant aspects of Canada, but for different reasons—which Laczko explicates.

David A. Smith reviews world-systems studies of urbanization. Not surprisingly, he argues that urbanization cannot be understood without careful attention to each urban area's role in the world-system. As the world-system changes, so do processes of urbanization. Here we have the first example of how social processes in the core work differently from the analogous processes in the periphery. Failure to see these differences from a world-systems perspective has led to a great deal of confusion. This is a theme that recurs in many chapters.

Debra Straussfogel highlights the "system" in world-systems analysis (see, too, Straussfogel 1997a, 1997b, 1997c, 1998). She explains how it relates to general systems theory and to its rapidly expanding progeny, chaos theory and complexity theory. She notes that the explanation of world-system evolution presented by Chase-Dunn and me (1997a) is a form of chaos theory that combines explanations of gradual change embedded in episodic rapid changes. Chaos models are relatively new in general, and quite recent in the social sciences. It seems that a major part of the explanation for the pulsation of all world-systems is found in the self-organizing tendency of all complex systems. This is an exciting new area.

Albert Bergesen provides a world-systems explanation of cultural change. He shows how trends in art and theory reflect the hegemonic cycles of the world-system. In it he explains the rise, and now apparent waning, of postmodernism. This is an insightful essay that is sure to raise more than a few hackles. But it shows a reasonable way out of the muddle where postmodern analyses all too often end up: claiming everything is relative, and there is no absolute truth. None of this gainsays postmodernism's fundamental insight that much of social science and history has been told from the core's point of view (see, too, Blaut 1993).

Part IV, "Gender, Urbanism, Cultures, Indigenous Peoples, and Ecology," provides more detailed case studies. For those who eschew the abstract and want concrete analyses of real people, these chapters present it. Wilma Dunaway's account of Cherokee women shows in considerable detail how participation in—incorporation into—the European-based world-system transformed the lives of Cherokees, especially the lives of women, and degraded their environment. Because there are so few documents that even mention women, she is unable to demonstrate how Cherokee women resisted these changes. But work on the nineteenth century shows that they did resist, with a modicum of success (see Dunaway 1996a, 1996b, 1997). Her work illustrates how analysis of the colonizers' accounts of the world can be used to see the world from the point of view of the colonized.

Carol Ward, Elon Stander, and Yodit Solomon follow similar issues into the twentieth century with Cheyenne women in Montana. They build their analysis on an old anthropological insight that drinking among contemporary Native Americans can be a form of resistance to Euroamerican domination—resistance to incorporation in world-systems terms. They show how contemporary Cheyenne women manipulate and transform the ubiquitous twelve-step programs into Cheyenne cultural efforts to preserve their identity and culture in the face of overwhelming pressure to assimilate. Along the way, they illustrate a point Joya Misra raises, that feminism for third world women is a very different matter than feminism for first world women.

In chapter 13 I draw on the previous two essays, my own work on that part of northern New Spain that became the southwestern United States, and much other work to develop a more general account of incorporation of indigenous peoples into expanding world-systems. I also begin to develop a world-systems account of frontier formation and transformation. While more abstract than the preceding two chapters, I seek to analyze the interaction of world-systems with indigenous peoples in ways that highlight their efforts to resist incorporation and show them as proactive participants, not just passive victims of the European juggernaut. I also try to cast these processes in a wider context of world-systems expansion and evolution.

Alvin So and Stephen Chiu argue that development in East Asia cannot be understood fully on a state-by-state basis or even by comparative analyses. Rather, a world-systems perspective is vital to understand the different paths taken by various East Asian states. They, too, illustrate the dialectic relations between an incorporating world-systems and incorporated regions. In this case the scale is larger—at the level of states—but the interactions share a broad similarity. In a separate book, Chiu, Ho, and Liu (1997) take the analysis to the city level and show how Singapore and Hong Kong are very different cities, largely because of their very different roles in the world-system.

Part V presents two visions of the future of the world-system and the analysis of it. Terry Boswell and Christopher Chase-Dunn summarize their arguments about the possibilities of global democracy oriented toward maximizing collective rationality—what they call "socialism." They explain why their "socialism" has at best a faint resemblance to the former Soviet Union, contemporary China, or Cuba. They, too, highlight the roles of current actors by suggesting where and how people might push on the current world-system to begin building a more just and humane system that does *not* abandon either individual freedom or technological productivity. This is a very provocative chapter that shows how even the seemingly most abstruse world-systems analysis is relevant to our everyday lives.

The last chapter by Albert Bergesen and Tim Bartley addresses the issue of ecology. They argue that without embedding human systems in their larger ecological contexts, our understanding of them is inherently limited. They suggest that the world-system is not the highest level of social relations but a subsystem of the larger ecosystem. Thus, true world-system, or global system, or ecosystem analysis must entail placing the human world-system within the structural framework of the more encompassing ecosystem. This is one of the most challenging analyses presented in this collection. Recently Wallerstein (1998) argued that world-systems analysis may eventually be succeeded by something else. Bergesen and Bartley show us what that might be.

These essays exemplify the exciting work being done within a world-systems perspective. Yet they are only a small sample from a large universe. Many puzzles remain to be solved, many preliminary findings require further exploration, many theories need further empirical testing, and many issues need further theoretical development. Still, they illustrate how a world-systems perspective can address culture, gender, race, ethnicity, and states. It is useful to recall Kuhn's (1970, 1977) criterion for evaluating a paradigm: that it leads us to ask questions that, when answered, tell us more about the world than we knew before we asked them. These chapters are an invitation to join the work of asking such questions and the fun of pursuing the answers from a very broad world-systems paradigm.

NOTES

Thanks to University of Texas Press for permission to reprint parts of this essay, which first appeared in *Sociological Inquiry* (Hall 1996). I also wish to thank the many commentators who helped me update the bibliography and check my interpretations of theirs and others' works. As is ever the case, they are not to be blamed for my failure to heed their sage advice.

1. I eschew the term *tribe* here both because of derogatory connotations some people associate with it and because of its lack of precision. I discussed this in more detail in my book (Hall 1989a, chaps. 2 and 3) and my chapter (13) on incorporation.

2. I am indebted to Terry Boswell and Christopher Chase-Dunn for sharing their excellent summary of these processes with me (Boswell and Chase-Dunn 1999). Their arguments are summarized in chapter 15. The various introductory accounts mentioned in the first paragraph of this section suggest similar trends and cycles. See, too, Hopkins and Wallerstein (1979).

3. *Kondratieff* is sometimes spelled *Kondratiev,* as it is by Grimes in the next chapter. The two spellings are used interchangeably.

4. There are two world-systems journals. The Fernand Braudel Center at the State University of New York, Binghamton, publishes the major journal *Review.* It has its own Web site: http://fbc.binghamton.edu/. Libraries often catalog it *Fernand Braudel Center Review* under the center's name, so it is found under "F" as often as "R." The *Journal of World-Systems Research* is an electronic journal available without cost via the World Wide Web (http://csf.colorado.edu/wsystems/jwsr.html) or via gopher and ftp (csf.colorado.edu/wsystems/jwsr). In addition to these journals there is a listserv, World-System Network (WSN; subscribe by sending: subscribe wsn "your name" to listproc@csf.colorado.edu), which discusses a variety of world-systems topics. The American Sociological Association section, Political Economy of the World-System (PEWS), has a newsletter (available via WSN), holds its own conference each spring, and produces an edited collection annually. Appendix 1 lists all the existing or planned PEWS annuals. PEWS also gives an annual book prize. Winners through 1999 are listed in appendix 2.

5. By "precapitalist" I mean before approximately 1500 C.E., the rise of Europe in the long sixteenth century. The prefix *pre-* is intended to convey a notion not of inevitability but of chronology.

6. We discuss this in Chapter 6 of *Rise and Demise.* We have since elaborated this account (see Chase-Dunn and Hall 1997b).

7. As this volume reached completion, Bentley's new world history text (Bentley and Ziegler 2000) appeared in print. This is an excellent entry point into world history for the novice.

REFERENCES

Abu-Lughod, Janet. 1989. *Before European Hegemony: The World System A.D. 1250–1350.* New York: Oxford University Press.

———. 1993. "Discontinuities and Persistence: One World System or a Succession of World Systems?" Pp. 278–290 in *The World System: Five Hundred Years or Five Thousand?* edited by Andre Gunder Frank and Barry K. Gills. London: Routledge.

Adas, Michael. 1993. "'High' Imperialism and the 'New' History." *Essays on Global and Comparative History.* Washington, D.C.: American Historical Association.

Agnew, John A. 1982. "Sociologizing the Geographical Imagination: Spatial Concepts in the World-System Perspective." *Political Geography Quarterly* 1:2(April):159–166.

———. 1987. *The United States and the World-Economy.* New York: Cambridge University Press.

Arrighi, Giovanni. 1994. *The Long Twentieth Century: Money, Power and the Origins of Our Times.* London: Verso.

Aston, T. H., and C. H. E. Philpin, eds. 1985. *The Brenner Debate: Agrarian Class Structure and Economic Development in Pre-Industrial Europe.* Cambridge: Cambridge University Press.

Barr, Kenneth. 1979. "Long Waves: A Selected Annotated Bibliography." *Review* 2:4(Spring):671–718.

Bartley, Tim, and Albert Bergesen. 1997. "World-System Studies of the Environment." *Journal of World-Systems Research* 3:369–380 (electronic journal; http://csf.colorado.edu/ wsystems/jwsr.html; ftp and gopher: csf.colorado.edu/wsystems/journals/).

Baugh, Timothy G. 1991. "Ecology and Exchange: The Dynamics of Plains-Pueblo Interaction." Pp. 107–127 in *Farmers, Hunters, and Colonists: Interaction between the Southwest and the Southern Plains,* edited by Katherine A. Spielmann. Tucson: University of Arizona Press.

Bentley, Jerry H. 1993. *Old World Encounters: Cross-Cultural Contacts and Exchanges in Pre-Modern Times.* Oxford: Oxford University Press.

———. 1996. "Shapes of World History in Twentieth-Century Scholarship." *Essays on Global and Comparative History.* Washington, D.C.: American Historical Association.

———. 1998. "Hemispheric Integration, 500–1500 C.E." *Journal of World History* 9:2(Fall):237–254.

Bentley, Jerry H., and Herbert F. Ziegler. 2000. *Traditions and Encounters: A Global Perspective in the Past.* Boston: McGraw-Hill.

Bergesen, Albert. 1990. "Turning World-System Theory on Its Head." *Theory, Culture & Society* 7:67–81.

———. 1991. "The Semiotics of New York's Artistic Hegemony." Pp. 121–132 in *Cities in the World-System,* edited by Resat Kasaba. New York: Greenwood.

———. 1992a. "Communism's Collapse: A World-System Explanation." *Journal of Military and Political Sociology* 20 (Summer):133–151.

———. 1992b. "Godzilla, Durkheim and the World-System." *Humboldt Journal of Social Relations* 18:1:195–216.

———. 1992c. "Pre- vs. Post-1500ers: Who Is Right?" *PEWS News* Summer:2–4.

———. 1995a. "Eco-Alienation." *Humboldt Journal of Social Relations* 21:1:111–126.

———. 1995b. "Deep Ecology and the Moral Community." Pp. 193–213 in *Rethinking Materialism: Perspectives on the Spiritual Dimension of Economic Behavior,* edited by Robert Wuthnow. Grand Rapids, Mich.: Erdmans.

———. 1995c. "Postmodernism: A World System Explanation." *Protosoziologie* 7:54–59, 304–305.

———. 1995d. "Let's Be Frank about World History." Pp. 181–191 in *Civilizations and World-Systems: Two Approaches to the Study of World-Historical Change,* edited by Stephen K. Sanderson. Walnut Creek, Calif.: Altamira.

———. 1996. "The Art of Hegemony." Pp. 259–278 in *The Development of Underdevelopment: Essays in Honor of Andre Gunder Frank,* edited by Sing Chew and Robert Denemark. Newbury Park, Calif.: Sage.

Bergesen, Albert, and Ronald Schoenberg. 1980. "Long Waves of Colonial Expansion and Contraction, 1415–1969." Pp. 231–277 in *Studies of the Modern World-System,* edited by Albert Bergesen. New York: Academic Press.

Blanton, Richard, and Gary Feinman. 1984. "The Mesoamerican World System." *American Anthropologist* 86:3(Sept.):673–682.

Blanton, Richard, Peter Peregrine, Deborah Winslow, and Thomas D. Hall. 1997. *Economic Analysis beyond the Local System.* Lanham, Md.: University Press of America.

Blaut, J. M. 1993. *The Colonizer's Model of the World: Geographical Diffusionism and Eurocentric History.* New York: Guilford.

Boli, John, and George M. Thomas. 1997. "World Culture in the World Polity." *American Sociological Review* 62:2(April):171–190.

Boswell, Terry, ed. 1989a. *Revolutions in the World-System.* Greenwich, Conn.: Greenwood.

———. 1989b. "Colonial Empires and the Capitalist World-System: A Time Series Analysis of Colonization, 1640–1960." *American Sociological Review* 54:2(April):180–196.

Boswell, Terry, and Christopher Chase-Dunn. 1999. *The Spiral of Capitalism and Socialism: The Decline of State Socialism and the Future of the World-System.* Boulder, Colo.: Reinner.

Boswell, Terry, and William Dixon. 1993. "Marx's Theory of Rebellion: A Cross-National Analysis of Class Exploitation, Economic Development, and Violent Revolt." *American Sociological Review* 58:5(Oct.):681–702.

Boswell, Terry, and Joya Misra. 1995. "Cycles and Trends in the Early Capitalist World-Economy: An Analysis of Leading Sector Commodity Trades, 1500–1600/50–1750." *Review* 18:3(Summer):450–485.

Boswell, Terry, and Mike Sweat. 1991. "Hegemony, Long Waves, and Major Wars: A Time Series Analysis of Systemic Dynamics, 1496–1967." *International Studies Quarterly* 35:2(June):123–149.

Bosworth, Andrew. 1995. "World Cities and World Economic Cycles." Pp. 192–213 in *Civilizations and World-Systems: Two Approaches to the Study of World-Historical Change,* edited by Stephen K. Sanderson. Walnut Creek, Calif.: Altamira.

Bradley, Candice, Carmella Moore, Michael Burton, and Douglas White. 1990. "A Cross-Cultural Historical Study of Subsistence Change." *American Anthropologist* 92:2(June):447–457.

Bunker, Stephen. 1987. *Peasants against the State.* Urbana: University of Illinois Press. (Reprinted 1991, Chicago: University of Chicago Press.)

Chase-Dunn, Christopher. 1992a. "The Spiral of Capitalism and Socialism." *Research in Social Movements: Conflict and Change* 14:165–187, edited by Louis Kriesberg. Greenwich, Conn.: JAI.

———. 1992b. "The Comparative Study of World-Systems." *Review* 15:3(Summer):313–333. (Introduction to special issue on precapitalist world-systems.)

———. 1996. "World-Systems: Similarities and Differences." Pp. 246–258 in *The*

Development of Underdevelopment: Essays in Honor of Andre Gunder Frank, edited by Sing C. Chew and Robert A. Denemark. Thousand Oaks, Calif.: Sage.

————. 1998. *Global Formation: Structures of the World-Economy.* 2d ed. Boulder, Colo.: Rowman & Littlefield (Originally 1989, London: Blackwell).

Chase-Dunn, Christopher, and Peter Grimes. 1995. "World-Systems Analysis." *Annual Review of Sociology* 21:387–417.

Chase-Dunn, Christopher, and Thomas D. Hall, eds. 1991. *Core/Periphery Relations in Precapitalist Worlds.* Boulder, Colo.: Westview.

————. 1993. "Comparing World-Systems: Concepts and Working Hypotheses." *Social Forces* 71:4(June):851–886.

————. 1994. "The Historical Evolution of World-Systems." *Sociological Inquiry* 64:3(Summer):257–280.

————. 1995. "Cross-World-Systems Comparisons: Similarities and Differences." Pp. 95–121 in *Civilizations and World-Systems: Two Approaches to the Study of World-Historical Change,* edited by Stephen K. Sanderson. Walnut Creek, Calif.: Altamira.

————. 1997a. *Rise and Demise: Comparing World-Systems.* Boulder, Colo.: Westview.

————. 1997b. "Ecological Degradation and the Evolution of World-Systems." *Journal of World-Systems Research* 3(Fall):403–431 (electronic journal: http://csf.colorado.edu/wsystems/jwsr.html).

————. 1998. "World-Systems in North America: Networks, Rise and Fall and Pulsations of Trade in Stateless Systems." *American Indian Culture and Research Journal* 22:1:23–72.

Chase-Dunn, Christopher, Susan Manning, and Thomas D. Hall. In press. "Pulsations in the Afro-Eurasian System: Indic City and Empire Growth and Decline." *Social Science History.*

Chase-Dunn, Christopher, and Kelly M. Mann. 1998. *The Wintu and Their Neighbors: A Very Small World-System in Northern California.* Tucson: University of Arizona Press.

Chase-Dunn, Christopher, and Kenneth O'Reilly. 1989. "Core Wars of the Future." Pp. 47–64 in *War and the World-System,* edited by Robert Schaeffer. Greenwich, Conn.: Greenwood.

Chase-Dunn, Christopher, and Bruce Podobnik. 1995. "The Next World War: World-System Cycles and Trends." *Journal of World-Systems Research* 1,6 (unpaginated electronic journal: http://csf.colorado.edu/wsystems/jwsr.html).

Chew, Sing C. 1992. *Logs for Capital: The Timber Industry and Capitalist Enterprise in the Nineteenth Century.* Westport, Conn.: Greenwood.

————. 1995a. "On Environmental Degradation: Let the Earth Live." *Humboldt Journal of Social Relations* 21:1:9–13.

————. 1995b. "Environmental Transformations: Accumulation, Ecological Crisis, and Social Movements." Pp. 201–215 in *A New World Order? Global Transformation in the Late Twentieth Century,* edited by David A. Smith and József Böröcz. Westport, Conn.: Greenwood.

Chirot, Daniel. 1977. *Social Change in the Twentieth Century.* New York: Harcourt, Brace, Jovanovich.

————. 1986. *Social Change in the Modern Era.* New York: Harcourt, Brace, Jovanovich.

Chiu, Stephen W. K., K. C. Ho, and Tai-Lok Lui. 1997. *City-States in the Global Economy: Industrial Restructuring in Hong Kong and Singapore.* Boulder, Colo.: Westview.

Clemens, Elizabeth S., Walter W. Powell, Kris McIlwaine, and Dina Okamoto. 1995.

"Careers in Print: Books, Journals, and Scholarly Reputations." *American Journal of Sociology* 101:2(Sept.):433–494.

Dassbach, Carl H. A. 1993. "Enterprises and B Phases: The Overseas Expansion of U.S. Auto Companies in the 1920s and Japanese Auto Companies in the 1980s." *Sociological Perspectives* 36:4:359–375.

Dassbach, Carl H. A., Nurhan Davutyan, Jianping Dong, and Barry Fay. 1995. "Long Waves Prior to 1790: A Modest Contribution." *Review* 18:2(Spring):305–325.

Denemark, Robert A. 1992. "Core-Periphery Trade: The Debate with Brenner over the Nature of the Link and its Lessons." *Humboldt Journal of Social Relations* 18:1:119–145.

———. 1995. "Toward a Theory of Ethnic Violence." *Humboldt Journal of Social Relations* 20:2:95–120.

Denemark, Robert A., and Kenneth P. Thomas. 1988. "The Brenner-Wallerstein Debate." *International Studies Quarterly* 32:1(March):47–65.

Deyo, Frederic C., ed. 1987. *The Political Economy of the New Asian Industrialism*. Ithaca, N.Y.: Cornell University Press.

Dunaway, Wilma. 1994. "The Southern Fur Trade and the Incorporation of Southern Appalachia into the World-Economy, 1690–1763." *Review* 18:2(Spring):215–242.

———. 1996a. *The First American Frontier: Transition to Capitalism in Southern Appalachia, 1700–1860*. Chapel Hill: University of North Carolina Press.

———. 1996b. "Incorporation as an Interactive Process: Cherokee Resistance to Expansion of the Capitalist World-System, 1560–1763." *Sociological Inquiry* 66:4(Fall):455–470.

———. 1996c. "The Incorporation of Mountain Ecosystems into the Capitalist World-System." *Review* 19:4(Fall):355–381.

———. 1997. "Rethinking Cherokee Acculturation: Women's Resistance to Agrarian Capitalism and Cultural Change, 1800–1838." *American Indian Culture and Research Journal* 21:1:231–268.

Faiman-Silva, Sandra L. 1997. *Choctaws at the Crossroads: The Political Economy of Class and Culture in the Oklahoma Timber Region*. Lincoln: University of Nebraska Press.

Fitzpatrick, John. 1922. "The Middle Kingdom, the Middle Sea and the Geographical Pivot of History." *Review* 15:3(Summer):477–521.

Flint, Colin, and Fred Shelley. 1996. "Structure, Agency and Context: The Contributions of Geography to World-Systems Analysis." *Sociological Inquiry* 64:4(November):496–508.

Foran, John. 1993. *Fragile Resistance: Social Transformation in Iran from 1500 to the Revolution*. Boulder, Colo.: Westview.

Frank, Andre Gunder. 1966. "The Development of Underdevelopment." *Monthly Review* September:17–31. (Reprinted in *Latin America* 1969 and *Monthly Review* 1988.)

———. 1995. "The Modern World System Revisited: Rereading Braudel and Wallerstein." Pp. 149–180 in *Civilizations and World-Systems: Two Approaches to the Study of World-Historical Change*, edited by Stephen K. Sanderson. Walnut Creek, Calif.: Altamira.

———. 1998. *ReOrient: Global Economy in the Asian Age*. Berkeley: University of California Press.

Frank, Andre Gunder, and Barry K. Gills. 1992. "The Five Thousand Year World System: An Interdisciplinary Introduction." *Humboldt Journal of Social Relations* 18:1:1–79.

———, eds. 1993. *The World System: Five Hundred Years or Five Thousand?* London: Routledge.

Fukuyama, Francis. 1992. *The End of History and the Last Man.* New York: Free Press.

Gereffi, Gary, and Miguel Korzeniewicz, eds. 1994. *Commodity Chains and Global Capitalism.* Westport, Conn.: Praeger.

Goldfrank, Walter L. 1987. "Socialism or Barbarism? The Long-run Fate of the Capitalist World-Economy." Pp. 85–82 in *America's Changing Role in the World-System,* edited by Terry Boswell and Albert Bergesen. New York: Praeger.

———. 1994. "Fresh Demand: The Consumption of Chilean Produce in the USA." Pp. 267–280 in *Commodity Chains and Global Capitalism,* edited by Gary Gereffi and Miguel Korzeniewicz. Westport, Conn.: Greenwood.

———. 1999. "Beyond Cycles of Hegemony: Economic, Social, and Military Factors." Pp. 66–77 in *Future of Global Conflict,* edited by Volker Bornschier and Christopher Chase-Dunn. London: Sage.

Goldfrank, Walter L., and Sergio Gomez. 1991. "World Market and Agrarian Transformation: The Case of Neoliberal Chile." *International Journal of Food and Agriculture* I:143–150.

Goldfrank, Walter, David Goodman, and Andrew Szasz, eds. 1999. *Ecology and the World-System.* Greenwich, Conn.: Greenwood.

Goldfrank, Walter L., Roberto P. Korzeniewicz, and Miguel Korzeniewicz. 1995. "Vines and Wines in the World-Economy." Pp. 113–138 in *Food and Agrarian Orders in the World-Economy,* edited by Philip McMichael. Westport, Conn.: Greenwood.

Goldfrank, Walter L., and Cynthia Maki. 1995. "Lessons from the Gulf Wars: Hegemonic Decline and Semiperipheral Turbulence." Pp. 57–70 in *A New World Order? Global Transformation in the Late Twentieth Century,* edited by David A. Smith and József Böröcz. Westport, Conn.: Greenwood.

Goldstein, Joshua. 1988. *Long Cycles: Prosperity and War in the Modern Age.* New Haven, Conn.: Yale University Press.

Hall, Thomas D. 1986. "Incorporation in the World-System: Toward a Critique." *American Sociological Review* 51:3(June):390–402.

———. 1987. "Native Americans and Incorporation: Patterns and Problems." *American Indian Culture and Research Journal* 11:2:1–30.

———. 1989a. *Social Change in the Southwest, 1350–1880.* Lawrence: University Press of Kansas.

———. 1989b. "Is Historical Sociology of Peripheral Regions Peripheral?" Pp. 349–372 in *Studies of Development and Change in the Modern World,* edited by Michael T. Martin and Terry R. Kandal. New York: Oxford University Press.

———. 1991. "The Role of Nomads in Core/Periphery Relations." Pp. 212–239 in *Core/Periphery Relations in Precapitalist Worlds,* edited by Christopher Chase-Dunn and Thomas D. Hall. Boulder, Colo.: Westview.

———. 1994. "The Case for a World-Systems Approach to Civilizations: A View from the 'Transformationist' Camp." *Comparative Civilizations Review* 30(Spring):30–49.

———. 1996. "The World-System Perspective: A Small Sample from a Large Universe." *Sociological Inquiry* 66:4(Fall):440–454.

———. 1997. "Finding the Global in the Local." Pp. 95–107 in *Economic Analysis beyond the Local System,* edited by Richard Blanton, Peter Peregrine, Deborah Winslow, and Thomas D. Hall. Lanham, Md.: University Press of America.

———. 1998. "The Effects of Incorporation into World-Systems on Ethnic Processes: Lessons from the Ancient World for the Contemporary World." *International Political Science Review* 19:3(July):251–267.

———. 1999. "World-Systems and Evolution: An Appraisal." Pp. 1–25 in *World-Systems Theory in Practice: Leadership, Production, and Exchange,* edited by P. Nick Kardulias. Lanham, Md.: Rowman & Littlefield.

Hall, Thomas D., and Christopher Chase-Dunn. 1993. "The World-Systems Perspective and Archaeology: Forward into the Past." *Journal of Archaeological Research* 1:2:121–143.

———. 1994. "Forward into the Past: World-Systems before 1500." *Sociological Forum* 9:2:295–306.

Harris, Betty J. 1990. "Ethnicity and Gender in the Global Periphery: A Comparison of Basotho and Navajo Women." *American Indian Culture and Research Journal* 14:4:15–38.

Himmel, Kelly D. 1999. *The Conquest of the Karankawas and the Tonkawas: A Study in Social Change, 1821–1859.* College Station: Texas A&M University Press.

Hopkins, Terence, and Immanuel Wallerstein. 1979. "Cyclical Rhythms and Secular Trends of the Capitalist World-Economy." *Review* 2:3(Winter):483–500.

Hugill, Peter J. 1993. *World Trade since 1431: Geography, Technology, and Capitalism.* Baltimore, Md.: Johns Hopkins University Press.

Kandal, Terry R. 1990–91. "Revolution, Racism and Sexism: Challenges for World-Systems Analysis." *Studies in Comparative International Development* 25:4(Winter):86–102.

Kardulias, P. Nick. 1990. "Fur Production as a Specialized Activity in a World System: Indians in the North American Fur Trade." *American Indian Culture and Research Journal* 14:1:25–60.

———, ed. 1999. *World-Systems Theory in Practice: Leadership, Production, and Exchange.* Lanham, Md.: Rowman & Littlefield.

Kasaba, Resat, ed. 1991. *Cities in the World System.* New York: Greenwood.

Kiser, Edgar, and Kriss A. Drass. 1987. "Changes in the Core of the World-System and the Production of Utopian Literature in Great Britain and the United States, 1883–1975." *American Sociological Review* 52:2(April):286–293.

Knox, Paul, and John Agnew. 1994. *The Geography of the World Economy.* 2d ed. London: Arnold.

Knox, Paul, and Peter J. Taylor, eds. 1995. *World Cities in a World-System.* New York: Cambridge University Press.

Kohl, Philip L. 1987. "The Use and Abuse of World Systems Theory: The Case of the Pristine West Asian State." *Advances in Archaeological Method and Theory* 11:1–35.

Korzeniewicz, Roberto P. 1996. "Challenging World Inequalities: Beyond Uneven Development?" *Political Power and Social Theory* 10:321–331.

Korzeniewicz, Roberto P., and Kimberley Awbrey. 1992. "Democratic Transitions and the Semiperiphery of the World-Economy." *Sociological Forum* 7:4:609–640.

Korzeniewicz, Roberto P., and Timothy Moran. 1997. "World-Economic Trends in the Distribution of Income, 1965–1992." *American Journal of Sociology* 102:4(Jan.):1000–1039.

Korzeniewicz, Roberto P., and William C. Smith, eds. 1996. *Latin America in the World-Economy.* Westport, Conn.: Greenwood and Praeger.

Kowalewski, David. 1991. "Periphery Revolutions in World-System Perspective, 1821–1985." *Comparative Political Studies* 24:1(April):76–99.

Kuhn, Thomas S. 1970. *The Structure of Scientific Revolutions.* 2d ed. Chicago: University of Chicago Press.

———. 1977. "Second Thoughts on Paradigms." Pp. 293–319 in T. S. Kuhn, *The Essential Tension: Selected Studies in Scientific Traditions and Change.* Chicago: University of Chicago Press.

Markoff, John. 1999. "From Center to Periphery and Back Again: Reflections on the Geography of Democratic Innovation." Pp. 229–246 in *Extending Citizenship, Reconfiguring States,* edited by Michael Hanagan and Charles Tilly. Lanham, Md.: Rowman & Littlefield.

Martin, Michael T., and Terry R. Kandal, eds. 1989. *Studies of Development and Change in the Modern World.* New York: Oxford University Press.

Martin, William G. 1994. "The World-Systems Perspective in Perspective: Assessing the Attempt to Move beyond Nineteenth Century Eurocentric Conceptions." *Review* 17:2(Spring):145–185.

Mathien, Frances Joan, and Randall McGuire, eds. 1986. *Ripples in the Chichimec Sea: Consideration of Southwestern-Mesoamerican Interactions.* Carbondale: Southern Illinois University Press.

McMichael, Philip. 1984. *Settlers and the Agrarian Question: Foundations of Capitalism in Colonial Australia.* Cambridge: Cambridge University Press.

———. 1996. *Development and Social Change: A Global Perspective.* Thousand Oaks, Calif.: Pine Forge.

McNeill, William H. 1995. "*The Rise of the West* after Twenty-Five Years." Pp. 289–306 in *Civilizations and World-Systems: Two Approaches to the Study of World-Historical Change,* edited by Stephen K. Sanderson. Walnut Creek, Calif.: Altamira.

Melko, Matthew. 1994. "World Systems Theory: A Faustian Delusion, I & II." *Comparative Civilizations Review* 30(Spring)8–21.

———. 1995. "The Nature of Civilizations." Pp. 11–31 in *Civilizations and World-Systems: Two Approaches to the Study of World-Historical Change,* edited by Stephen K. Sanderson. Walnut Creek, Calif.: Altamira.

Meyer, Melissa L. 1990. "Signatures and Thumbprints: Ethnicity among the White Earth Anishinaabeg, 1889–1920." *Social Science History* 14:3(Fall):305–345.

———. 1991. "'We Can Not Get a Living as We Used To': Dispossession and the White Earth Anishinaabeg, 1889–1920." *American Historical Review* 96:2:368–394.

———. 1994. *The White Earth Tragedy: Ethnicity and Dispossession at a Minnesota Anishinaabe Reservation, 1889–1920.* Lincoln: University of Nebraska Press.

Modelski, George. 1987. *Long Cycles in World Politics.* London: Macmillan.

Modelski, George, and William R. Thompson. 1988. *Seapower and Global Politics, 1494–1993.* London: Macmillan.

———. 1996. *Leading Sectors and World Powers: The Coevolution of Global Politics and Economics.* Columbia: University of South Carolina Press.

Morrissey, Marietta. 1989. *Slave Women in the New World: Gender Stratification in the Caribbean.* Lawrence: University Press of Kansas.

Pailes, Richard A., and Joseph W. Whitecotton. 1975. "Greater Southwest and Mesoamerican World-System." Paper presented at the Southwestern Anthropological Association meeting, Santa Fe, N.M., March.

———. 1979. "The Greater Southwest and the Mesoamerican 'World' System: An Exploratory Model of Frontier Relationships." Pp. 105–121 in *The Frontier: Comparative Studies,* vol. 2, edited by William W. Savage, Jr., and Stephen I. Thompson. Norman: University of Oklahoma Press.

Peregrine, Peter N. 1992. *Mississippian Evolution: A World-System Perspective.* In *Monographs in World Archaeology No. 9.* Madison, Wisc.: Prehistory Press.

————. 1995. "Networks of Power: The Mississippian World-System." Pp. 132–143 in *Native American Interactions,* edited by M. Nassaney and K. Sassaman. Knoxville: University of Tennessee Press.

Peregrine, Peter, and Gary Feinman, eds. 1996. *Pre-Columbian World-Systems.* Madison, Wisc.: Prehistory Press.

Rasler, Karen, and William R. Thompson. 1994. *The Great Powers and Global Struggle, 1490–1990.* Lexington: University of South Carolina Press.

Robinson, William I. 1996. *Promoting Polyarchy: Globalization, U.S. Intervention, and Hegemony.* New York: Cambridge University Press.

Sanderson, Stephen K. 1994. "Expanding World Commercialization: The Link between World-Systems and Civilizations." *Comparative Civilizations Review* 30(Spring):91–103.

————, ed. 1995. *Civilizations and World-Systems: Two Approaches to the Study of World-Historical Change.* Walnut Creek, Calif.: Altamira.

Santley, Robert S., and Rani T. Alexander. 1992. "The Political Economy of Core-Periphery Systems." Pp. 23–59 in *Resources, Power, and Interregional Interaction,* edited by Edward M. Schortman and Patricia Urban. New York: Plenum.

Schneider, Jane. 1977. "Was There a Pre-Capitalist World-System?" *Peasant Studies* 6:1(Jan.):20–29. (Reprinted in *Core/Periphery Relations in Precapitalist Worlds,* edited by C. Chase-Dunn and T. D. Hall, pp. 45–66. Boulder, Colo.: Westview, 1991.)

Schortman, Edward M., and Patricia A. Urban. 1994a. "Living on the Edge: Core/Periphery Relations in Ancient Southeast Mesoamerica." *Current Anthropology* 35:4(Aug.–Oct.):401–430.

————. 1994b. "Reply." *Current Anthropology* 35:4(Aug.–Oct.):421–426.

Shannon, Thomas R. 1996. *An Introduction to the World-System Perspective.* 2d ed. Boulder, Colo.: Westview.

Silver, Beverly. 1995. "World Scale Patterns of Labor-Capital Conflict: Labor Unrest, Long Waves and Cycles of Hegemony." *Review* 18:1(Winter):155–192.

Smith, David A. 1996. *Third World Cities in Global Perspective: The Political Economy of Uneven Urbanization.* Boulder, Colo.: Westview.

Smith, David A., and József Böröcz. 1995. *A New World Order? Global Transformation in the Late Twentieth Century.* Westport, Conn.: Greenwood.

Smith, Joan, Jane Collins, Terence K. Hopkins, and Akbar Muhammad, eds. 1988. *Racism, Sexism and the World-System.* New York: Greenwood.

So, Alvin. 1986. *The South China Silk District: Local Historical Transformation and World-System Theory.* New York: State University of New York Press.

————. 1990. *Social Change and Development: Modernization, Dependency, and World-System Theory.* Newbury Park, Calif.: Sage.

So, Alvin, and Stephen Chiu. 1995. *East Asia and the World-Economy.* Newbury Park, Calif.: Sage.

Sokolovsky, Joan. 1985. "Logic, Space, and Time: The Boundaries of the Capitalist World-Economy." Pp. 41–52 in *Urbanization in the World-Economy,* edited by Michael Timberlake. New York: Academic Press.

Straussfogel, Debra. 1997a. "World-Systems Theory: Toward a Heuristic and Conceptual Tool." *Economic Geography* 73:1(Jan.):118–130.

———. 1997b. "A Systems Perspective on World-Systems Theory." *Journal of Geography* 96:2(March/April):119–126.

———. 1997c. "Redefining Development as Humane and Sustainable." *Annals of the Association of American Geographers* 87:2(June):280–305.

———. 1998. "How Many World-Systems? A Contribution to the Continuationist/ Transformationist Debate." *Review* 21:1:1–28.

Suter, Christian. 1992. *Debt Cycles in the World-Economy: Foreign Loans, Financial Crises, and Debt Settlements, 1820–1990*. Boulder, Colo.: Westview.

Taylor, Peter J., ed. 1993. *Political Geography of the Twentieth Century: A Global Analysis*. London: Belhaven.

———. 1996. *The Way the Modern World Works: World Hegemony to World Impasse*. New York: Wiley.

———. 1997. "Modernities and Movements: Antisystemic Reactions to World Hegemony." *Review* 20:2(Winter):1–17.

Teggart, Frederick J. 1918. *The Processes of History*. New Haven, Conn.: Yale University Press.

———. 1925. *Theory of History*. New Haven, Conn.: Yale University Press. (Both Teggart works cited here have been reprinted together twice: University of California Press, 1942; Peter Smith, 1972.)

———.1939. *Rome and China: A Study of Correlations in Historical Events*. Berkeley: University of California Press.

Terlouw, Cornelis P. 1992. *The Regional Geography of the World-System: External Arena, Periphery, Semiperiphery, Core*. Utrecht, Netherlands Geographical Studies.

Thompson, William R., ed. 1983a. *Contending Approaches to World System Analysis*. Beverly Hills: Sage.

———. 1983b. "Introduction: World System Analysis with and without the Hyphen." Pp. 7–24 in *Contending Approaches to World System Analysis*, edited by William R. Thompson. Beverly Hills: Sage.

———. 1988. *On Global War: Historical-Structural Approaches to World Politics*. Columbia: University of South Carolina Press.

Timberlake, Michael, ed. 1985. *Urbanization in the World-Economy*. New York: Academic Press.

Tomich, Dale W. 1990. *Slavery in the Circuit of Sugar: Martinique and the World Economy, 1830–1848*. Baltimore, Md.: Johns Hopkins University Press.

Troillot, Michel-Rolph. 1988. *Peasants and Capital: Domenica in the World Economy*. Baltimore, Md.: Johns Hopkins University Press.

Wagar, W. Warren. 1999. *A Short History of the Future*. 3d ed. Chicago: University of Chicago Press.

Waller, David V. 1998. "Anticipating China's Future: Geopolitical and World System Considerations." Unpublished manuscript, University of Texas, Arlington, summer 1998.

Wallerstein, Immanuel. 1974a. "The Rise and Future Demise of the World Capitalist System: Concepts for Comparative Analysis." *Comparative Studies in Society and History* 16:4(Sept.):387–415. (Also in Wallerstein 1979, chap.1.)

———. 1974b. *The Modern World-System: Capitalist Agriculture and the Origins of European World-Economy in the Sixteenth Century*. New York: Academic Press.

———. 1979. *The Capitalist World-Economy*. Cambridge: Cambridge University Press.

———. 1980. *The Modern World-System II: Mercantilism and the Consolidation of the European World-Economy, 1600–1750.* New York: Academic Press.

———. 1983. "An Agenda for World-Systems Analysis." Pp. 299-308 in *Contending Approaches to World System Analysis,* edited by William R. Thompson. Beverly Hills: Sage.

———. 1984. *The Politics of the World-Economy: The States, the Movements, and the Civilizations.* Cambridge: Cambridge University Press.

———. 1989. *The Modern World-System III: The Second Era of Great Expansion of the Capitalist World-Economy, 1730–1840s.* New York: Academic Press.

———. 1991. *Unthinking Social Science: The Limits of Nineteenth-Century Paradigms.* Oxford: Polity.

———. 1992. "The West, Capitalism, and the Modern World-System." *Review* 15:4(Fall):561–619.

———. 1993a. "World System vs. World-Systems." Pp. 291–296 in *The World System: Five Hundred Years or Five Thousand?* edited by Andre Gunder Frank and Barry K. Gills. London: Routledge.

———. 1993b. "The Timespace of World-Systems Analysis: A Philosophical Essay." *Historical Geography* 23:172:5–22.

———. 1995a. "Hold the Tiller Firm: On Method and the Unit of Analysis." Pp. 225–233 in *Civilizations and World-Systems: Two Approaches to the Study of World-Historical Change,* edited by Stephen K. Sanderson. Walnut Creek, Calif.: Altamira.

———. 1995b. "The Modern World-System and Evolution." *Journal of World-Systems Research* 1:19 (electronic journal; Web: http://csf.colorado.edu/wsystems/jwsr.html; ftp and gopher: csf.colorado.edu/wsystems/journals/).

———. 1997. "Social Science and the Quest for a Just Society." *American Journal of Sociology* 102:5(March):1241–1257.

———. 1998. "The Rise and Future Demise of World-Systems Analysis." *Review* 21:1(Winter):103–112.

Ward, Kathryn B. 1984. *Women in the World-System: Its Impact on Status and Fertility.* New York: Praeger.

———, ed. 1990. *Women Workers and Global Restructuring.* Ithaca, N.Y.: ILR.

———. 1993. "Reconceptualizing World-System Theory to Include Women." Pp. 43–68 in *Theory on Gender/Feminism on Theory,* edited by Paula England. New York: Aldine.

Wilkinson, David. 1995. "Civilizations *Are* World Systems!" Pp. 234–246 in *Civilizations and World-Systems: Two Approaches to the Study of World-Historical Change,* edited by Stephen K. Sanderson. Walnut Creek, Calif.: Altamira.

Wolf, Eric R. 1990. "Distinguished Lecture: Facing Power—Old Insights, New Questions." *American Anthropologist* 92:3(Sept.):586–596.

2

Recent Research on World-Systems

Peter Grimes

Today the terms *world-economy* and *world-market* are routine, appearing in the media sound bites of politicians, executives, and unemployed workers alike. But in the background, beyond the reach of media interest, social scientists working in the area are trying to better understand the history and evolution of the *whole system,* as well as how local and national regions have been integrated into it. This current research has required broadening our perspective to include ever larger periods of historical time and geographical space. For example, some recent research has compared the modern Europe-centered world-system of the last five hundred years with earlier, smaller intersocietal networks that have existed for millennia (Chase-Dunn and Hall 1997). Other work attempts to use the knowledge of cycles and trends that has grown out of world-systems research to anticipate events likely in the future with a precision impossible before the advent of the theory. This is still a new field, and much remains to be done.

This survey of current research will focus on three main topics:

- how different definitions of the world-system concept imply different ways of breaking down history into discrete periods (e.g., when did the *current* system begin, and what are the relevant criteria for separating it from *other systems* in either time or space?);
- how the patterned changes introduced by trends and cycles structure the reproduction of the modern world-system; and
- how the global hierarchy of wealth and power reproduces itself by the constraints it imposes on the range of policy options for most nations.

World-systems theory brings a new and unique perspective to the *big* questions of history, such as whether it reveals any repeating patterns or sense of "direction." What makes the world-system perspective distinctive is that it is:

- geographically *holistic* by studying the development of all human societies on the planet at any given time;
- aggressively *transdisciplinary* by considering all relevant evidence, whether it comes from anthropology, geology, economics, or climatology; and
- emphatically *structuralist* in its focus on how the habitual routines followed by individuals within societies both reflect and perpetuate social power hierarchies across many generations spanning centuries of time, thereby faithfully (but unintentionally) reproducing complex social structures intact over long periods of history.

WHAT DEFINES A WORLD-SYSTEM

The way we define a world-system reflects both our understanding of history and the questions we try to answer. For example, how connected must peoples be before we call them a "system"? How does a "world"-system differ from a "society"? Have there been several such systems in history, and, if so, what distinguishes one from another in time or space? While these questions may at first seem abstract and scholastic, our answers define and channel our understanding of history and thereby directly affect our understanding of the present. How we respond to the challenges of our times reflects our analysis of history. Some of the most important issues are surveyed here.

The modern world-system is understood as a set of nested and overlapping interaction networks that link all units of social analysis—from individuals and households up to international regions and global structures. It is all of the economic, political, social, and cultural relations among the people of the Earth. Thus, the world-system is *not* just "international relations" or the "world market." It is the *whole* interactive system, where the whole is greater than the sum of the parts (Chase-Dunn and Hall 1997). Within this system, all boundaries and identities (individuals, ethnic groups, nations) are socially defined and reproduced. Trade in bulk goods is spatially restricted by transport costs to a small region, political/military interactions occur over a larger territory, and trade in prestige goods is the largest important interaction network. Interaction inside the system is routinized so that the connected actors come to depend and form expectations based on the habitual repetition of their connections.

The most important structure of the current world-system is a power hierarchy between core and periphery in which powerful and wealthy "core" societies dominate and exploit weak and poor "peripheral" societies. The peripheral countries, instead of developing along the same paths taken by core countries in earlier periods (the assumption of "modernization" theories), are structurally *constrained* by trade relations and geopolitical pressures to reproduce their weak and impoverished status, while the countries in the core are *enabled* by those same coercive structures to benefit from the surplus received from the periphery. Put simply, it is *the whole system* that develops, not just the national societies that are its parts.

In this moving context core and peripheral countries generally retain their positions relative to one another over time, although there are individual cases of upward and downward mobility in the core/periphery hierarchy. Between the core and the periphery is an intermediate layer of countries referred to as the *semiperiphery*. These countries combine features of both the core and the periphery and are located in intermediate or mediating positions in larger interaction networks.

RESEARCH ON WORLD-SYSTEM BOUNDARIES

The use of data for spatially bounding world-system interactions is still in its infancy. The best work to date has been done by David Wilkinson (1987, 1991, 1992, 1993), who studies the boundaries of political/military interaction networks (PMNs). Wilkinson conceptualizes "world systems/civilizations" primarily in terms of military alliances and conflicts among a group of states in a region. Using this type of interconnection, Wilkinson produced a spatiotemporal map of the expansion of "Central Civilization," the world-system that was formed by the merging of the Mesopotamian and Egyptian world-systems in the fifteenth century B.C.E. (see Chase-Dunn and Grimes 1995; Chase-Dunn and Hall 1997; Sanderson 1999).

SURPLUS ACQUISITION

Another basic conceptual issue yet to be resolved concerns the *temporal* bounding of world-systems and the question of similarities or qualitative differences in systemic logic. The debate around this issue revolves around the *logic of accumulation*.[1] Immanuel Wallerstein (1974, 1984a, 1984b) asserts that the modern world-system is unique because it is capitalist. He thought that the transition from feudalism to capitalism happened for the first and only time in sixteenth-century Europe. Subsequent transitions in the mode of accumulation in the rest of the world have resulted from the global expansion of and conquest by the formerly regional Europe-centered world-system. He argues that it was only in the post-sixteenth-century capitalist world-economy that powerful actors focus primarily on the goal of "ceaseless accumulation." Samir Amin (1993) also agrees that a transformation in systemic logic happened in the way Wallerstein describes, while at the same time acknowledging that precapitalist world-systems also had important core/periphery dimensions and processes of uneven development in which old core areas were superseded by new ones.

However, Frank and Gills (1993) and Ekholm and Friedman (1982) argue that capitalism has been an important aspect of the Eurasian world-system for millennia.

Documentary evidence confirms the existence of commodified forms of wealth and property and exchange in the early states and empires of the Near East. Ekholm and Friedman (1982) assert the existence of a "capital-imperialist" mode of accumulation in which core states and wealthy families exploited peasants and peripheral regions by means of a combination of capitalist accumulation and state-organized extraction and conquest. They argue that the core regions of both ancient *and* modern world-systems oscillate back and forth between state-based and capitalist accumulation.

Frank and Gills likewise argue that there has been a continuity in systemic logic in a single five thousand–year–old world-system that emerged out of the Near East and eventually expanded to the whole globe. Frank (1993) contends that there was no transition from feudalism to capitalism in Europe in the sixteenth century, because the world-system has had the same capital-imperialist mode of accumulation for five millennia. Frank and Gills also contend that China was the core of the Afroeurasian world system until at least the eighteenth century C.E. Europe remained a peripheral backwater exporting bullion to China in exchange for silk and porcelain.

Both sides in this debate agree on the facts: capital accumulation has appeared throughout history, and also Europe used imperialist expansion/conquest to leap from a peripheral backwater to become the contemporary source of core accumulation during the sixteenth century. Their differences lie in their respective estimates of the degree of influence that capitalist logic had on state administrators before that time.

Chase-Dunn and Hall (1997) have expanded the scope of world-system research in yet another direction to include small stateless and classless systems. This allows them to study earlier examples of qualitative transformations in systemic logic from kin-based to state-based modes of accumulation. They also note the existence and growing importance of commodified wealth, land, exchange and labor in the precapitalist world-systems, but they agree with Wallerstein and Amin that the Europe-centered subregion of the Afroeurasion world-system was the first region to experience a *predominantly* capitalist regional system. Many commodified institutional forms developed inside the tributary empires, especially Rome and Han China. The Sung Empire in particular nearly underwent a transformation to capitalism in the tenth century C.E. But the only states to be *controlled* by capitalists before the European transformation in the seventeenth century were semiperipheral capitalist *city-states* such as the Phoenician cities, Venice, Genoa, and Malacca. These operated in the interstices between the tributary states and empires, and though they were agents of commodification, they existed within larger systems in which the logic of state-based coercion remained dominant. The first capitalist *nation-state* was the Dutch Republic in the seventeenth century. This coming to state power by capitalists in an emerging core region signaled the triumph of regional capitalism in the European subsystem.

One refreshing new approach to these issues is offered by Debra Straussfogel (1998; see also chap. 9). She recasts the entire debate in terms of energy and thermodynamics. When societies are viewed as systems that, like organisms, survive only by virtue of

- acquiring energy from the environment and then;
- channeling that energy in a controlled way so as to maintain their structure against their entropic tendency to fall apart, and, finally;
- expelling the energy as pollution, garbage, and heat;

then they fit into the definition created by Prigogine (e.g., 1984, 1996) of a *dissipative structure*. The utility of this definition lies in that much work has already been done on how such structures work (see Straussfogel 1998 and chap. 9; Holland 1995; Waldrop 1994; Gleick 1991). This research shows that when a variable affecting a dissipative structure changes enough, the structure either reorganizes itself or collapses. This crisis point is called a "bifurcation." When collapse is avoided, the reorganization process takes pieces of the earlier structure and changes their functions and relative importance so as to allow the system itself to continue, although now in a new form. An example is the stress caused by a rising population, which ultimately compels a society to expand and become more hierarchical or else collapse into fragments (Tainter 1988; Sanderson 1999; Chase-Dunn and Hall 1997). Thus can be understood the various transitions from kinship-based hunter-gatherers into kinship-based horticultural chiefdoms; followed in turn by state-based agrarian empires.

When recast this way, 1600 can be thought of as the approximate time of a system bifurcation, prior to which capitalist accumulation existed but was a less important piece of the system structure than it was after. But if one rejects that there was a bifurcation at all, the problem remains.

This point is also consistent with the observation made originally by Elton (1927) that the food chain is arranged as a pyramid of trophic levels (plants, herbivores, carnivores), whose numbers shrink with each level out of thermodynamic necessity: each level can access only about 10 percent of the energy available to the level beneath it (Colinvaux 1977). The social pyramid of classes is likewise constrained by thermodynamic necessity, so expansion of the numbers at the top (and thereby the achievement of social complexity) can only be attained at the expense of increased work by all (especially those below). Population pressure at all levels must eventually either be met with revolutionary advances in energy acquisition and/or efficiency involving social reorganization or else by a reduction in population (collapse).

HIERARCHY

One of the main structures of all state-based world-systems is a hierarchy linking core, peripheral, and semiperipheral societies. In the modern world-system this structure has been reproduced over centuries despite the upward and downward mobility of a few national societies.

Before the modern capitalist world-system, in the era of world-empires built around the tributary mode of accumulation, peripheral societies gave up a portion of their surplus to the core in return for nominal independence or were absorbed altogether (in which event they gave up an even larger portion of their surplus to the imperial core). The maintenance of peripheral status during those times relied exclusively on the credible threat of military force by the core. Today, the reproduction of the international power hierarchy is achieved more subtly through market mechanisms, and force is used only when the market "rules" (which act to sustain the dominance of the core) are challenged by open insurrection. The hypothesis of "unequal exchange" (Emmanuel 1972; Mandel 1975; Raffer 1987) contends that the central mechanism by which the global market acts to gather together the global surplus and channel it to the core is through price inequality, in which the political and military suppression of wages in the periphery allows the products of peripheral labor to be much cheaper than those of the core. Put simply, an hour of labor in the periphery costs capital only a fraction of its costs in the core, so that a commodity produced there is much cheaper than the same commodity produced in the core. When core and periphery come together to exchange products in the world market, the exchange results in a net transfer of value from the periphery to the core. Hence, the market masks a process of exploitation, a process backed up by the military power of the states in the periphery and, behind them, the military power of the core.

The chronic impoverishment of the periphery prevents the typical peripheral state from being able to finance programs of public welfare or infrastructural improvement (even if they wished to), so its popular legitimacy is low. It is thus always vulnerable to coups or popular insurrection. Yet despite these barriers, some states have managed gradually to improve their infrastructure and to combine these improvements with policies that encourage key industries that seem the most promising in the world market.

Occasionally, such policies pay off in upward mobility. The most recent examples are the Asian "tigers" (i.e., Hong Kong, Taiwan, Singapore, and South Korea), while Japan has achieved an upward trajectory since 1880. Japanese mobility was enabled by protectionist state policies that allowed local industries to remain profitable and grow just as in the United States during the same period (Baran 1957), but the "tigers" grew by hitching their fortunes to becoming recipients of the capital flight from the core in the 1970s and 1980s. They managed this by suppressing their own labor forces as fiercely as in the rest of the periphery but also by guiding their state revenue (itself buoyed by loans and grants from the United States and World Bank) into financing an advanced infrastructure that could power and support the sophisticated factories built by capital fleeing the militant unions of the core. Unfortunately, this strategy relies on an ever-expanding core market, the periodic saturation of which will necessarily lead to recurring "meltdowns," bankruptcies, and political instability (e.g., Thailand and Korea in 1997, Indonesia and Russia in 1998) (see chap. 14).

The only sure way to *permanent* upward mobility in the modern world-system is the development of a well-paid workforce that can consume its own products. This is extremely difficult for even the most efficient and well-intentioned state today.

Perhaps the most spectacular case of upward mobility has been the United States, a region that went from being peripheral to semiperipheral and then core (1880) and finally hegemony within the core (1945). This case highlights an important distinction among former European colonies that explains much about their differential paths of mobility and also underscores the exceptional obstacles to mobility today.

The distinction is between colonies of European "acquisition" versus those of European "settlement" (Amin 1976). The former were areas that were already densely populated by a preexisting tributary state (Egypt, Syria, Persia, India, Inca, Aztec, Indonesia, China). In these areas, Europeans used a combination of diplomacy and military force to redirect the surplus generated from the existing tributary class structure away from the local elite toward the conquering European country (typically England, although initially Portugal, Spain, and the Netherlands). This left the original class structure undisturbed but raised the level of exploitation for all. Braudel, for example, reports that the income of the Indian peasantry fell by half between 1700 and 1900, as the resulting surplus was expropriated by England and thereby fueled its own dynamic growth during that period (Braudel 1984, vol. 3). During the nineteenth century, the British systematically sought to increase both labor discipline and thereby also surplus extracted by monetizing the labor forces. By imposing taxes that had to be paid in cash, the colonial authorities compelled workers to move out of self-supporting village agriculture and into plantations that paid wages (Beckford 1972). This created a type of labor force that was "semiproletarianized"— part wage labor, part peasant, in which peasant production continued (often run by women) as a subsidy to the wages received. This subsidy was, on a social level, a transfer of surplus from the tributary landlord-peasant relation to the international capitalist relation, while on an ecological level it was a "free" subsidy from nature to the construction of industrial capitalism in the core.

One of the results was the spread of plantation economies throughout the tropics, with the plantations occupying the best alluvial land surrounded by a peasantry on marginal land. The employment of swidden technology by the marginal peasantry is a major cause of contemporary deforestation.

In contrast to the colonies of acquisition were the colonies of European settlement: the United States, Canada, northern Mexico, Argentina, Australia, New Zealand, South Africa, and (much later) Israel. In general (Israel aside), the aboriginal peoples in these areas were either nomadic hunters or horticulturists with low population density and little means of resistance. Furthermore, their climates were temperate and thus conducive to European farming technology. The immigrants were easily able to diffuse out into the countryside, indirectly raising urban wages (by creating a viable and low-cost alternative to wage labor). This had the effect of creating a *mass consumer market* made up of peasant farmers selling food for profit and a relatively high-wage urban labor force. In the United States, the legacy of this process of a growing rural consumer market is reflected in the names of railroads established during that era: "Baltimore and *Ohio*," "Chesapeake and *Ohio*," as eastern cities raced each other to establish infrastructural access to the expanding consumer base of farmers in the Ohio river valley.

These unusual conditions of cheap land hiking up wages explains why the colonies of European settlement were able to support local industries that would later emerge as serious competitors on the world market, allowing these ex-colonies to move from peripheral to semiperipheral and even (for the United States) hegemonic core status.

However, in the southern United States and throughout Latin America, the interests of the plantation economies were in opposition to those of the urban industries—most acutely in tariff policy. The former wanted free trade while the latter wanted protectionism, a conflict that led to series of civil wars, of which the one in the United States was the last and the *only* one where the industrialists won. In the rest of the Americas, the victory of the planters condemned the entire region to peripheral status, with the sporadic exceptions of Argentina and Brazil (Frank 1978). A similar conflict separates the Boers and English in South Africa. A structural legacy of plantation culture and class structure is worship of the military interwoven with and justified by religious ideology, strict patriarchy, and caste stratification (Beckford 1972).

EMPIRICAL MEASURES OF WORLD-SYSTEM POSITION

Some excellent work has tried to measure empirically the placement of states in the core/periphery hierarchy. It started yielding publications in the 1980s and continues today.

The first work was done by David Snyder and Edward Kick (1979), followed by Roger Nemeth and David Smith (1985), and expanded upon by Smith and White (1992). Each of these initial forays relied on computer programs using network analysis. Network analysis originated in the research done in the 1930s and 1940s on community power within cities (Grimes 1996). The idea was to ask who contacts whom and how often when a problem arose within a city. When these contacts were carefully noted, the most important power brokers would be revealed by their frequency and breadth of contacts, and a chart of the hierarchy of power within the community generated. Computer programs were created to produce such charts, using as their input a spreadsheet-style table in which each cell was marked with a one when there was a contact or a zero when no contact was made. These world-system workers creatively changed the units of analysis from individuals to states and filled in the spreadsheet inputs not with telephone calls but measures such as diplomats exchanged, trade volume, and military interventions. The graphic results showed a rat's nest of ties, with the most connected states as a "core," the least as "periphery," and intermediate groups as presumed "semiperipheries" or, in some versions, "semicores" or even remoter outer "peripheries" (typically the countries of Africa, which were documented by Terlouw [1992] as having dropped far behind all others in the 1980s). The results of these measures necessarily reflected the choice of input variables, leaving them vulnerable to the critics of those results (Grimes 1996). Further, the limited availability of relevant data has limited the application of the method to only the most recent decades. However, the method is itself quite promising and worthy of further refinement when and as new data become ready.

Another approach during the same time was taken by Arrighi and Drangel (1986), who looked at changes over several decades in the distributions of GNP per capita of large numbers of states. By avoiding the need for data only recently available, they could go much further back in time. This kind of time depth is quite important for analyzing structural change in the world-system. They found a rather stable trimodal (again presumably "core-semiperiphery-periphery") distribution of national GNPs. But the use of GNP per capita by itself led to strange conclusions, such as the inclusion of Libya in the core because of its high revenues due to oil resources. A more recent study using the same method by Korzeniewicz and Martin (1994) uses data on more countries and more time points to confirm the findings of Arrighi and Drangel (1986).

More recently, Terlouw (1992) used six variables per nation to create an index of core/periphery position for the mid-1980s. His index includes two kinds of economic measures (production and trade), political measures (diplomats and diplomatic missions), and military power. Once again, the sophistication of his data constrains his application to just the past two decades.

After Terlouw, Van Rossen (1996) proposed that GNP alone was sufficient to locate world-system position. In arguing that economic size was the sole determinant of power, Van Rossen was able to simplify the method potentially to allow for a very broad time period to be included.

My own work (Grimes 1996) has sought a compromise between the virtues of temporal longevity and maximal data. The former requirement constricts the input data to only those items that are available for long time periods, while the latter compels one to be creative about the type of data used. My particular compromise was to use GDP, population, value of trade, and the identity of trading partners to compose an index of economic power, so that the available data could generate a measure for many countries stretching back to 1800. While yet incomplete, my analysis of my data seems to indicate that my measure correlates strongly with GNP, thereby affirming the results of Van Rossen.

As of this writing, Kentor (1998) has proposed yet another measure that compromises between data longevity and availability. He has added military power and foreign investment to my data set, and thereby constructed a more delicately sensitive gauge than mine, but it is again limited by the investment information from

Table 2.1   Percent Share of Global Income Going to Richest and Poorest

Year	Share of Richest 20%	Share of Poorest 20%	Ratio Rich/Poor
1960	70.2	2.3	30:1
1970	73.9	2.3	32:1
1980	76.3	1.7	45:1
1989	82.7	1.4	59:1

1940 forward. Manning (1998) is additionally considering portfolio investments as a mechanism of both assessing position and extracting resources from the periphery.

This brief survey obscures the commonality of the collective results: regardless of how measured, all of these workers have demonstrated that power, position, and therefore income in the world-economy is distributed grossly unevenly, and the gap is growing. In 1992, the United Nations published the *Human Development Report* containing data tracking the recent history of the distribution of global income (see table 2.1).

In a supplement to that report issued in 1994, the ratio of rich/poor had risen in only two years from 1989 to 1991 to *61 to 1* (United Nations 1994). These data show that the global polarization of income is increasing at an *increasing rate,* which bodes ill for long-term system stability. Another "bifurcation" between reorganization and collapse may be due.

TRENDS AND CYCLES IN THE MODERN SYSTEM

All scholars agree that world-systems exhibit both cycles and long-term trends. The simplest cycles are weather-dependent, but longer ones (population density, city size) seem to have more complex roots in trade and the organization of production. Trends may either prompt or reflect such cycles and include population, deforestation, extension of irrigation, or general intensification of production effort. Described in the following subsections are some of the most important trends and cycles characterizing recent history.

Trends

Population

While population growth has always been both an important cause and consequence of the historical evolution of world-systems, the pace of population growth has increased dramatically in the modern world. Research has indicated that fertility is mainly affected by economic factors (Folbre 1977; Gimenez 1977; Grimes 1981; Mamdani 1972; Seccombe 1983, 1992). An important implication of this finding is that the structure of economic incentives and costs accompanying each type of world-system structure varies across both classes and accumulation modes. For example, universal and state-subsidized education among the contemporary states of the core, combined with wage scales that (until recently) made possible the support of children by their parents, has removed the economic incentive to have large families. But in the periphery, the persistence of a semiproletarianized, semipeasant coerced workforce suspended between the current capitalist and "tributary" modes has until recently sustained the high value of children as valuable economic assets and provide a motivation for high rates of fertility. Now that global agricultural production is once again being reorganized to reduce family-owned farms (McMichael 1996; Magdoff, Buttel, and Foster 1998), the fertility rate in the periphery is at last plunging in a delayed echo of the drop in rates within the core a century ago—albeit for very different reasons.

Technological Change

Technological change has also accelerated in modern capitalism. In earlier systems, the implementation and diffusion of new approaches and techniques was slow because such novelties were evaluated against not just their effect on production efficiency but also on their potential effect on the structures of social stratification (Anderson 1974a). For a new technique to displace an old one, it had to be compatible with the preexisting "occupational structure" so that traditional status hierarchies were not challenged. Under the current capitalist mode, considerations of the social effects of technical change are largely ignored, because the producers are firms whose guiding principle is their individual profit and the suppression of their production costs, not the effect of technological change on the broader social structure. Furthermore, interfirm competition compels firms to be on a constant and aggressive search for new techniques that hold the promise of reducing production costs. Hence, the rate of technical change has blossomed within contemporary capitalism in a way that has been historically unprecedented. Meanwhile, the energy required to drive that technology has risen far faster than even the geometric growth of the population (Singer 1970; Star 1971; Roberts and Grimes 1997; Grimes 1999; Podobnik, n.d.).

Commodification

Another trend that modern capitalism has accelerated is that of "commodification" (Wallerstein 1984b). By this is meant the assignment of a market "price" to an ever-expanding percentage of the products of human activity. In the early stages of capitalism, money complemented barter as a method of payment, while only those products created specifically for trade were created simply with an end market in mind (e.g., textiles). In this early phase, the population was largely agricultural, and food and clothing were individually produced as needed by each family or community

(Braudel 1984, vol. 1). As late as the second half of the nineteenth century, workers in the United States were still expected to supplement their wages with private gardens (Braverman 1974). In the age of McDonald's, this view seems surprising, which is itself an indicator of just how far the trend toward commodification has progressed.

Proletarianization and Capital Intensity

Among the most important subjects of the trend toward commodification has been human labor itself. The wage relation between employers and employees is widespread in both the core and the periphery. Because this means that human labor is a precisely priced cost of production, the cost of labor can more easily be compared with machinery, and where the latter "wins," labor is expelled. In this way labor costs are a constant inducement toward cheaper and more capable machinery, along with cheaper energy to run them. This has meshed with, and largely powered, the drive toward faster technological change. The overall pattern has been the displacement of labor by machinery, which can as well be described as the substitution of "inorganic" for "organic" energy. While an increase in economic efficiency (more items produced per wage dollar spent) is the motivation for this substitution, it entails environmental costs which are undervalued by the calculus of capitalism (e.g., global warming, ozone depletion, top soil loss, and deforestation—all linked in various ways to mechanization and automation).

Increasing Size of Firms

Despite oscillations, the largest firms have grown by every measure in the last two hundred years: production capacity, number of employees, amount of capital controlled, and size of the market. This "monopoly" sector, characterized by a small number of huge firms, was originally born out of the broader "competitive" sector of small-scale companies, and the competitive sector continues to be reproduced (O'Connor 1973). But the emergence of the monopoly sector has changed the face of modern capitalism most obviously in the realm of international trade, notwithstanding the recent literature on "postmodern" flexible accumulation (Harvey 1989). Among the key advantages accruing to large size are cheap access to large amounts of capital. This access allows for ready reinvestment in upgrading older production facilities or building new ones, enabling the development and implementation of new technologies. This benefit applies not just to production but also to marketing and the substitution of new materials. Access to capital also allows for financing new bouts of automation with new waves of investment, thereby propelling the secular trend toward greater capital intensity.

Geographically, the physical area that a company's decisions could affect because of the range of local markets they supply, or the area from which they draw their employees, or the number of governments toward whom they pay taxes, has always expanded, albeit unevenly and cyclically.

State Formation

Tilly (1989) argues that the European nation-state arose to facilitate the prosecution of war. This may be true, but another important function of the modern state has been to control the economy and the surplus it generates, because it depends on this surplus for its basic existence.

The power of states over the lives of their citizens has been expanding since the beginning of the first states, but this expansion has grown explosively in the last two centuries (Foucault 1980; Boli-Bennett 1980). While the "absolutist state" emerging during the end of feudalism asserted arrogant claims about the power of the monarch over the lives, bodies, and property of subject citizens (Anderson 1974b), only today has technology enabled some of these claims to become almost literally true. This expansion of the state has taken on two forms: a geographical expansion of the power of the central government ever-farther away from the capital city combined with a deepening of its power over the daily life of its citizens.

Each bout of corporate and state expansion seems to have followed the same general course: a wave of technological change enabled an expansion in corporate size and control, stimulating popular demands for compensatory regulation on a corresponding governmental scale. Corporate attempts to evade regulation have often involved crossing governmental boundaries, in turn leading to bilateral and multilateral agreements between governments on regulations. Each round has led to an expanded state chasing after an expanding corporate size. International political integration has increased the level of global governance such that some world-system scholars predict the eventual emergence of a world state (e.g., Chase-Dunn 1990; Chase-Dunn and Podobnik 1995; Arrighi 1990, 1994).

Limits

As with population growth, there are ceilings to each of these trends. Technical change often relies on energy sources that are not infinite while producing effluents that are deadly; commodification requires the equal expansion of the ability to pay; proletarianization is curtailed by automation; both corporate and governmental size must stop at the global level.

Each ceiling has the capacity to throw the modern capitalist world-economy into either acute crisis or systemic collapse. For example, much of our technology relies on the use of fossil fuels, which is the leading source of global warming. If we continue using these fuels as we have been, global warming and air pollution will kill increasing numbers of people, threatening the viability of the entire world-system on both biological and political levels. Also, as we run out of these fuels (an inevitable event; see "Special Report" 1998), prices will rise dramatically, introducing rates of inflation that threaten system stability (yet another portent of system bifurcation).

An example of the effect of trend limits in commodification and proletarianization is provided by every recession, which is largely due to the failure by the working classes to consume an adequate volume of the products created by their own work. This is always attributable to inadequate wages (the inverse side of excessive commodification) and/or automation (the necessary by-product of proletarianization). The recent depressions in the world-economy between 1988 and 1992 and the current one starting in 1998 illustrate these limits. Automation (fossil fuel–dependent) is generating a vast and growing number of unemployed lacking the ability to buy. In the United States the employed remainder brings home a real wage that has been falling since 1965 (Economic Policy Institute 1996). This has placed a restraint on the economic recovery and may eventually block growth altogether, pending a redistribution of income. Such a situation is already creating a crisis of political legitimacy and economic viability evident throughout the core in rising rates of unemployment and political extremism.

These examples should highlight just how vulnerable the current world-system is and how close to its limits we have already come. Current rates of increase in population and energy use combined with capitalism as it exists today simply cannot be sustained, so a major breakdown is as inevitable as earthquakes in California, yet equally difficult to pin down to a particular moment. This should become clearer after the review of research on cycles below. Different cycle phases either accelerate or retard trend developments, and each cycle acts along its own time scale. The sum of the respective influences of all of these cycles and trends affects the likely trajectory of our collective future.

Cycles

Cycles in production have always existed. In societies closely tied to agricultural production, the cycle of the seasons imposes an annual periodicity on planting and harvests. Longer-term fluctuations in economic output in the precapitalist era have also been noted (Chase-Dunn and Willard 1993, 1994; Frank and Gills 1993).

In the current period, economic cycles appear to be endemic to modern capitalism, and several have been identified, differing mainly in length. The most important among these are the Juglar cycle of six to ten years, the Kuznets cycle of twenty to twenty-five years, and the Kondratiev, thought to last about fifty years (Kleinknecht et al. 1992; Modelski and Thompson 1996; Mandel 1975, 133–161).

Regardless of length, each has a similar underlying logic: a new set of products are introduced that sell well, the market expands, and related employment swells, allowing for an expansion of worker/"consumer" spending. The market eventually becomes saturated, sales drop, income contracts, and workers are laid off. The effect of the contraction is prolonged by the extended feedback loop through those firms producing capital goods. These manufacturers of the means of production take orders in advance, which means that they are producing machinery for constructing the end product long after the slump in sales of that product has started. This long feedback only prolongs the downturn. But eventually, the excess inventory is sold out, production resumes, and renewed growth is possible (Mandel 1968, 1975; Marx 1967).

Juglar Cycles

Otherwise known as the "normal" or "classic" business cycle, these cycles last six to ten years (Gordon 1986, 522; Maddison 1982, 77). Even though discussions of these cycles can be traced as far back as Marx (1967), the National Bureau of Economic Research has conferred upon them the name of the much more recent Jacek Juglar.[2] The causes of these particular cycles are thought to lie in the average life span of capital equipment: after about eight years, the machinery of production tends to wear out or depreciate to the point that replacement becomes necessary (Moore and Zarnowitz in Gordon 1986, 738). The fact that the periodization of equipment purchases and factory construction *within* firms becomes also synchronous *across* firms throughout the economy as a whole reflects how the firms are themselves customers of one another. Recent examples of Juglar downturns in the United States are the recessions of 1998, 1989–92, 1981–83, and 1973–76, 1968, and 1957.

Kuznets Cycles

In works published at various times through the period 1930–1960, Simon Kuznets traced what appeared to be cycles of about twenty to twenty-five years in the records of a variety of indicators for several core states and one semiperipheral state (Maddison 1982, 262, note 15). These cycles have been the subject of many different investigators (e.g., Grimes 1993; Thomas 1954; Solomou 1990), and several hypotheses have been put forward for their cause, although none have been uniformly accepted (Maddison 1982, 73–77). One possible explanation may be to link the timing of the Kuznets oscillation to generational turnover in the demand for housing and other buildings, but this linkage is still speculative (Solomou 1990, 157–159). I have extended the coverage of nations to include all countries having GDP data back until 1790 and have found that a Kuznets-length cycle exists for all countries measured to the present (discussed later) (Grimes 1993).

Kondratiev Cycles

A Soviet economist writing in the 1920s, Kondratiev observed that the historical record of prices then available to him appeared to indicate a cyclic regularity of phases of gradual price increases followed by phases of decline (Kondratiev, reprinted in 1979). The period of these apparent oscillations seemed to him to be about fifty years.

Kondratiev's controversial concept of a forty- to sixty-year-long cycle has faded in and out of academic favor often since he wrote. Most recently, the onset of persistent sluggishness in the world economy in the 1970s has brought about a revival (Berry 1991; Freeman 1984; Goldstein 1988; Gordon 1978; Kleinknecht et al. 1992; Mandel 1975, 1980; Matteo et al. 1989; Van Duijn 1983).

The basic argument is that periodically (approximately every forty to sixty years), factories, means of production, communication, and transport are all rebuilt incorporating new technologies. A list of specific historical examples of such "revolutionary" new technologies could include the construction of paved roads (Braudel 1984, vol. 3, 316–317), the centralized factory system of manufacture (Marglin 1974), the early machinery employed in the textile mills, the infrastructure for the railroad system, the discovery and application of electric power, automobiles and the application of internal combustion engines (Baran and Sweezy 1966), electric home appliances and factory machinery, and microprocessors and robotics (Mandel 1975, chap. 4). Throughout the period of initial implementation of these new technologies, the "normal" business cycles (both Juglar and Kuznets) proceed unimpeded. However, recessions are shorter and recoveries surer and longer (Mandel 1975, chap. 4).

Eventually, however, the diffusion of the new technologies becomes general. The market becomes saturated in precisely the same fashion as occurs in the Juglar cycle. Parallel with this diffusion comes a decline in profits. Once more, investment shifts away from production and into speculation, while unemployment mounts and effective demand decays. Yet because the whole process is spread out over the entire infrastructure of the economy, its decline also stretches over several Juglar cycles, during which the depressive phases grow longer and the recoveries more anemic (Mandel 1975, chap. 4).

Recent Cycle Research

Most of the work on Kondratiev cycles has used price data (e.g., Berry 1991), although researchers have also examined inventories, consumption patterns, labor strikes, profit rates, and international debt (Gordon 1986; Kowalewski 1994; Poletayev 1993; Shaikh 1992; Silver 1992; Suter 1992). Until very recently (e.g., Kleinknecht 1992), less empirical work has been done with data on production, even though there are sound theoretical reasons for thinking that the essential engine of the manifold cycles of capitalism lie ultimately in the dynamics of production and accumulation (Mandel 1975, 1980). But that deficit has started to be filled. Metz (1992), in a reanalysis of GDP production data on eight European countries from 1850 to 1979, and a "world" production index originally presented by Bieshaar and Kleinknecht (1984) for 1780–1979, seems to have found strong evidence of long waves. His contribution lay in omitting the data distorted by the two world wars and substituting in their place interpolated values, combined with a creative application of spectral analysis.

The entire area of research into long waves is still young, so there remain important disagreements over relevant data and methods. One of these disagreements is over the best approach to detecting cycles within time series data. One school (exemplified by the works found in Kleinknecht 1992) advocates use of procedures that are variations on spectral analysis, while another (typified by Goldstein [1988], Solomou [1990], and Grimes [1993, 1996]) examines data series by using percentage change between data values (typically referred to as taking "first differences"). Put simply, the first approach starts by taking out any long-term trends in the data series, and then ignores fluctuations outside of the target cycle length. The second does not throw out any information but assumes that any cycles (regardless of length) will be revealed by the percentage change data. Unfortunately, the method chosen appears to influence the results—those using spectral analysis typically find Kondratiev-length cycles more often than those using first differences (e.g., Solomou 1990).

An example of the results available from the application of percentages is in some of my own recent work (Grimes 1993, 1996).[3] Most of the research literature surveyed in this chapter has restricted itself to analysis of data from core countries, usually those in Europe. Missing has been the inclusion of data from countries beyond these few. Until recently, such data were not easily available. However, over the 1980s an important new source of data was presented in a series of historical statistics compiled and published by Mitchell (1980, 1983, 1985). Using these new data, I computed the average percentage growth rate for available measures of national product at five-year intervals over the period 1790–1988.[4]

My calculations revealed a regular cycle of twenty to thirty years in the growth rates of GDP per capita across the entire period, despite a considerable variation in the number of cases (Chase-Dunn and Grimes 1995). *This* method of approaching the data shows no obvious support for a Kondratiev-length pattern, but there *is* strong support for a Kuznets-length oscillation. Further work with these data is needed, especially its submission to the procedures of spectral analysis. So at this point the jury on the existence of Kondratievs in production is still out.

It easy to imagine, if there are regular cycles lasting at least one generation (such as the Kuznets) or possibly two (like the Kondratiev), that these oscillations might express themselves socially and politically. Particularly in the case of the latter cycle length, where either phase ("up" or "down") is capable of lasting twenty to thirty years, entire generations could come of age during either generalized depression or prosperity, which would have political consequences (Berry 1991). Research has already shown that general attitudes about the relative importance of economic "well-being" as against economic "equity" is detectably different by cohort and is powerfully affected by the economic circumstances of one's youth (Inglehart 1971). This linkage between cohort attitude and the economic circumstances of childhood has also been demonstrated cross-nationally. Berry (1991), Gordon (1978, 1986), and Mandel (1975, 1980) have all used this observation to suggest that the Kondratiev cycle shapes the timing of "key" elections, as well as the form and timing of class conflict. Yet efforts to test this seemingly clear corollary to cycle research have yielded inconsistent results.

Berry (1991) found strong support for linking key elections in the United States to K-waves. However, research on strikes (Silver 1992, 1995) has been less clear. While there was mild support for a long wave influence on class conflict, it was weak. Instead, there was an unexpected match between such conflict and the decentralization phase of the "hegemonic sequence" (described later). Another example of similar research is that of Kowalewski (1991a, 1991b, 1991c). Like Silver, he used the *New York Times* and *The Times* (London) to gather reports of political events to check against the pattern of the Kondratiev for the period 1821–1985. But in his case, he looked for reports of revolutionary movements in the periphery, core interventions to suppress those movements, and the frequency of military coups. For the frequency of revolutionary movements and also that of coups, he found a clear upward trend throughout, while for core interventions there appeared to be a gentle upside-down U curve, peaking near 1920. In looking for connections to the generally agreed-on timing of the Kondratiev, like Silver, he found weak support, but more apparent correlation with the hegemonic sequence—again unexpectedly. As is necessarily the case in empirical research, the results echo the data used. Presumably elections, strikes, coups, and core interventions are each only glimpses through a glass darkly, true enough as measured yet still mere glints from the depths of a murky social whole.

Finally, in other work on Kondratiev waves, Suter's (1992) important study of debt cycles shows that the debt crisis of the 1980s in the periphery and semiperiphery is only the most recent instance of a recurrent cycle of foreign indebtedness characterizing the world economy as a whole. Suter identifies three repeating phases (expansion, crisis, and settlement) that have accompanied four major periods of global debt-servicing incapacity among sovereign borrowers: the late 1830s, the mid-1870s, the early 1930s, and the early 1980s. Suter concludes that debt cycles are regularly related to *both* Kondratiev waves *and* Kuznets cycles.

HEGEMONIC SEQUENCES

One of the cyclical features that is common to all hierarchical world-systems is an oscillation between centralization and decentralization of political organization. In systems of chiefdoms, states, and empires the size of the largest polity increases and then decreases, in a sequential sequence of rise and fall. But in the modern world-system, capitalism has changed the dynamics and the form of the political centralization/decentralization sequence (Chase-Dunn and Hall 1997).

In the modern world-system the sequence takes the form of *the rise and fall of hegemonic core powers*. This is analytically similar to the rise and fall of empires, but the differences are important. State-based world-systems prior to the modern one oscillated back and forth between corewide tributary empires and interstate systems in which the core region contained several states. In some regions the decentralization trend went so far as to break up into ministates. Thus feudalism may be understood as a very decentralized form of a state-based tributary system.

The simplest structural difference between a corewide empire and a hegemonic core state is with regard to the degree of the concentration of political/military power in a single state. It is in this sense that Wallerstein's distinction between world-empires and world-economies points to an important structural difference between the modern capitalist system and earlier state-based tributary systems. But this is not only a difference in the degree of peak political concentration. The rise and fall of hegemonies have occurred in a very different way from the rise and fall of empires. Empire formation was a matter of conquering and exploiting adjacent core states by means of plunder, taxation, or tribute. In contrast to these earlier blunt methods, modern hegemons have sought to control international trade, especially oceanic trade, that linked cores with peripheries. Upward mobility to core status (e.g., Amsterdam, London, New York, Tokyo) appears to entail gaining a disproportionate share of global surplus by successively producing cheaper

- consumer goods,
- consumer durables,
- producer goods,
- infrastructural goods,
- financial control, and
- military control.

This is why the modern world-system is resistant to empire formation. The most powerful state in the system acts to block empire formation and to *preserve the interstate system*. Thus, the cycle of political centralization/decentralization takes the form of the rise and fall of hegemonic core powers. This important difference is primarily due to the relatively great importance that capitalist accumulation has for the modern system.

There is general agreement that certain core states have been able to access disproportionate amounts of political, military, and economic power for prolonged periods averaging about one century. These unusual core states are variously called either "leading" or "hegemonic," because during their dominion they can promulgate and militarily enforce international rules of trade and international relations that favor their status. With the most dynamic economy and often the largest military, the hegemon also disseminates its language, culture, and currency as "global" standards (Cha 1991; Chase-Dunn 1998; Goldstein 1988; Modelski and Thompson 1988, 1996; Thompson 1992). It is also accepted that at some point the transition between the hegemony of one country and that of another is punctuated and affirmed by global war. The key linkage between the hegemonic cycle and war that all students of these issues accept is that the agreements ending any global war ratify the global power hierarchy existing at war's end. Meanwhile, in the generations following the war, uneven investment and development gradually work to reorder the hierarchy in ways unanticipated by the earlier international agreements. Eventually, the disparity between the "actual" hierarchy formed by long-term economic investment patterns and the "formal" hierarchy ratified by the last war grows large enough—and the aging hegemon weak enough—that another member of the core seeks to unseat and replace it by initiating a new war (Boswell and Misra 1995; Goldstein 1988; and Thompson 1988, 1996).[5]

Other important questions remain unsettled: how many core states have been hegemonic and for how long? What are the links between economic cycles, wars, and the hegemonic sequence?

Braudel (1984, vol. 3) provides one list of dominant economic "centers" focused on cities: Venice (1378–1498), Antwerp (1500–1569), Genoa (1557–1627), Amsterdam (1585–1773), London (1773–1929), and New York (1929–present). For him, these centers were not "hegemonic" in the sense described earlier. Rather, they were the sites of the most dynamic economies in their day, locations that acted as midwives to the most important institutional inventions facilitating capitalist development, and also the trading nodes through which the most advanced products were routed. Others writing since have imposed criteria for hegemony that delete some of Braudel's candidates because they conceptualize hegemony differently. Wallerstein (1984a, 1984b) understands hegemony as a combination of economic power based on the most profitable leading industries and military power. Modelski and Thompson (1988) originally focused on the importance of naval forces that allowed the "leader" to exercise global reach. Their study of changes in the distribution of naval capacity among the "great powers" remains one of the most important contributions to our understanding of the hegemonic sequence. In their more recent work, Modelski and Thompson (1996) see global leadership as based on economic innovations that allow the leader to be central in the newest and most profitable industries in the world economy. Their empirical work on the distribution of shares of the most profitable sectors over long periods of time is very valuable, though the sectors chosen remain somewhat controversial (see chap. 5 for more details).

Modelski and Thompson (1988, 1996) argue that there have been four "global leaders": Portugal (1516–1609), United Provinces of the Netherlands (1609–1714), United Kingdom (1714–1945), United States (1945–?), and they contend that the United Kingdom managed to lead through two "leadership cycles," one in the eighteenth and one in the nineteenth centuries. Wallerstein (1984a) is even more restrictive, admitting into the club of hegemonies only the United Provinces, United Kingdom, and the United States.

The cyclical nature of the rise and fall of these powers strongly implies a linkage with the economic cycles surveyed previously. But the nature of this linkage has yet to yield a consensus. Most tie the hegemonic sequence to the Kondratiev cycle, despite its empirical ambiguity. Modelski and Thompson (1988, 1996) are the most explicit in this regard, arguing that each hegemonic cycle encloses two Kondratievs. Goldstein (1988) and I (Grimes 1996) are more flexible, asserting only that a global war and birth of a new hegemon coincides with a Kondratiev upswing but that not all upswings are accompanied by war. Boswell and Misra (1995) assert that the relevant cycle governing the hegemonic sequence is not the Kondratiev at all but instead a "logistic" wave, a wave postulated to have the same dynamic as the

Kondratiev—infrastructural innovation eventually giving way to market saturation—but whose period is thought to be roughly twice as long (about 120 years).

But despite these disagreements, there is a consensus that the engine driving a rising core state and contender for hegemonic status is that it is the geographical home for the emergence of some revolutionary. This new technology (and its derivative products) requires a change in basic infrastructure to allow their full use. Hence, it takes more than one generation to saturate the market. The novelty of these products allows for disproportionate accumulation via technological rents, enabling the aspiring new core to gain rapidly the capital required both for continued investment and military buildup.

CONCLUSIONS

Thirty years of research by workers inspired by the broad perspective provided by world-systems theory has generated exciting new insights into the underlying patterns of global history. By exposing these patterns in a scientifically disciplined way, world-system researchers have shown that a science of history is possible and that the superficial vagaries of human social experience and change fit into a logical sequence consistent with the evolution of all other life forms.

This potential for world-systems analysis to enable us to *understand* social change and act in a collectively rational way to avoid *predictable* disasters (such as global war and environmental collapse) is the most alluring aspect encouraging research in the area. Already we know that our current system is wobbling dangerously close to the edge of growth limits.

As the best unit of analysis for studying social change on a global level, world-systems theory provides social science with a new purchase that can produce a robust and scientific theory explaining both the past and likely future. Today the study of world-systems promises to wrest our expectations about the future away from theology and into the realm of science.

NOTES

1. Here the term *mode of accumulation* is used as approximately equivalent to the Marxist *mode of production*, but without the implication that only production processes are important for systemic logic.

2. This despite Maddison's (1982, 77) assertion that "Juglar never claimed to have discovered the existence of an eight- to nine-year rhythm."

3. A more thorough report of this research can be found in Grimes (1993).

4. Data were collected at five-year intervals on GDP and population for 104 of the largest countries from 1790 to 1990. For the period 1790 to 1945, data were obtained from Mitchell (1981, 1982, 1983) and Maddison (1982, 1983). For the period 1950–1990, the

GDP and population data are drawn from the World Bank. Although data on GDP are not available for every country for all years, the countries for which data exist in the period 1790–1950 include the United States, Mexico, most of Europe, and the largest countries in Latin America and Asia.

5. The threat of yet another global war in an age of nuclear and biological weapons represents an additional reason to presume that the current world-system is transitory.

REFERENCES

Amin, Samir. 1976. *Unequal Development.* New York: Monthly Review Press.

———. 1993. "The Ancient World-Systems versus the Modern Capitalist World-System." Pp. 247–277 in *The World System: Five Hundred Years or Five Thousand?* edited by Andre Gunder Frank and Barry K. Gills. London: Routledge.

Anderson, Perry. 1974a. *Passages from Antiquity to Feudalism.* London: New Left Books.

———. 1974b. *Lineages of the Absolutist State.* London: New Left Books.

Arrighi, Giovanni. 1990. "The Three Hegemonies of Historical Capitalism." *Review* 13:3(Summer):365–408.

———. 1994. *The Long Twentieth Century: Money, Power and the Origins of Our Times.* London: Verso.

Arrighi, Giovanni, and Jessica Drangel. 1986. "The Stratification of the World-Economy: An Exploration of the Semi-peripheral Zone." *Review* 10:1(Summer):9–74.

Baran, Paul A. 1957. *The Political Economy of Growth.* New York: Monthly Review Press.

Baran, Paul A., and Paul Sweezy. 1966. *Monopoly Capital.* New York: Monthly Review Press.

Beckford, George. 1972. *Persistent Poverty: Underdevelopment in Plantation Economies of the Third World.* New York: Oxford University Press.

Berry, Brian J. L. 1991. *Long Wave Rhythms in Economic Development and Political Behavior.* Baltimore, Md.: Johns Hopkins University Press.

Bieshaar, H., and A. Kleinknecht. 1984. "Kondratieff Long Waves in Aggregate Output? An Econometric Test." *Konjunkturpolitik* 30:5:279–303.

Boli-Bennett, John. 1980. "Global Integration and the Universal Increase of State Dominance, 1910–1970." Pp. 77–107 in *Studies of the Modern World-System,* edited by Albert J. Bergesen. New York: Academic Press.

Boswell, Terry, and Joya Misra. 1995. "Cycles and Trends in the Early Capitalist World-Economy: An Analysis of Leading Sector Commodity Trades, 1500–1600/50–1750." *Review* 18:3(Summer):450–485.

Braudel, Fernand. 1981–84. *Civilization and Capitalism, 15th–18th Century.* 3 vols. New York: Harper & Row.

Braverman, Harry. 1974. *Labor and Monopoly Capital.* New York: Monthly Review Press.

Cha, Yun-Kyung. 1991. "Effect of the Global System on Language Instruction, 1850–1986." *Sociology of Education* 64:19–32.

Chase-Dunn, Christopher. 1990. "World State Formation: Historical Processes and Emergent Necessity." *Political Geography Quarterly* 9:2(April):108–130.

———. 1998. *Global Formation: Structures of the World-Economy.* Boulder, Colo.: Rowman & Littlefield. (Originally published 1989, Cambridge, Mass.: Blackwell.)

Chase-Dunn, Christopher, and Peter Grimes. 1995. "World-Systems Analysis." *Annual Review of Sociology* 21:387–417.

Chase-Dunn, Christopher, and Thomas D. Hall. 1997. *Rise and Demise: Comparing World-Systems.* Boulder, Colo.: Westview.

Chase-Dunn, Christopher, and Bruce Podobnick. 1995. "The Next World War: World-System Cycles and Trends." *Journal of World-Systems Research* 1:6 (electronic journal: http://csf.colorado.edu/wsystems/jwsr.html).

Chase-Dunn, Christopher, and Alice Willard. 1993. "Systems of Cities and World-Systems: Settlement Size Hierarchies and Cycles of Political Centralization, 2000 BC to 1988 AD." Paper presented at the annual meeting of the International Studies Association, Acapulco, March 25. Data appendix is available from an electronic archive at Boulder, Colo.: csf.colorado.edu/wsystems/datasets/citypop/civilizations/citypops_2000bc-1988ad

———. 1994. "Cities in the Central Political/Military Network since CE1200: Size Hierarchy and Domination." *Comparative Civilizations Review* 30(Spring):104–132.

Colinvaux, Paul. 1978. *Why Big Fierce Animals Are Rare.* Princeton, N.J.: Princeton University Press.

Economic Policy Institute. 1996. *The State of Working America 1996–97.* Washington, D.C.: Author.

Ekholm, Kasja, and Jonathan Friedman. 1982. "'Capital' Imperialism and Exploitation in the Ancient World-Systems." *Review* 6:1(Summer):87–110.

Elton, Charles S. 1927. *Animal Ecology.* New York: Macmillan.

Emmanuel, Arghiri. 1972. *Unequal Exchange: A Study of the Imperialism of Trade.* New York: Monthly Review Press.

Folbre, Nancy. 1977. "Population Growth and Capitalist Development on Zongolica, Veracruz." *Latin American Perspectives* 4:4:41–55.

Foucault, Michel. 1980. *The History of Sexuality. Vol. 1: An Introduction.* New York: Vintage.

Frank, Andre Gunder. 1978. *World Accumulation 1492–1789.* New York: Monthly Review Press.

———. 1993. "Transitional Ideological Modes." Pp. 200–220 in *The World System: Five Hundred Years or Five Thousand?* edited by Andre Gunder Frank and Barry K. Gills. London: Routledge.

Frank, Andre Gunder, and Barry K. Gills, eds. 1993. *The World System: Five Hundred Years or Five Thousand?* London: Routledge.

Freeman, Christopher, ed. 1984. *Long Waves in the World Economy.* Dover, N.H.: Pinter.

Gimenez, Martha E. 1977. "Population and Capitalism." *Latin American Perspectives* 4:4:5–40.

Gleick, James. 1987. *Chaos: Making a New Science.* New York: Viking Penguin.

Goldstein, Joshua. 1988. *Long Cycles: Prosperity and War in the Modern Age.* New Haven, Conn.: Yale University Press.

Gordon, David M. 1978. "Up and Down the Long Roller Coaster." *URPE U.S. Capitalism in Crisis.* New York: URPE, Economic Education Project.

———. 1986. "What Makes Epochs? A Comparative Analysis of Technological and Social Explanations of Long Economic Swings." Pp. 267–304 in *Technological and Social Factors in Long Term Fluctuations,* edited by Massimo di Matteo, Richard M. Goodwin, and Alessandro Verceli. Proceedings, Siena, Italy, December 1989. Vol. 321 of *Lecture Notes in Economics and Mathematical Systems.* New York: Springer.

Gordon, Robert J., ed. 1986. *The American Business Cycle: Continuity and Change.* Studies in Business Cycles, vol. 25. National Bureau of Economic Research. Chicago: University of Chicago Press.

Grimes, Peter. 1981. "Poverty, Exploitation, and Population Growth: Marxist and Malthusian Views on the Political Economy of Childbearing in the Third World." Unpublished master's thesis, Michigan State University, East Lansing.

———. 1993. "Harmonic Convergence? Frequency of Economic Cycles and Global Integration." Paper presented at the Social Science History Association meeting, November 4–7, Baltimore, Md.

———. 1996. "Economic Cycles and International Mobility in the World-System: 1790–1990." Unpublished Ph.D. dissertation, Johns Hopkins University, Baltimore, Md.

———. 1999. "The Horsemen and the Killing Fields: The Final Contradiction of Capital." Pp. 13–42 in *Ecology and the World-System,* edited by Walter Goldfrank, David Goodman, and Andrew Szaz. Westport, Conn.: Greenwood.

Harvey, David. 1989. *The Condition of Postmodernity.* Oxford: Blackwell.

Holland, John H. 1995. *Hidden Order: How Adaptation Builds Complexity.* New York: Addison-Wesley.

Inglehart R. 1971. "The Silent Generation in Europe: Intergenerational Change in Post-Industrial Society." *American Political Science Review* 65:4(Aug.):991–1017.

Kentor, Jeffrey D. 1998. "Wealth and Power: An Examination of the Economic and Military Dimensions of the Core/Periphery Hierarchy in the World-Economy 1820–1990." Unpublished Ph.D. dissertation. Johns Hopkins University, Baltimore, Md.

Kleinknecht, Alfred, Ernest Mandel, and Immanuel Wallerstein, eds. 1992. *New Findings in Long-Wave Research.* New York: St. Martin's.

Kondratiev, Nikolai D. 1979. "The Long Waves in Economic Life." *Review* 2:4(Spring):519–562.

Kowalewski, David. 1991a. "Periphery Revolutions in World-System Perspective, 1821–1985." *Comparative Political Studies* 24:1(April):76–99.

———. 1991b. "Core Intervention and Periphery Revolution, 1821–1985." *American Journal of Sociology* 97:1(July):70–95.

———. 1991c. "Periphery Praetorianism in Cliometric Perspective 1855–1985." *International Journal of Comparative Sociology* 32:3–4:289–303.

———. 1994. "The African Periphery in the World System: Kondratieff Longwaves of Exports and Indebtedness." *Journal of the Third World Spectrum* 1:1(Spring):17–32.

Korzeniewicz, Roberto, and William Martin. 1994. "The Global Distribution of Commodity Chains." Pp.67–92 in *Commodity Chains and Global Capitalism,* edited by Gary Gereffi and Miguel Korzeniewicz. Westport, Conn.: Praeger.

Maddison, Angus. 1982. *Phases of Capitalist Development.* New York: Oxford University Press.

Magdoff, Fred, Frederick H. Buttel, and John Bellamy Foster, eds. 1998. *Hungry for Profit: Agriculture, Food, and Ecology.* Special issue of *Monthly Review* 50:3.

Mandel, Ernest. 1968. *Marxist Economic Theory.* New York: Monthly Review Press.

———. 1975. *Late Capitalism.* London: New Left Books.

———. 1980. *Long Waves of Capitalist Development: The Marxist Interpretation.* New York: Cambridge University Press.

Mamdani, Mahmood. 1972. *The Myth of Population Control: Family, Caste, and Class in an Indian Village.* New York: Monthly Review Press.

Manning, Susan E. 1998. "Finance Capital and International Development: A Study of Post–World War II Portfolio Investment Dependence." Unpublished Ph.D. dissertation proposal. Johns Hopkins University, Baltimore, Md.

Marglin, Stephan A. 1974. "What Do Bosses Do? The Origins and Functions of Hierarchy in Capitalist Production." *Review of Radical Political Economics* 6:2(Summer):33–60.

Marx, Karl. 1967. *Capital*. 3 vols. New York: International House.

Matteo, Massimo Di, Richard M. Goodwin, and Alessandro Verceli, eds. 1989. *Technological and Social Factors in Long Term Fluctuations*. Proceedings, Siena, Italy, December 1986, Vol. 321 of *Lecture Notes in Economics and Mathematical Systems*. New York: Springer.

McMichael, Philip. 1996. *Development and Social Change: A Global Perspective*. Thousand Oaks, Calif.: Pine Forge.

Metz, Rainer. 1992. "A Re-examination of Long Waves in Aggregate Production Series." Pp. 80–119 in *New Findings in Long-Wave Research,* edited by Alfred Kleinknecht, Ernest Mandel, and Immanuel Wallerstein. New York: St. Martin's.

Mitchell, Brian R. 1980. *European Historical Statistics: 1750–1975*. London: Macmillan.

———. 1983. *International Historical Statistics: The Americas and Australasia*. Detroit: Gale Research.

———. 1985. *International Historical Statistics: Africa and Asia*. New York: New York University Press.

Modelski, George, and William R. Thompson. 1988. *Seapower in Global Politics, 1494–1993*. Seattle: University of Washington Press.

———. 1996. *Innovation, Growth and War: The Coevolution of Global Economics and Politics*. Columbia: University of South Carolina Press.

Moore, Geoffrey H., and Victor Zarowitz. 1986. "The Development and Role of the National Bureau of Economic Research's Business Cycle Chronologies." Appendix A, pp. 735–781 in *The American Business Cycle: Continuity and Change,* edited by Robert J. Gordon. Chicago: University of Chicago Press.

Nemeth, Roger S., and David A. Smith. 1985. "International Trade and World-System Structure: A Multiple Network Analysis." *Review* 8:4(Spring):517–560.

O'Connor, James. 1973. *The Fiscal Crisis of the State*. New York: St. Martin's.

Podobnick, Bruce. n.d. "Global Energy Shifts: Future Possibilities in Historical Perspective." Unpublished Ph.D. dissertation, Johns Hopkins University, Baltimore, Md.

Poletayev, Andrey V. 1993. "Long Waves in Profit Rates in Four Countries." Pp.151–167 in *New Findings in Long-Wave Research,* edited by Alfred Kleinknecht, Ernest Mandel, and Immanuel Wallerstein. New York: St. Martin's.

Prigogine, Ilya. 1984. *Order Out of Chaos: Man's New Dialogue with Nature*. New York: Bantam.

———. 1996. *The End of Certainty: Time, Chaos, and the New Laws of Nature*. New York: Free Press.

Raffer, Kunibert. 1987. *Unequal Exchange and the Evolution of the World System*. New York: St. Martin's.

Roberts, J. Timmons, and Peter Grimes. 1997. "Carbon Intensity and Economic Development 1962–1991: A Brief Exploration of the Environmental Kuznets Curve." *World Development* 25:2(Feb.):191–198.

Sanderson, Steven K. 1999. *Social Transformations: A General Theory of Historical Development*. Lanham, Md.: Rowman & Littlefield. (Originally published 1995, London: Blackwell.)

Seccombe, Wally. 1983. "Marxism and Demography." *New Left Review* 137(Jan.–Feb.):22–47.

———. 1992. *A Millennium of Family Change: Feudalism to Capitalism in Northwestern Europe*. London: Verso.

Shaikh, Anwar. 1992. "The Falling Rate of Profit as the Cause of Long Waves: Theory and Empirical Evidence." Pp. 174–194 in *New Findings in Long-Wave Research*, edited by Alfred Kleinknecht, Ernest Mandel, and Immanuel Wallerstein. New York: St. Martin's.

Silver, Beverly. 1992. "Class Struggle and Kondratieff Waves, 1870 to the Present." Pp. 279–295 in *New Findings in Long-Wave Research*, edited by Alfred Kleinknecht, Ernest Mandel, and Immanuel Wallerstein. New York: St. Martin's.

———. 1995. "World-Scale Patterns of Labor-Capital Conflict: Labor Unrest, Long Waves & Cycles of Hegemony." *Review* 18:1(Winter):155–187.

Singer, Fred. 1970. "Human Energy Production as a Process in the Biosphere." Pp.163–172 in *Energy*, edited by Fred Singer. New York: Scientific American/Freeman.

Smith, David A., and Douglas White. 1992. "Structure and Dynamics of the Global Economy: Network Analysis of International Trade, 1965–1980." *Social Forces* 70:4(June):857–893.

Snyder, David, and Edward Kick. 1979. "Structural Position in the World-System and Economic Growth, 1955–1970: A Multiple-Network Analysis of Transnational Interactions." *American Journal of Sociology* 84:5(March):1096–1126.

Solomou, Solomos. 1990. *Phases of Economic Growth: 1850–1973*. London: Cambridge University Press.

"Special Report: The End of Cheap Oil." 1998. *Scientific American* (March).

Star, Chauncey. 1971. "Energy and Power." Pp. 8–21 in *Energy*, edited by Fred Singer. New York: Scientific American/Freeman.

Straussfogel, Debra. 1998. "How Many World-Systems? A Contribution to the Continuationist/Transformationist Debate." *Review* 21:1:1–29.

Suter, Christian. 1992. *Debt Cycles in the World-Economy: Foreign Loans, Financial Crises and Debt Settlements, 1820–1990*. Boulder, Colo.: Westview.

Tainter, Joseph A. 1988. *The Collapse of Complex Societies*. New York: Cambridge University Press.

Terlouw, Cornelius Peter. 1992. *The Regional Geography of the World-System: External Arena, Periphery, Semiperiphery, Core*. Utrecht: Faculteit Ruimtelijke Wetenschappen, Rejksuniversiteit Utrecht.

Thomas, Brinley. 1954. *Migration and Economic Growth*. London: Cambridge University Press.

Thompson, William R. 1992. "Long Cycles and the Geohistorical Context of Structural Transitions." *World Politics* 43:127–152.

Tilly, Charles. 1989. *Coercion, Capital and European States, AD 990–1990*. Cambridge, Mass.: Blackwell.

United Nations. 1992. *Human Development Report 1992*. New York: Oxford University Press.

———. 1994. *Human Development Report 1994*. New York: Oxford University Press.

Van Duijn, Jaap. 1983. *The Long Wave in Economic Life*. London: Allen Unwin.

Van Rossen, Ronan. 1996. "The World-System Paradigm as a General Theory of Development: A Cross-National Test." *American Sociological Review* 61:3(April):508–527.

Waldrop, M. Mitchell. 1992. *Complexity: The Emerging Science at the Edge of Order and Chaos.* New York: Simon & Schuster.

Wallerstein, Immanuel. 1974. *The Modern World-System: Capitalist Agriculture and the Origins of European World-Economy in the Sixteenth Century.* New York: Academic Press.

———. 1984a. "The Three Instances of Hegemony in the History of the Capitalist World-Economy." Pp. 100–108 in *Current Issues and Research in Macrosociology, International Studies in Sociology and Social Anthropology,* vol. 37, edited by Gerhard Lenski. Leiden: Brill.

———. 1984b. *Historical Capitalism.* London: Verso.

Wilkinson, David. 1987. "Central Civilization." *Comparative Civilizations Review* 17(Fall):31–59.

———. 1991. "Cores, Peripheries and Civilizations." Pp. 113–166 in *Core/Periphery Relations in Precapitalist Worlds,* edited by Christopher Chase-Dunn and Thomas D. Hall. Boulder, Colo.: Westview.

———. 1992. "Cities, Civilizations and Oikumenes: I." *Comparative Civilizations Review* 27(Fall):51–87.

———. 1993. "Cities, Civilizations and Oikumenes: II." *Comparative Civilizations Review* 28(Spring):41–72.

Part II

From Many Disciplines

3

Archaeology and World-Systems Theory

Peter N. Peregrine

I hope to make a simple point in this article: that sociologists stand to gain a lot by learning something about non-Western and noncapitalist societies, particularly as they are understood by archaeologists. I make this point in the framework of world-systems theory because I think it is one of the few, perhaps the only, common theoretical grounds the social sciences have, a unifying framework that "pushes social science toward an understanding of change in which Western and non-Western, traditional and modern peoples are subject, if not to similar outcomes, then at least to similar laws" (Schneider 1977, 26). Unfortunately, world-systems theory has not yet attained its unifying potential. One reason is that many social scientists are not closely familiar with scholarship on non-Western and noncapitalist societies, and scholarship on such societies has not been actively drawn into the wider development and application of world-systems theory. I suggest that archaeology holds a key to unlocking the potential of world-systems theory and expanding it to a truly encompassing theoretical framework.

Archaeology has tremendous potential for developing world-systems approaches to non-Western and noncapitalist societies because it has the ability both to explore non-Western and noncapitalist societies with the sophistication of anthropology and to explore societies in existence long before the capitalist world-system began to evolve. The idea that archaeology has a special role in developing world-systems theory has been recognized by a number of archaeologists who have produced a series of valuable studies of prehistoric world-systems (for reviews see Chase-Dunn and Hall 1991b, 36–37; Hall and Chase-Dunn 1993, 1994; Sanderson 1991, 173–175; Schortman and Urban 1992b; and volumes by Champion 1989; Chase-Dunn 1992; Chase-Dunn and Hall 1991a; Peregrine and Feinman 1996; Rowlands, Larsen, and Kristiansen 1987; and Schortman and Urban 1992a). Perhaps more importantly, a number of archaeologists have begun the difficult task of reshaping world-systems theory to make it more applicable to noncapitalist and non-Western societies. This reformulation has been proceeding along two general lines: (1) redefining the nature of geographic differentiation within world-systems and (2) redefining the nature of economic interdependence among differentiated regions.

REDEFINING CORE/PERIPHERY RELATIONS

Under the first of these two research agendas, most work has proceeded by a rethinking of the specific roles or functions of cores and peripheries. Ekholm and Friedman (1985) suggest that social scientists should consider all societies, whether capitalist or noncapitalist, as part of "global systems" based on both modes of production and modes of social reproduction, rather than as parts of world-systems based solely on production. Within the concept of global systems, Ekholm and Friedman (1985, 113–115) define cores ("centers" in their terminology) and peripheries in terms similar to the way most sociologists do, but they add two additional structures: (1) dependent structures, which are dependent on interaction with the global system for their reproduction, but which are neither cores nor peripheries (and which they admit can be similar to semiperipheries); and (2) independent structures, which can reproduce themselves outside the global system, but whose reproduction can affect or be affected by the system. Independent structures may include "predatory" groups such as militaristic nomads (Hall 1991) or groups of "refugee" peoples surviving, untouched, in remote areas of the system.

These independent structures may be related to Upham's (1992) idea of "empty spaces" within world-systems. He defines these spaces simply as "those areas between major population centers: areas without obtrusive or distinctive evidence of population" (Upham 1992, 141). While empty spaces may seem devoid of population, Upham (1992) argues that they profoundly affect world-systems by posing "significant obstacles to travel, trade, and communication" (149). More importantly (and perhaps in more direct relation to the idea of independent structures), Upham (150) explains that empty spaces provide living areas for people whose social, political, and economic organization is different than that in other areas of the system. Maintaining access to or control over empty spaces requires societies in both the core and the periphery to develop mechanisms for interrelating with these peoples, mechanisms that are often maintained with difficulty (the need for interrelation with peoples of different sociopolitical organization seems to be related, in turn, to Hall's [1989] conception of contact peripheries in the capitalist world-system).

Kristiansen (1987) employs Ekholm and Friedman's conception of independent structures in world-systems to conduct a more substantial redefinition of core/periphery relations. Through his analysis of the world-system in Bronze Age Scandinavia he suggests that a two-tiered hierarchy of core/periphery relations may exist in some world-systems: (1) at the higher level based on organization at a regional scale, but lacking formal exploitive relationships, and (2) at the lower level based on organization between local centers and attached communities, with formal exploitive relationships (Kristiansen 1987, 82). In this model, the independent structures are themselves small world-systems, with local cores exploiting local peripheries. However, these independent world-systems also exist within a larger world-system based on the dependent, but not necessarily exploitive, trade of elite symbols between independent polities. The independent world-systems could physically survive without linkage to the larger system, but elites could not socially reproduce without this connection.

In a further extension of the general conception of independent structures in world-systems, Kohl (1987, 20–21) argues that peripheries themselves may not be dependent on cores in some world-systems. He explains that for Near Eastern Bronze Age societies, "peripheries situated between cores were far from helpless in dictating the terms of exchange; they could develop or terminate relations depending upon whether or not these relations were perceived to be in their best interest" (20). Kohl also argues (21–23) that in situations where technology is fairly simple and transportable, peripheries can be technologically sophisticated and even innovative. He concludes that "relations between ancient cores and peripheries were not . . . analogous to those that underdevelopment theorists postulate are characteristic . . . today" (21), but that comparing prehistoric core/periphery relations with historic ones is important, since it leads us to ask why and in what way those relatives are different (29).

Chase-Dunn and Hall (1991b, 1991c, 1992, 1993) have spent considerable effort attempting to answer questions similar to Kohl's: Why and in what way were prehistoric, precapitalist core/periphery relations different from modern ones? They explain that, in order to answer these kinds of questions, "what is needed is an explicit effort to formulate competing notions of coreness and peripherality with an eye to a comparative study of intersocietal inequalities" (Chase-Dunn and Hall 1991, 7). As a first step toward the comparative study of world-systems, they have developed a number of guiding principles for examining cores and peripheries. First, they argue that there is a fundamental difference between core/periphery differentiation and core/periphery hierarchy. All world-systems have differentiated cores and peripheries, they argue, but not all have a hierarchical or exploitive relationship between them (Chase-Dunn and Hall 1991b, 18–19), and it is actually important to understand how and why exploitive relationships develop. With this in mind, Chase-Dunn and Hall (1991b, 21–32) develop a "typology" of world-systems—kin based, tributary, and capitalist—and provide hypotheses for the kinds of core/periphery relations that are present in each under different sociopolitical conditions. They suggest these are useful categories through which we can begin the comparative study of world-systems, both prehistoric and historic, capitalist and noncapitalist (Chase-Dunn and Hall, 1991b, 33; 1993).

Frank (1993) (see also Frank and Gills 1993) has put forward a very different conception of core/periphery relations, and indeed of world-systems themselves. He argues that there has only been one world system (without the hyphen) in world history that has been continuously developing for 5,000 years. Wallerstein's modern world-system (with the hyphen) is only the most recent transformation of the ancient one. The logic of cores and peripheries in Frank's conception is simple: Cores are constant accumulators of goods, information, capital, or whatever and peripheries are constant suppliers. This is an intriguing idea that Frank and others are already beginning to formulate into a provocative model of world history; however, there are problems when their ideas are applied to non-Western societies. For example, since the core of Frank's world system began in Mesopotamia and spread from there, it is unclear how the presence of multiple prehistoric world-systems in Oceania (Knapp 1993) and the Americas (Peregrine and Feinman 1996) fit into Frank's conception of a singular world system centered in the eastern Mediterranean.

What all these redefinitions of core/periphery relations seem to have in common is the notion that world-systems did exist in noncapitalist situations but that standard definitions of core/periphery relations are too strict to be directly applied to them (Chase-Dunn and Hall 1992, 88–89). Most of the scholars discussed have argued in one way or another that dependency or exploitation is a basic characteristic of the capitalist world-system but may not have been for precapitalist world-systems. Models of core/periphery relations in the absence of this dependency open world-systems theory to a wide variety of precapitalist and noncapitalist situations (see Chase-Dunn and Hall 1994 for an application of this idea).

REDEFINING ECONOMIC INTERDEPENDENCE

Under the second of the two research agendas discussed earlier, work has generally proceeded by rethinking the nature of the goods involved in trade within world-systems, and particularly by considering the possibility of an economic interdependence based on luxury goods. Wallerstein (1974a, 398) insists that world-systems must be based on trade in staples rather than luxury goods. He tells us bluntly that "in the long run, staples account for more of men's economic thrusts than luxuries" (Wallerstein 1974b, 42). The distinction between luxuries and staples is never clearly defined by Wallerstein, although he indicates that within trade relations luxuries are goods that are highly valued by one society but not by another, creating "rich trades" rather than the systemic interdependence of a world-system (Wallerstein 1982, 100; 1989, 130–132), and that staples are goods valued by all and used by individuals to maintain and reproduce themselves both physically and socially (Wallerstein 1974b, 302). The idea that social reproduction can require access to certain goods, goods valued by all members of a trade relation, has been the basis for a wide-ranging critique of Wallerstein's exclusion of luxury goods from the processes of world-systems.

Schneider (1977) argues Wallerstein's insistence that staples and staples alone could create a systemic interdependence between polities contradicts the reality of many noncapitalist economic systems. As Schneider (1977) plainly states:

Before the emergence of the capitalist world-system . . . most exchanges between distant places involved the movement of luxuries. . . . For many authorities . . . [luxuries and staples] are implicitly categorized as opposites: preciosities versus essentials or utilities. I suggest that this dichotomy is a false one which obscures the systemic properties of the luxury trade. (21)

Schneider (1977) goes on to consider a variety of prehistoric and historic cases that support her proposition that luxury trade can have systemic properties (23–27). She concludes that, for Europe and Asia, "it is possible to hypothesize a *pre-capitalist* world-system, in which core areas accumulated precious metals while exporting manufactures, whereas peripheral areas gave up these metals . . . against an inflow of finished goods" (25).

Much theoretical work in the archaeology of prehistoric world-systems has followed Schneider's lead, predicting and modeling situations in which luxury trade is a systemic process. Blanton and Feinman (1984), for example, examine luxury trade in pre-Hispanic Mesoamerica. They tell us that the exchange and consumption of particular categories of luxury goods was vital to the maintenance of political power in many pre-Hispanic Mesoamerican polities—so vital that it was a major source of both internal conflict and external conquest (676–677). Indeed, they argue that "a consequence of the growth of powerful core states in ancient Mesoamerica was a widespread stimulation of trade, a reorienting of priorities in many places toward production and exchange in the world-system arena" (678). Blanton and Feinman argue further that, through the demands of tribute in the form of labor-intensive luxury goods that could be exchanged in the world-system, cores actively peripheralized outlying regions in ancient Mesoamerica (678–679).

Schneider's argument that luxury trade can have systemic impact in noncapitalist worlds has been more directly tied to the idea of social reproduction through the concept of prestige-goods systems. As defined in two seminal articles, by Friedman and Rowlands (1977) and Frankenstein and Rowlands (1978), prestige-goods systems exist when important aspects of political alliance or social reproduction are tied to the consumption or exchange of specific exotic preciosities that can be obtained only through foreign trade. Frankenstein and Rowlands (1978) lucidly explain the economic logic of prestige-goods systems:

The specific economic characteristics of a prestige-goods system are dominated by the political advantage gained through exercising control over access to resources that can only be obtained through external trade. However, these are not the resources required for general material well-being or for the manufacture of tools and other utilitarian items. Instead, emphasis is placed on controlling the acquisition of wealth objects needed in social transactions, and the payment of social debts. Groups are linked to each other through the competitive exchange of wealth objects as gifts and feasting in continuous cycles of status rivalry. Descent groups reproduce themselves in opposition to each other as their leaders compete for dominance through differential access to resources and labour power. (76)

While never explicitly linked to world-systems theory, it is clear that there is economic interdependence, geographic differentiation, and competition between independent polities, so it is reasonable to conceive of prestige-goods systems as world-systems (Peregrine 1991a, 1992, 5–6). Indeed, Frankenstein and Rowlands (1978, 80–81) discuss economic interdependence in terms of core/periphery relations, and in terms of cores dominating peripheries by controlling their access to prestige goods.

A number of scholars have applied the concept of prestige-goods systems in an explicitly world-system perspective. Kristiansen (1987) employed the concept in the article on Bronze Age Scandinavia discussed earlier, where he argued that much of the interdependence at the higher level of the world-system, the level of dependent structures, was based on the exchange of prestige goods needed by political leaders to display and maintain their power. Friedman (1982) applied the idea of prestige-goods systems in an attempt to explain the different evolutionary trajectories evident in Oceanic prehistory. He argued that different social and political strategies of control over access to and distribution of prestige goods led to different patterns of political centralization and decentralization. McGuire (1987) used the idea to examine core/periphery relations in the southwestern United States. He argued that a systemic interdependence existed between elites in the Phoenix Basin, who needed shell from the western Papagueria to produce prestige goods for their social reproduction, and the peoples of the Papagueria, who relied on the Phoenix Basin core for agricultural products when their unreliable method of obtaining irrigation water from rainfall runoff failed.

In my own work on prehistoric world-systems (Peregrine 1991a, 1991b, 1992, 1995), I have attempted to reformulate the idea of prestige-goods systems into a broadly applicable model of noncapitalist world-systems. Through a cross-cultural analysis of prestige-goods systems I defined the nature of prestige goods themselves and demonstrated that the character of these goods changes as societies become more politically centralized. Specifically, prestige goods tend to become more ornate, requiring extensive labor or sophisticated techniques to manufacture, and I posit that this change allows elites to better control production of these goods (Peregrine 1991b; 1992, 47–67). I also demonstrated that prestige-goods systems can exist in a variety of sociopolitical contexts, from simple, egalitarian societies to states, and that core/periphery relations in prestige-goods systems can also range from simple differentiation to domination (Peregrine 1992, 9–26).

In addition, I developed a model of social evolution in prestige-goods systems that predicts the conditions under which political centralization will change (Peregrine 1992, 27–46). I argue that political power in prestige-goods systems is inversely related to the density of prestige goods in the system. Density, in turn, is the product of three interacting variables: (1) population, (2) the volume of trade (i.e., the number of prestige goods obtained per transaction), and (3) the system's boundedness (i.e., the number of transactions for prestige goods that take place over a given time). An alteration in any of these variables can lead to changes in the density of prestige goods in the system, and hence to changes in political power. I suggest that most political strategy, and indeed much of elite activity, in prestige-goods systems focuses on controlling these variables (Peregrine 1991a, 1995). Because they are so readily comprehensible, and because so much work has already been done on them, prestige-goods systems seem to be a particularly useful "test case" for exploring political strategy and the evolution of political authority in noncapitalist world-systems.

It seems apparent that work under this second research agenda is more limited in scope than under the first. Most research has focused on addressing the single problem of demonstrating that trade in luxury goods can have a systemic impact and on creating models that explore the impact of luxury trade. More specifically, most of this work has been focused on a single kind of systemic luxury trade—that taking place to use to help develop theory, few scholars have considered other potential reformulations of the nature of trade, and many questions remain. Are there other systemic forms of luxury trade outside of prestige-goods systems? Can, for example, the exchange of information or ritual create a systemic interdependence?

CONCLUSIONS

The two research agendas discussed in this article are not as disparate as they may seem. Both attempt to expand the realm of world-systems theory from a focus on a mode of production (particularly capitalist) to one that includes or focuses on a mode of reproduction (see Chase-Dunn and Hall 1992, 101–105). In noncapitalist worlds most societies are largely self-sufficient in terms of subsistence production, but they often rely on interaction with other societies for social reproduction (Blanton et al. 1996). Reshaping our conceptions about world-systems to accept the idea that core/periphery relations often involve social reproduction, not subsistence, seems to be a focal point of both agendas. In reformulating world-systems theory to make it more applicable to non-Western and noncapitalist economies, both agendas are, in their own way, focusing on interdependence that is necessary for social reproduction rather than physical reproduction. Certainly social reproduction may be more vital to survival in the kinds of group-oriented and kin-based societies archaeologists typically study than in our individualistic society, but I suggest that the examination of social reproduction and its systemic impact should not be limited to non-Western, noncapitalist societies, but should also be integrated into the study of the capitalist world-system.

To conclude, it seems clear that sociology, and indeed the social sciences as a whole, have a lot to gain from the work of archaeologists. Not only are new insights into the importance of social reproduction obtained, but by engaging the data and perspectives of archaeology, a number of problems that often confound world-systems analyses can be addressed. For example, research on the transformation of luxury goods into staple commodities (see Mintz 1985), the role of nonstate societies in the expansion of the modern world-system (e.g., Hall 1989), and the social transformations and disruptions that seem central to the operation of the modern world-system (e.g., Ferguson and Whitehead 1992; Wolf 1982) have all benefited from the active inclusion of archaeological data and perspectives. Frank's provocative perspective on the 5,000-year evolution of the modern world system could not have even been conceived of without insights from archaeology. Surely other problems in world-systems analysis will similarly benefit, and I suggest that actively engaging archaeology can only help unlock the potential world-systems theory has for being an encompassing framework for the social sciences.

REFERENCES

Blanton, Richard, and Gary Feinman. 1984. "The Mesoamerican World-System." *American Anthropologist* 86:673–682.

Blanton, Richard, Gary Feinman, Stephen Kowalewski, and Peter Peregrine. 1996. "A Dual-Processual Theory for the Evolution of Mesoamerican Civilization." *Current Anthropology* 37(1):1–14.

Champion, Timothy, ed. 1989. *Centre and Periphery: Comparative Studies in Archaeology.* London: Unwin Hyman.

Chase-Dunn, Christopher, ed. 1992. "Comparing World-Systems." Special issue. *Review* 15(3).

Chase-Dunn, Christopher, and Thomas Hall. 1994. "The Historical Evolution of World-Systems." *Sociological Inquiry* 64:257–280.

———. 1993. "Comparing World-Systems: Concepts and Working Hypotheses." *Social Forces* 71:851–886.

———. 1992. "World-Systems and Modes of Production: Toward the Comparative Study of Transformations." *Humbolt Journal of Social Relations* 18:81–117.

———, eds. 1991a. *Core/Periphery Relations in Precapitalist Worlds.* Boulder, CO: Westview.

———. 1991b. "Conceptualizing Core/Periphery Hierarchies for Comparative Study." Pp. 5–44 in *Core/Periphery Relations in Precapitalist Worlds,* edited by C. Chase-Dunn and T. Hall. Boulder, CO: Westview.

———. 1991c. "Epilogue." Pp. 277–290 in *Core/Periphery Relations in Precapitalist Worlds,* edited by C. Chase-Dunn and T. Hall. Boulder, CO: Westview.

Ekholm, Kasja, and Jonathan Friedman. 1985. "Towards a Global Anthropology." *Critique of Anthropology* 5:97–119.

Ferguson, R. Brian, and Neil Whitehead, eds. 1992. *War in the Tribal Zone.* Santa Fe: School of American Research Press.

Frank, Andre Gunder, and Barry Gills, eds. 1993. *The World System: Five Hundred Years or Five Thousand?* London: Routledge.

Frankenstein, Susan, and Michael Rowlands. 1978. "The Internal Structure and Regional Context of Early Iron Age Society in South-Western Germany." *Bulletin of the Institute of Archaeology of London* 15:73–112.

Friedman, Jonathan. 1982. "Catastrophe and Continuity in Social Evolution." Pp. 175–196 in *Theory and Explanation in Archaeology,* edited by C. Renfrew, M. Rowlands, and B. Segraves. New York: Academic.

Friedman, Jonathan, and Michael Rowlands. 1977. "Notes Towards an Epigenetic Model of the Evolution of 'Civilisation'." Pp. 201–276 in *The Evolution of Social Systems*, edited by J. Friedman and M. Rowlands. London: Duckworth.

Hall, Thomas. 1991. "The Role of Nomads in Core/Periphery Relations." Pp. 212–239 in *Core/Periphery Relations in Precapitalist Worlds*, edited by C. Chase-Dunn and T. Hall. Boulder, CO: Westview.

———. 1989. *Social Change in the Southwest, 1350–1880*. Lawrence: University of Kansas Press.

Hall, Thomas, and Christopher Chase-Dunn. 1994. "Forward into the Past: World-Systems before 1500." *Sociological Forum* 9:295–306.

———. 1993. "World-Systems Perspective and Archaeology: Forward into the Past." *Journal of Archaeological Research* 1:121–143.

Knapp, A. Bernard. 1993. "Comment on Bronze Age World System Cycles." *Current Anthropology* 34:413–414.

Kohl, Philip. 1987. "The Use and Abuse of World-Systems Theory: The Case of the Pristine West Asian State." Pp. 1–36 in *Advances in Archaeological Method and Theory*, vol. 11, edited by M. B. Schiffer. San Diego: Academic.

Kristiansen, Kristian. 1987. "Centre and Periphery in Bronze Age Scandinavia." Pp. 75–85 in *Centre and Periphery in the Ancient World*, edited by M. Rowlands, M. Larsen, and K. Kristiansen. Cambridge: Cambridge University Press.

McGuire, Randall. 1987. "The Papaguerian Periphery: Uneven Development in the Prehistoric U.S. Southwest." Pp. 123–139 in *Polities and Partitions: Human Boundaries and the Growth of Complex Societies,* edited by K. M. Trinkaus. Tucson: Arizona State University, Anthropological Research Papers, Number 37.

Mintz, Sidney W. 1985. *Sweetness and Power*. New York: Viking.

Peregrine, Peter N. 1995. "Networks of Power: The Mississippian World-System." Pp. 132–143 in *Native American Interactions,* edited by M. Nassaney and K. Sassaman. Knoxville: University of Tennessee Press.

———. 1992. *Mississippian Evolution: A World-System Perspective*. Madison, WI: Prehistory Press.

———. 1991a. "Prehistoric Chiefdoms on the American Midcontinent: A World-System Based on Prestige Goods." Pp. 193–211 *Core/Periphery Relations in Precapitalist Worlds*, edited by C. Chase-Dunn and T. Hall. Boulder, CO: Westview.

———. 1991b. "Some Political Aspects of Craft Specialization." *World Archaeology* 23:1–11.

Peregrine, Peter, and Gary Feinman, eds. 1996. *Pre-Columbian World-Systems*. Madison, WI: Prehistory Press.

Rowlands, Michael, Mogens Larsen, and Kristian Kristiansen, eds. 1987. *Centre and Periphery in the Ancient World*. Cambridge: Cambridge University Press.

Sanderson, Stephen. 1991. "The Evolution of Societies and World-Systems." Pp. 167–192 in *Core/Periphery Relations in Precapitalist Worlds*, edited by C. Chase-Dunn and T. Hall. Boulder, CO: Westview.

Schneider, Jane. 1977. "Was There a 'Pre-Capitalist' World-System?" *Peasant Studies* 6:20–29.

Schortman, Edward, and Patricia Urban, eds. 1992a. *Resources, Power, and Interregional Interaction*. New York: Plenum.

———. 1992b. "The Ancient World-System." Pp. 17–21 in *Resources, Power, and Interregional Interaction*, edited by E. Schortman and P. Urban. New York: Plenum.

———. 1987. "Modeling Interregional Interaction in Prehistory." Pp. 37–95 in *Advances in Archaeological Method and Theory*, vol. 11, edited by M. B. Schiffer. San Diego: Academic.

Upham, Steadman. 1992. "Interaction and Isolation: The Empty Spaces in Panregional Political and Economic Systems." Pp. 139–152 in *Resources, Power, and Interregional Interaction*, edited by E. Schortman and P. Urban. New York: Plenum.

Wallerstein, Immanuel. 1989. *The Modern World-System*, vol. 3. New York: Academic.

———. 1982. "World-Systems Analysis: Theoretical and Interpretive Issues." Pp. 91–203 in *World-Systems Analysis: Theory and Methodology*, edited by T. K. Hopkins and I. Wallerstein. Beverly Hills, CA: Sage.

———. 1974a. *The Modern World-Systems*, vol. 1. New York: Academic.

———. 1974b. "The Rise and Future Demise of the Capitalist World-System: Concepts for Comparative Analysis." *Comparative Studies in Society and History* 16:387–415.

Wolf, Eric. 1982. *Europe and the People Without History*. Berkeley: University of California Press.

4

Geography, Place, and World-Systems Analysis

Fred M. Shelley and Colin Flint

A theme of fundamental interest in geography is that of place. Throughout the history of the discipline, geographers have focused on analyzing changes in places and relationships between them. Because world-systems analysis is explicitly concerned with social change, it has inherent appeal to geographers concerned with explaining the development of places and their relationships with one another. Not only has a considerable volume of geographical research been influenced by world-systems analysis in recent years, but geographers have made important contributions to world-systems analysis itself. Some of the major contributions are listed in the "Suggested Readings" section at the end of this chapter, and this literature is reviewed in greater detail in an earlier paper (Flint and Shelley 1996).

In our earlier paper, we reviewed some of the important contributions of geographers to the world-systems analysis literature, identified the potential for future research linking geography and world-systems analysis, and described some of the pitfalls associated with uncritical application of world-systems analysis to geographical research. We identified two lines of inquiry: geographical research based on world-systems analysis as its theoretical framework, and other research by both geographers and nongeographers that attempts explicitly to incorporate aspects of the geographical perspective into world-systems analysis. Within these bodies of literature, we identified and examined three themes: regions, hegemony, and place. Each of these themes involves research perspectives that are central to the discipline of geography and within which world-systems analysis plays a central role. In this chapter, we examine the role of place in greater detail.

Over the past few years, geographers—whose discipline has long been concerned about the description and understanding of places—have achieved a deeper and more sophisticated understanding of the relationship between place and the world-system. In developing this relationship between places and the world-system, we review some of the recent literature linking place to the world economy, and we examine the potential impacts of the profound changes currently taking place within the global economy on places within the world-system. Such an examination may encourage a future research agenda that would encourage further cross-fertilization between geography and world-systems analysis. As in our earlier paper, we use the term *world-systems analysis* to refer to the body of work generated or inspired by Immanuel Wallerstein and other scholars at the Fernand Braudel Center and summarized in the introduction to this collection.

GEOGRAPHICAL REFORMULATIONS OF WORLD-SYSTEMS ANALYSIS

Wallerstein (1979) conceptualizes the structure of the capitalist world-economy as a hierarchy consisting of the core, periphery, and semiperiphery. This conceptualization is inherently geographical (Flint and Shelley 1996). Throughout history, whether a particular society is regarded as having core, peripheral, or semiperipheral status has often been linked to its locational characteristics. Core status is often associated with a favorable endowment of resources or with a favorable location relative to other societies. For example, the Middle Atlantic colonies within what is now the United States quickly achieved core status relative to New England and the South because of their central location and high-quality agricultural land and other natural resources (Agnew 1987; Meinig 1986). The theoretical basis of how location, resources bases, and other factors have influenced core and peripheral status outside the modern European world-economy has been enriched in far more detail by Chase-Dunn and Hall (1997).

The spatial relationships between core, periphery, and semiperiphery soon formed the basis for a more explicit spatialization of world-systems analysis. Taylor (1981, 1982) argues that world-systems analysis provides a theoretical basis for the development of the hierarchy of spatial scales, from global to state to local, that has long formed an important component of geographical research. The location and development of states on the world political map has long been an important concern of political geographers. By reformulating this traditional hierarchy of spatial scales in world-systemic terms, however, Taylor illustrates the utility of examining states and localities as components of the world economy. This reformulation provided a fresh perspective on analysis of spatial differentiation in state forms and functions as well as in understanding the historical development of individual cities, communities, regions, localities, and other places.

In developing this world-systemic reformulation of spatial hierarchies, geographers also expressed concern about the dangers associated with treating categories such as core, periphery, and semiperiphery or global, state, and local scale as actors. Categories themselves are not actors; rather, it is human activity that creates cores and peripheries which in turn mediate people's choices (Taylor 1988, 1991). The concepts of core and periphery related to processes of the capitalist world-economy and not to territorial entities. Core processes involve high-wage, high-value-added production processes coupled with high consumption, while peripheral processes are the exact opposite. A region or place may be regarded as "core-like" if core processes predominate there, but the territory itself can be neither core nor peripheral. Similar to Swyndegouw's (1997, 141) contribution, we urge that the starting point for analyses of social change is the definition of economic, political, and cultural processes that in turn transform and transgress scales and places. World-systems analysis provides a framework with which to identify these processes. Geography illustrates how scale and place are created as part of social conflict and then, in turn, mediate future conflicts.

World-systems analysis has been criticized as characterized by excessive structural determinism (Skocpol 1977; Agnew 1982). Sensitive to local-scale differences between places, geographers working from a world-systemic perspective have recognized that place-specific institutions and circumstances generally mediate social interaction with larger-scale structures, including the world economy itself (Flint and Shelley 1996). Such analysis has occurred at a variety of spatial scales. At a national and global scale, the world-systemic perspective has influenced analysis of geopolitics, colonialism, world trade, and other macroscale social processes. Other analysis, which is examined in more detail in this chapter, emphasizes cities or communities and how these places influence, and in turn are influenced by, the changing world-economy.

GEOGRAPHICAL ANALYSIS OF PLACES WITHIN THE WORLD-ECONOMY

The description, analysis, and understanding of local-scale geographical units, or places, has a long tradition within geography. In recent years, the infusion of world-systems analysis into geography has enhanced geographical analysis and explanation of places. Places influence and in turn are influenced by the world-system. Each place is characterized by a different location, mode of production, transportation system, resource endowment, and population. Any place can be conceptualized and analyzed from a variety of perspectives that emphasize linkages between the local and the global. For example, Knox and Taylor (1995) have developed an analysis of cities as nodes within global commodity chains. This conceptualization may be especially useful in a contemporary world-economy characterized by production for global markets, worldwide distribution systems, and global transportation and production networks.

For centuries, individual communities have specialized in the production of particular goods and services. The very identities of many communities—for example, the association of Detroit with the automobile industry, California's Silicon Valley with computers, and Pittsburgh with steel production—are the result of historical and contemporary concentrations of particular industries in particular communities. Other communities specialize in trade and exchange. Prior to the development of modern aviation and telecommunications, this function was generally undertaken by port cities. Today, cities that are favorably located with respect to telecommunications, aviation, or previously impenetrable political boundaries take advantage of this location to enhance their status within the global economy. Perhaps even more fundamental is the association of particular communities with the circulation of capital. Cities such as New York, London, Frankfurt, and Tokyo owe their importance in the contemporary global economy to a considerable degree to their importance as financial centers and nodes managing globally dispersed economic activity (Sassen 1991).

As Taylor (1996) has pointed out, the nation-state as a social structure is seldom questioned. Terms such as "America's interest" or "Japanese influence" commonly appear in ordinary discourse. In short, much of modern mainstream social science treats societies as bounded by state borders. The world-systemic reformulation of spatial scale illustrates, however, understanding the nation-state must be linked to analysis of processes operating at the global scale as well as at the local scale.

These linkages are illustrated with the example of Detroit, which has been a center for automobile production since the beginning of the twentieth century (Halberstam 1986). Detroit was chosen as a center for automobile production in part because its banking community was more willing to invest in fledgling companies than were bankers in older, larger, and more conservative industrial cities. The rapid development of the automobile industry brought unprecedented growth to Detroit, whose population increased from approximately 100,000 in 1900 to 1.8 million in 1950. Many of Detroit's new arrivals were immigrants from Eastern Europe, whites from Appalachia, and African-Americans from the rural South. These migrants, coming as they did from very different rural cultures, established ethnic communities located in uneasy proximity to one another. Ethnic communities were maintained by racial bias in the housing market (Kodras 1997, 52).

In the 1970s, dramatic increases in the cost of petroleum had profound effects on the Detroit area itself, within the United States and neighboring Canada, and within the world-economy as a whole. The decline in automobile sales following the energy crisis of the 1970s cost the Detroit area many thousands of jobs. Those losing their jobs moved in large numbers to other places, such as the Sunbelt. The energy crisis influenced public policy within the United States and other countries, and it contributed to a dramatic increase in sales of imported automobiles, contributing to the rise of Japan's economy (Halberstam 1986). Thus, understanding changes in Detroit is closely related to a fundamental change in the global economy as well as to state and local responses to that change.

The simple example of Detroit illustrates the importance of analyzing linkages between the local, state, and global scale in examining places. Yet geographers examining places within the changing world economy must pay particular attention to the warnings about structural determinism articulated by Skocpol (1977), Agnew (1982), and others. Such concern is especially important at the local scale because it is at the local scale where persons lead their lives, make decisions, and in short exercise human agency most directly. Moreover, critics of world-systems analysis in geography have argued that the world-systems perspective is characterized not only by structural determinism but also by economic reductionism. Following Giddens (1985), Painter (1995, 112) argues that "where politics and culture are examined [in world-systems analysis], they tend to be explained in terms of the economy." In combination with the traditional reliance of political geographers on large-scale international forces affecting world politics, economic reductionism has meant that noneconomic influences on human agency are often downplayed or ignored.

Exclusive reliance on economic explanations, of course, eliminates the potential effects of culture, race, religion, gender, ethnicity, politics, and other factors that influence choices made by people within localities. These factors frequently affect social movements that originate within localities and yet impact on persons at much larger spatial scales. As Painter points out, however, political geographers who have embraced the world-systems paradigm have sometimes neglected such social movements and other local influences. He states that "social movements which political geographers study more frequently than any other are those associated with regionalism and nationalism" (Painter 1995, 24). Ignoring social movements associated with other social and cultural factors, he argues, runs the risk of overlooking the rich variety of local experience that both influences and is affected by global forces.

Painter's comments can be regarded as both prescriptive and prophetic. His concern that geographers risk neglecting local social and cultural influences on human agency reinforces the warnings articulated by Skocpol, Agnew, and others. At the same time, recent scholarship applying world-systems analysis to human geography has responded to these concerns through even more explicit focus on human agency as mediated by social and cultural forces. Such analysis is especially evident in recent literature focusing on particular places within the world economy. Here, we examine three themes: the geography of specific places; the geography of interacting places, and the actual and potential influence of current and impending changes in the global economy on places in the twenty-first century.

GEOGRAPHY, PLACE, AND WORLD-SYSTEMS ANALYSIS

Throughout the history of the discipline, geographers have been concerned with analyzing the characteristics of specific places. In recent years, this analysis has been enhanced by incorporating linkages between individual places and the global economy.

As we indicated in our earlier paper, a series of books known as the "World Cities Series" is being produced by John Wiley and Sons. This series includes books about such world cities as Tokyo (Cybriwsky 1995) and Mexico City (Ward 1995).

The authors of these and other books in the series have explicitly traced and developed linkages between the respective cities and the world economy. In short, it is useless to try to understand the geography of Tokyo, Mexico City, or other world cities without reference to the development of their changing positions within the world economy.

Agnew's (1995) analysis of Rome provides an effective illustration. The "Eternal City's" position within the modern world economy as well as within preindustrial social orders has changed many times over the course of the past three millennia. Founded as a city-state on the banks of the Tiber River, Rome emerged as the world's most populous and dominant city during the heyday of the Roman Empire. As the Roman Empire began to decay, Rome's population and influence declined. Indeed, the lessening importance of the city of Rome is a textbook example of the instability of world-empires such as the Roman Empire described by Wallerstein (1979).

Once the Roman Empire collapsed, Rome remained an important city because of its association with the Roman Catholic Church. By the Middle Ages, the pope was regarded simultaneously as the leader of the church worldwide and as the bishop of Rome itself—a dual role continuing to the present day. The very name "Roman Catholic" in fact associates the universality of the Catholic faith with the city of Rome. As Agnew points out, this association has invested the city with great mythic power within Western civilization. This mythic power in turn gave Rome political authority within reunified Italy during the late nineteenth and early twentieth centuries. Industrialized northern Italy has been the economic center of the country since it became independent in 1871, yet Italian leaders recognizing Rome's history as the seat of the Roman Empire and the church made the Eternal City its political capital. In examining contemporary Rome, Agnew pays considerable attention to the interface between Rome's mythic power as it affects Rome's residents as well as the city's interaction with the rest of the world-economy.

Leyshon and Thrift (1997) provide another effective analysis of locality, place, and the world economy in their analysis of world financial markets in London, New York, and Tokyo. As Leyshon and Thrift document, changes in the global economy have affected the status of London as one of the world's great financial centers. This in turn has had profound effects on life in London itself and in nearby southeast England.

In examining the impact of the financial sector on London, Leyshon and Thrift identify changes in the global economy that have changed both the location and the importance of major world markets. Over the course of the late twentieth century, they illustrate the increasing strength of international financial capital relative to control by the nation-state. As the state's power to control the international mobility declined, national stock markets declined and major financial markets consolidated. London, New York, and Tokyo emerged as the world's three major stock markets. The concentration of international financial capital markets in these three cities was attributed to the dominance of each center over regional financial markets in Europe, North America, and East Asia, respectively, as well as to their locations in widely separated time zones, enabling transactions to take place at all hours. As early as 1985, these three cities were responsible for half of all of the world's international banking activity and were the venues of the world's three largest stock exchanges (Leyshon and Thrift 1997, 126). Indeed, London had branch offices of all of the world's hundred largest banks as of 1994 (Clark 1996, 152).

This concentration of financial capital has had major implications not only for the global economy but for the cities themselves. Examining London and its hinterland in detail, Leyshon and Thrift document how the development of London as a financial center has influenced a variety of areas of life in southeastern England, including the region's employment base, housing market, retail sector, and so on. Thus, an important shift with the world economy resulted in a dramatic change in the location of an important commodity flow (that of money), which in turn had considerable local impact, as Leyshon and Thrift point out.

Analyses of places with an implicit or explicit world-systemic perspective is by no means limited to large cities such as Rome and London. Many interesting and effective analyses of suburbs, rural communities, and peripheral places have been published (e.g., Hanna 1995). For example, Wood (1997) has examined the development of the Vietnamese-American community of northern Virginia. This community was established by refugees from urban Vietnam who moved to the United States in the mid-1970s following the collapse of the South Vietnamese government. Many of these refugees were educated government officials, military officers, and white-collar workers. Wood documents how these refugees established a variety of retail businesses in Arlington, Virginia. Arlington became a Vietnamese-American center because at the time it was a declining inner-city suburb, with "low rents and available space—and the disarray that was part and parcel of Washington's Metrorail construction" (Wood 1997, 65). The presence of the Vietnamese community revitalized the area but generated a new set of social problems associated with gang activity, overcrowding, and allegations of violations of American immigration law.

Of course, analyses of places in the global economy are by no means limited to entire cities or ethnic communities within cities. Much attention is being paid to the impacts of various aspects of identity construction on communities within larger places. Geographers have long been concerned with mapping and describing ethnic communities within cities and ethnic influences on landscapes (Allen and Turner 1996). In many cities, an increasing disparity between affluent middle- and upper-class communities and less affluent communities structured by gender, class, sexual orientation, ethnicity, and other factors has become evident (Dear and Flusty 1998). Fincher and Jacobs (1998) refer to these less affluent communities as "cities of difference." Cities of difference are structured in different ways in different places, and frequently these differences can be attributed to large-scale political and economic forces. For example, Dear and Flusty (1998) point to the role of global communications technology in creating a class of global economic managers and a large underclass handicapped by lack of access to information and electronic communication.

In a direct contribution to world-systems analysis, Jones (1998) has reemphasized the problem of conceptualizing the core, semiperiphery, and periphery in purely spatial terms. From a survey of the business elite in Santiago, Chile, Jones found evidence of a "social core"—the people employed and involved in core economic activities. Although "this core space does have a physical spatial form: a diaspora of business elites across the globe . . . the physical location of this group at any one time is not its defining feature" (Jones 1998, 304). In other words, the processes of the world-economy are driving forces behind the constructions of places, scales, and landscapes (Swyndegouw 1997).

PLACES IN THE TWENTY-FIRST CENTURY WORLD-ECONOMY

The examples of Rome, London, and Arlington illustrate the potential of integrating world-systemic concepts into the analysis of particular places. These analyses and many others are examples of how places are impacted by and in turn impact on the world economy. Such literature has revitalized urban geography, which has long relied on case studies of specific cities in analyzing processes of urbanization and community formation. In the years ahead, world-systemic concepts can and should continue to influence the study of places, and vice versa. As this literature develops, however, careful attention must be paid to the impacts of actual and impending changes in the world-economy on local places.

As indicated in our earlier paper, the history of the modern world economy can be interpreted in terms of ongoing conflict over hegemony. From that perspective, historical events can be interpreted in relation to the operation of large-scale economic and political forces within the world-economy. Taylor (1990, 1993) has traced the relationship between hegemonic cycles and Kondratieff waves over time to develop the concept of geopolitical orders and transitions. Geopolitical orders are defined by either periods of hegemonic dominance by a single power or relatively stable multipolar power relationships within the core. In recent world history, geopolitical orders include the European-dominated Victorian era of Great Power politics which ended abruptly with World War I, the period between World Wars I and II, and the cold war. Geopolitical orders are punctuated by abrupt transitions. For example, what Taylor (1990) calls the "long 1945" was a short but significant transition between the interwar world order and the cold war.

The cold war world order, which was dominated by conflict between the United States and the Soviet Union for hegemonic dominance, came to an abrupt end with the collapse of communism in the Soviet Union and its allies in "long 1989." Since that time, geographers and others have written much about the characteristics of the "new world order" that is replacing that of the cold war (Agnew and Corbridge 1995; Taylor 1996; Kegley and Raymond 1994). Of course, these and other scholars and journalists predicting changes in the global economy are far from united as to their predictions. Here, we identify a few of the major predictions and illustrate their actual and potential impacts on the analysis of place in the world economy.

One already evident transition is the increasing mobility of people, goods, and capital from place to place around the world. Increased mobility is associated with improved transportation and communication technology, the lowering or elimination of trade barriers in various parts of the world such as the European Union and the North American Free Trade Agreement, and the removal of the Iron Curtain and other political barriers that during the cold war had restricted or forbidden interaction between places. The elimination of such barriers has already had profound impacts on communities near the Iron Curtain in central Europe. Brunn (1998) examines the effects of the removal of these barriers on communities located near the borders. He documents increased opportunities for trade, cultural interchange, investment and migration in these border regions.

Examining local impacts of global changes in geopolitical orders such as these points out the need to reconceptualize the very meaning of a community. Traditional studies of places that have examined a community's economy, political structure, ethnic mix, demography, and other characteristics are based on an implicit assumption that the community's population is self-contained. Each community resident is assumed to work, shop, vote, socialize, and otherwise interact predominantly within that community. Studies of ethnic "homelands" populated by migrants from one culture area to another focus on the movement, re-creation, or transformation of various economic activities and cultural landscapes from the "old country" to a new location. For example, Ostergren (1988) describes the migration of entire communities from Dalarna, Sweden, to Isanti County, Minnesota, and the re-creation of Swedish cultural activities in the New World.

In an era characterized by increased capital mobility, the lowering of political barriers and lowered costs of long-distance transportation and communication, such a conceptualization of a community may be anachronistic. Not only is it impossible to understand a place without reference to global economic considerations that affect that place, but individuals and entire communities can and often do maintain active associations with two or more widely separated places. Brunn's analysis of day-to-day life in border communities in Eastern Europe calls attention to the fact that these communities' residents can now take advantage of their location relative to the current geopolitical transition.

The opportunity to take advantage of a border location is by no means limited to the countries located along the former Iron Curtain. Zelinsky and Lee (1998) describe this process as "heterolocalism." Physical proximity of home, work, shopping and other activities within one neighborhood or even one city or country is less important than was the case in the past. The concept of heterolocalism is clearly linked to world-systems analysis. The same global economic forces that have created a single world-economy have dramatically reduced the cost of movement within that economy. These changes have made it far easier for individuals and indeed entire communities to maintain and exploit enduring long-distance linkages between communities, and these linkages in turn have impacts on all communities involved.

The concept of heterolocalism is illustrated by comparing the experience of nineteenth-century immigrants into the United States with contemporary immigrants. Nineteeth-century immigrants from Europe to the United States such as the Swedish migrants to Minnesota described by Ostergren (1988) endured journeys of as much as several months to reach their new homes. Communication with friends and relatives in the "old country" might take even longer. Once an immigrant moved, it was difficult and expensive to return. Today, in contrast, an immigrant can travel around the world in a matter of hours or days. For this reason, immigrants need no longer cut ties with their former homes. Instead, they can exploit location within both old and new communities to individual and collective advantage.

The recent experience of European guest workers and recent migrants from Asia and Latin America to the United States illustrates this phenomenon. Millions of people move from origins in less developed countries to Europe and North America in search of employment opportunity. In contrast to their predecessors of centuries ago, such people do not need to leave their old homelands behind. Many regularly migrate back and forth between their birthplaces and work environments in the developed countries. Some retire to their birthplaces: the local economies of entire villages in Mexico are geared to the arrival of pension and social security checks in the United States. Remittances also contribute to economic development in areas that send large numbers of people abroad. One recent estimate suggests that upwards of $500 million annually is remitted from Vietnamese-Americans to their relatives in Vietnam, providing a major source of capital to the development of Vietnam's economy.

Heterolocalism is being reinforced by revolutionary changes in global communications. Recent technological advances have enabled instantaneous, cost-free global communication and capital transfer. This has spurred the rise of telecommuting, freeing many people from dependence on living near their workplaces. Popular resort communities, such as the mountain communities of Aspen, Vail, and Telluride in Colorado, are populated by an increasing number of persons who use the Internet and other communications technologies to telecommute while enjoying mountain scenery, skiing, and other recreational activities. Of course, the freedom of such people to reside in such resort communities while engaged in the global economy elsewhere has imperiled the fragile ecosystems of such environments.

The global diffusion of democracy illustrates the way in which political processes create struggles and decisions which in turn create scales and place-specific behavior (O'Loughlin et al. 1998). Democracy has diffused throughout the world in historic waves, and the diffusion of each wave has been constrained by resistance in particular regional contexts. An interpretation of these patterns suggests that democracy is a core process and that struggles for democracy are inseparable from the structural constraints of the world-economy. In a similar vein, the norms of political and economic liberalism have been disseminated by hegemonic powers (Taylor 1996) but are mediated in different ways across geographical settings. The result is the construction of new scales of political authority, through separatism and regionalism, as well as new forms of political behavior as people in different places interpret global trends in different ways.

What other changes in the contemporary global economy might affect the geographical analysis of places? In their description of a potential twenty-first century world order, Kegley and Raymond identify a number of issues affecting the global economy and its linkages to the international state system. These issues include the continued development of weapons of mass destruction, the continuing internationalization of the economy, the transfer of power from state governments to multinational corporations, the widening gap between rich and poor, environmental degradation, and the resurgence of ethnic conflicts in various parts of the world. In reviewing these issues, Painter (1995) adds three others: the growth of international terrorism and organized crime, the growth of religious fundamentalism, and the globalization of the media and communications technology.

How might these concerns affect geographical analysis of place from a world-systemic perspective? Geographers have already been identified and investigated some of these concerns. For example the "cities of difference" identified and described by Fincher and Jacobs (1998) are a local response to the widening gap between rich and poor. The growth of London as a financial center is a response to the internationalization of the economy and the declining power of the state relative to privately held corporations. Economic internationalization and the removal of barriers between places has also contributed to the emergence of heterolocalism and its impacts on individual communities.

Other issues associated with these changes in the world-economy have direct potential for such investigation for geographical analysis of places. For example, we are only beginning to understand the relationship between place, environmental degradation, and the global economy (Steinberg 1997). Community leaders, especially in less developed areas, have often expressed willingness to relax environmental protection standards to promote economic development. This view helps explain why some of the world's major cities of the less developed and newly industrializing countries—cities such as Cairo, Mexico City, Tokyo, and Seoul—are notorious for their air and water pollution problems. At the local level, analyses of the environmental justice movement relate communities, production, and environmental quality (Heiman 1996).

CONCLUSION

Understanding the dynamics of the changing world-system requires the understanding of individual places within the world-system. The discipline of geography is uniquely qualified to contribute to the understanding of places. In recent years, the geographical literature has been enriched by the infusion of world-systems analysis.

Geographical analysis of important and mundane places has in turn contributed to world-systems analysis by calling attention to the unique qualities of localities and thir characteristics at small spatial scales. In the twenty-first century, the role of place in the changing world-system will undoubtedly change, and geographers are well positioned to play an important role in understanding these changes.

SUGGESTED READINGS

Agnew (1987, 1995)
Agnew and Corbridge (1995)
Cox (1997)
Cybriwsky (1995)
Fincher and Jacobs (1998)
Flint and Shelley (1996)
Knox and Taylor (1995)
Leyshon and Thrift (1997)
O'Loughlin et al. (1998)
Shelley et al. (1996)
Staeheli et al. (1997)
Taylor (1994, 1995, 1996)

REFERENCES

Agnew, John A. 1982. "Sociologizing the Geographical Imagination: Spatial Concepts in the World-System Perspective." *Political Geography Quarterly* 1:2(April):159–166.
———. 1987. *The United States in the World Economy: A Regional Geography.* Cambridge: Cambridge University Press.
———. 1995. *Rome.* Chichester: Wiley.
Agnew, John A., and Stuart Corbridge. 1995. *Mastering Space: Hegemony, Territory and International Political Economy.* London: Routledge.
Allen, James P., and Eugene Turner. 1996. *The Ethnic Quilt: Population Diversity in Southern California.* New York: Macmillan.
Brunn, Stanley D. 1998. "When Walls Come Down and Borders Open: New Geopolitical Worlds at the Grassroots in Eastern Europe." In *Teaching Political Geography,* edited by Fiona M. Davidson, J. I. Leib, Fred M. Shelley, and G. R. Webster. Washington, D.C.: National Council for Geographic Education.
Chase-Dunn, Christopher, and Thomas D. Hall. 1997. *Rise and Demise: Comparing World-Systems.* Boulder, Colo.: Westview.
Clark, David. 1996. *Urban World/Global City.* London: Routledge.
Cox, Kevin R., ed. 1997. *Spaces of Globalization: Reasserting the Power of the Local.* New York: Guilford.
Cybriwsky, Roman. 1995. *Tokyo.* Chichester: Wiley.

Dear, Michael, and Steve Flusty. 1998. "Postmodern Urbanism." *Annals, Association of American Geographers* 88:1(March):50–72.

Fincher, Ruth, and Jane Jacobs. 1998. *Cities of Difference.* New York: Guilford.

Flint, Colin, and Fred M. Shelley. 1996. "Structure, Agency and Context: The Contributions of Geography to World-Systems Theory." *Sociological Inquiry* 66:4(Fall): 496–508.

Giddens, Anthony. 1985. *The Nation-State and Violence.* Cambridge: Polity.

Halberstam, David. 1986. *The Reckoning.* New York: Morrow.

Hanna, Stephen P. 1995. "Finding a Place in the World-Economy: Core-Periphery Relations, the Nation-State and the Underdevelopment of Garrett County, Maryland." *Political Geography* 14:5:451–472.

Heiman, Michael. 1996. "Race, Waste and Class: New Perspectives on Environmental Justice." *Antipode* 28:2:111–121.

Jones, Andrew. 1998. "Re-theorizing the Core: A 'Globalized' Business Elite in Santiago, Chile." *Political Geography* 17:3:295–318.

Kegley, Charles W., and Gregory Raymond. 1994. *A Multipolar Peace? Great Power Politics in the Twenty-first Century.* New York: St. Martin's.

Knox, Paul L., and Peter J. Taylor. 1995. *World Cities in a World-System.* Cambridge: Cambridge University Press.

Kodras, Janet E. 1997. "Globalization and Social Restructuring of the American Population: Geographies of Exclusion and Vulnerability." Pp. 41–59 in *State Devolution in America: Implications for a Diverse Society,* edited by Lynn A. Staeheli, Janet E. Kodras, and Colin Flint. Thousand Oaks, Calif.: Sage.

Leyshon, Andrew, and Nigel Thrift. 1997. *Money Space: Geographies of Monetary Transformation.* London: Routledge.

Meinig, Donald. 1986. *The Shaping of America: Atlantic America, 1492–1800.* New Haven, Conn.: Yale University Press.

O'Loughlin, John, Michael D. Ward, and Michael Shin. 1998. "The Diffusion of Democracy, 1946–1994." *Annals, Association of American Geographers* 88:4(Dec.):545–574.

Ostergren, Robert. 1988. *A Community Transplanted: The Trans-Atlantic Experience of a Swedish Immigrant Settlement in the Upper Middle West, 1835–1915.* Madison: University of Wisconsin Press.

Painter, Joe. 1995. *Politics, Geography and "Political Geography": A Critical Perspective.* London: Arnold.

Sassen, Saska. 1991. *The Global City: New York, London, Tokyo.* Princeton, N.J.: Princeton University Press.

Shelley, Fred M., J. Clark Archer, Fiona M. Davidson, and Stanley D. Brunn. 1996. *The Political Geography of the United States.* New York: Guilford.

Skocpol, Theda. 1977. "Wallerstein's World Capitalist System: A Theoretical and Historical Critique." *American Journal of Sociology* 82:5(March):1075–1090.

Staeheli, Lynn A., Janet E. Kodras, and Colin Flint, eds. 1997. *State Devolution in America: Implications for a Diverse Society.* Thousand Oaks, Calif.: Sage.

Steinberg, Phillip. 1997. "Political Geography and the Environment." *Journal of Geography* 96:2:113–118.

Swyndegouw, E. 1997. "Neither Global or Local: 'Globalization' and the Politics of Scale." In *Spaces of Globalization: Reasserting the Power of the Local,* edited by Kevin Cox. New York: Guilford.

Taylor, Peter J. 1981. "Political Geography and the World Economy." Pp. 157–172 in *Political Studies from Spatial Perspectives,* edited by Alan D. Burnett and Peter J. Taylor. Chichester: Wiley.

———. 1982. "A Materialist Framework for Political Geography." *Transactions, Institute of British Geographers* 7:15–34.

———. 1988. "World-Systems Analysis and Regional Geography." *Professional Geographer* 40:3:259–265.

———. 1990. *Britain and the Cold War: 1945 as Geopolitical Transition.* London: Pinter.

———. 1991. "A Theory and Practice of Regions: The Case of Europes." *Environment and Planning D: Society and Space* 9:2:183–195.

———. 1993. "The Last of the Hegemons: British Impasse, American Impasse, World Impasse." *Southeastern Geographer* 33:1(May):1–22.

———. 1994. "The State as Container: Territoriality in the Modern World-System." *Progress in Human Geography* 18:2:151–62.

———. 1995. "Beyond Containers: Inter-nationality, Inter-stateness, Inter-territoriality." *Progress in Human Geography* 19:1:1–15.

———. 1996. *The Way the Modern World Works: World Hegemony to World Impasse.* New York: Wiley.

Wallerstein, Immanuel. 1979. *The Capitalist World-Economy.* Cambridge: Cambridge University Press.

Ward, Peter. 1995. *Mexico City.* Chichester: Wiley.

Wood, Joseph. 1997. "Vietnamese American Place Making in Northern Virginia." *Geographical Review* 87:1(Jan.):58–72.

Zelinsky, W., and B. A. Lee. 1998. "Heterolocalism: An Alternative Model of the Sociospatial Behaviour of Immigrant Ethnic Communities." *International Journal of Population Geography* 4:1–18.

5

K-Waves, Leadership Cycles, and Global War: A Nonhyphenated Approach to World Systems Analysis

William R. Thompson

How does the world work? This is a basic question on which analysts from the entire spectrum of academic disciplines converge. In the pursuit of answers to this question, we have developed various divisions of labor—some of them understandable, others simply curious. Geologists study rock formations. Biologists look at animals of varying complexity. Historians restrict themselves to reconstructing how things have happened in the past. In the social sciences, economists specialize in markets and exchange processes, political scientists examine institutions and decision-making processes, and sociologists focus on stratification and inequality. Within both history and the social sciences, there is an additional tendency for analysts to specialize exclusively in either domestic or international economics, politics, or stratification/inequalities. Just why we approach our analyses in these fashions is a complicated story that need not be examined at the present time. Suffice it to say that some analysts rebel at the utility or wisdom of these disciplinary barriers and inside/outside orientations. No one is proposing that social scientists study rock formations and single-cell creatures. Some divisions of labor make perfect sense. But some questions demand that we examine economics, politics, and stratification/inequalities simultaneously. Nor is it clear that we can readily disentangle internal and external processes. There can be little doubt that they interact. And one important possibility is that domestic processes are shaped to some extent by external or systemic structures and processes. Moreover, all of these processes have pasts and variable tendencies toward path dependencies. That is, history matters. Accordingly, some analysts insist that political-economic and social processes need to be studied as mutually contingent phenomena; that their past, present, and future interactions are related systematically; and that these interactions are influenced in important ways by macrosystemic forces. People who share these fundamental analytical assumptions may be called world system analysts. Yet even though the questions raised by world system analysts are apt to overlap, one should not assume that these analysts embrace a single set of assumptions about how best to proceed in answering the questions.

For instance, some of the most central questions in the study of world systems include these:

- Why do preeminent states, and their associated political orders, rise and fall?
- Why are some types of economic activity more critical at some times than at others?
- Why do some parts of the world economy seem more central to economic operations at some times than at others?
- Why is economic growth intermittent rather than continuous, and what difference does it make?
- What relationships link intermittent processes of economic growth to such ostensibly political phenomena as war, domestic stability, and state making?

These are all important questions. They address the basic rhythms of political-economic life; they are very much about how the world works. These questions are also closely related because rise and fall dynamics, the shifting centrality of the world economy, and the discontinuities of growth share roots in a thousand-year-old, evolutionary process of long run economic growth and structural change.

Among the core dynamics of these processes are long, Kondratieff waves (K-waves) of innovation and even longer cycles of political-military preeminence and order (leadership long cycles). The shapes of these processes are not perfectly uniform throughout time. Their periodicities are less than precise. They have a specific historical genealogy in the sense that their operations and transitions can be traced back roughly over the last millennium in a continuous fashion and seemingly no further, at least in a continuous fashion. Nor did they emerge abruptly with all of the characteristics that K-waves and leadership long cycles possess currently. On the contrary, almost five hundred years went by before these processes began to assume the attributes associated with their contemporary manifestations. Similarly, we should not assume that these core dynamics must continue forever. They may but it is most unlikely that they will do so without undergoing substantial modification—just as they already have done in the past.

This chapter will focus on providing an overview of the interdependence between long-term economic growth, global political leadership, and global war and how that interdependence has evolved over the past millennium. The arguments that are advanced constitute a type of world systems analysis that does not embrace the central tenets of Wallersteinian "world-systems" analysis (e.g., Hopkins and Wallerstein 1996). While this interpretation shares an interest in macrostructures and their impact on microprocesses, the concern of the present focus is not oriented primarily toward the processes of capital accumulation, capitalism, and core-periphery divisions of labor. Rather, and perhaps betraying a disciplinary origin in political science as opposed to sociology, the focus is on waves of political-economic leadership, order, and large-scale violence closely linked to processes of long-term economic growth. Thus, in the end, we arrive at a classical world-system position: "economic" and "political" processes are tightly intertwined and reciprocally interdependent. We also share the notion that systemic processes are characterized by alternating periods of concentration and deconcentration. In periods of high concentration, one state, whether we call it the hegemon or the system leader, is predominant in leading sector production and capabilities of global reach. In periods of low concentration, states compete for leadership and leading sector production tends to be depressed. In this respect, we also share an imagery of phased sequences of high-low growth and interstate competition as the fundamental rhythm of world systemic processes. However, our initial assumptions are not the same as those associated with most world-systems analysis.[1]

At the same time, these assumptions are hardly congruent with most political science approaches to interstate conflict and international political economy. While there are certainly many different approaches to interpretation in the political science literature on international relations, mainstream approaches are reluctant to embrace the notions of systemic structures, high degrees of economic and military concentration, and historically patterned fluctuations in long-term economic growth and its variable impacts. The mainstream international relations prefers to stress the role of nation-states as the principal actors in world politics. Many analysts argue that nation-states are entirely free from structural constraints and opportunities. Movement to and from bipolarity and multipolarity are conceivable, even though the probable outcome is disputed, while unipolarity is deemed beyond the pale of likelihood. Short-term economic growth is the preferred province of economists and political economists alike. The operating attitude is that in the long term, we are all dead, so why worry about the ephemeral? The idea that some of those long-term patterns even exist and/or might influence behavior in the long and short term is more often ridiculed than examined empirically.

THE DOUBLE HELIX OF K-WAVES AND LONG CYCLES

K-waves represent surges of radical innovations that peak and decay over durations of approximately fifty to sixty years. Long cycles are periods of variable political-military leadership reflecting the initially ascending and then eroding primacy of a single world power over durations of approximately a hundred years. Each world power's order is tied to a specific pair of K-waves. The first wave in the set helps to propel a new systemic leader to preeminence. The second wave follows a period of intense conflict and is made more likely by the nature of the conflict. The period of intense conflict, in its own turn, is made more probable by the destabilizing, political-economic outcomes of the first K-wave, which catapults one or more states ahead of its competitors. The subsequent political-military leadership is very much dependent on the nature of the struggle and the outcome of the intensive competition. Each set of four shocks (two innovation surges, one period of intense conflict, and the development of political-military preeminence) thus define a distinctive era organized around the finite salience of a single state—the "world power" or the lead actor in the global politics and economy of its time.

Especially critical to this interpretation is the distinction between global and nonglobal activity. The *global* adjective is not used as a synonym for *world* or *international* political and economic activities. Global activities refer to interregional, long-distance transactions which, of course, have become increasingly salient. A principal question is when did a global system specializing in the political management of interregional transactions begin to emerge? To answer this question, it is not simply a matter of finding the first recorded instance of interregional transaction and commencing one's history of the global system from that point on. Interregional transactions have been around for quite a long time, but they were also characterized by a great deal of intermittence in ancient times. The question, then, is if one works back from our own time, how far back can one trace the perceived continuity of the contemporary system?

Modelski and Thompson (1996) find nine sequentially related sets of innovation-based leadership cycles between 930 and 1973: two Chinese (Northern and Southern Sung), two Italian (the first one Genoese led and the second one Venetian led), one Portuguese, one Dutch, two British (Britain I and II) and, so far, one American.[2] The scope of these instances of leadership have become increasingly "global" and, therefore, focused on the management of interregional transactions. For a variety of reasons, conditions (population scale, urbanization, marketization, and maritime commerce) came together propitiously in tenth-century China to begin a continuous sequence of innovation-driven, long-term surges of economic growth. The expansion of maritime trade (in the South China Sea and Indian Ocean) and the revived use of the Silk Roads on land facilitated the emergence of competing trading empires in the eastern Mediterranean that helped to transplant/transmit the growth surges and the innovation-based sequence of paired K-waves to the other end of Eurasia. At the end of the twentieth century and after several more geographical shifts in location, we appeared to be in the process of entering a tenth K-wave set that may or may not assume an American leadership identity. It is too soon to tell for sure. Too many degrees of freedom remain open to human agency.

Table 5.1 lists the leaders and the chronological pattern of political and economic leadership. Each long cycle is characterized by four phases that can be interpreted in two different ways. If one wishes to emphasize the rise of a new leader, the appropriate sequence of phases is agenda setting, coalition building, macrodecision, and execution. To emphasize the decline of an incumbent leader, the phase sequence is global war, world power, delegitimation, and deconcentration. Each phase deline-

Table 5.1 Long Cycles in Global Politics: Learning and Leadership Patterns

Long Cycle Mode	Phases			
Learning ("Rise") Leadership ("Decline")	Agenda Setting Delegitimation	Coalition Building Deconcentration	Macrodecision Global war	Execution World power
		Starting in		
		Chinese/Italian Renaissance		
N. Sung/S. Sung	930	990	1060	1120
Genoa/Venice	1190	1250	1300	1355
		Western European		
Portugal	1430	1460	1494	1516
Dutch Republic	1540	1560	1580	1609
Britain I	1640	1660	1688	1714
Britain II	1740	1763	1792	1815
		post-European		
United States	1850	1873	1914	1945
United States?	1973	2000	2030	2050

Source: Modelski and Thompson (1996). Note that this table can be read in two different modes. The dates are set for the rise mode phases ("agenda setting" through "execution") but can be read for the decline mode if one starts in the "world power" column and moves through to the following "global war" column.

ates different types of behavior. The macrodecision/global war phase determines the identity of the next system leader or world power. The postwar execution/world power phase is the period of peak economic and political-military leadership. Agenda setting/delegitimation is a period in which leadership decline and challenges to the existing world order become increasingly noticeable. New problems emerge that demand innovative responses. Finally, coalition building/deconcentration is a period marked by competitive preparations for developing new versions of world order and succession to systemic leadership.

The political processes of the leadership long cycle, it is argued, have coevolved with the economic processes of the K-wave. There are at least three reasons that this might be anticipated:

1. Political leadership and waging global wars are expensive propositions that depend on adequate economic resources. Economic fluctuations, therefore, are likely to influence the exercise of political leadership.
2. The world economy's activities are dependent on a minimal level of stability and security. Intensive conflicts within the political system are likely to influence the functioning of the world economy.
3. To emerge as the world's political-military leader requires technological leadership. Technological leadership, in turn, is predicated on the development of innovation in commerce and industry. Once some level of technological leadership is attained, political-military leadership, or its pursuit in global war, can be quite useful in expanding the edge associated with technological leadership and protecting the consequent accumulation of wealth.

The economic carriers of these interdependent processes are the rise and decline of leading sectors or clusters of basic innovations that periodically restructure economic activities. The innovations are Schumpeterian in the sense that they encompass new ways of production, the opening of new markets and sources of raw materials, as well as new forms of business organization. They are also Schumpeterian in the sense that many of the old ways of doing things tend to be destroyed by the ascendance of new leading sectors. The innovations that fuel these long waves of economic growth may be limited to one or more of the various types of change, but their implications for economic restructuring must be substantially more than merely routine increments to existing practices and activities. They revolutionize commerce and/or industrial production. They are not the only source of long-term growth (e.g., population growth is another important source), but new technology does constitute one of the more significant sources.

Why should this pattern characterize long-term economic growth? Innovations are responses to problems encountered by economic (and political) agents. As commercial routes or industrial profits become increasingly unpredictable or unattainable, the search for new routes and production possibilities is quite likely. For instance, Europeans began searching for a way around the Moslem/Mameluk lock on the east-west flow of spices some three hundred years before they finally found a route around Africa in the late fifteenth century. Or, looking for a way to deal with water in mines led to the development of steam engines with subsequent implications for railroads, the iron industry, and the transportation of agricultural goods to markets (and soldiers to battle).

Once a sufficiently radical innovation or complex of innovations is launched, its potential for facilitating economic growth is finite. Just how finite depends on the specific nature of the innovation, the rate of diffusion, the number of new competitors, demand elasticities, and impacts on supply. One good example is pepper and its long life as a leading sector of growth for Genoa, Venice, Portugal, and the Dutch—each of which discovered new ways of supplying this commodity to European markets. The various commercial innovations relating to the pepper trade each had their own growth curves. But in the process of innovating new ways of supplying pepper, each new leader managed also to increase the supply. Prices and profits fell as the pepper trade became routinized and lost its ability to function as a leading sector. Attention shifted, partially as a consequence but also in accordance with concomitant changes in supply and demand, to other drugs (in the seventeenth century, sugar and tobacco).

Not coincidentally, the shift in leading sector commodity emphasis favored the English who controlled a good number of what were to become the prime North American production sites over the Dutch who had been more successful in the Asian trades than in acquiring territory in Brazil and in North America. The paths to success in the world economy changed in the mid- to late seventeenth century. The overcommitment of the incumbent lead economy (in this case, the Dutch) to earlier paths created an opportunity for new sources of entrepreneurial innovation.

The rise and decline of leading sectors are thus concentrated in both time and space. Fifty to sixty years are needed for what is initially revolutionary to become either routine or to be superseded by some new way of doing things. The innovations tend also to be monopolized initially by entrepreneurs in one national economy. Gradually, the innovations spread to other (but not all) economies. The K-wave sources of long-term economic growth are thus highly concentrated spatially and subject to processes of diffusion in which some other economies catch up to the leading positions first established by the economic pioneers.

Keying on the timing of a rising leader (see table 5.1), we should expect these growth surges or K-waves to peak immediately before and after phases of global war if the political-military and economic processes are as truly interdependent as argued earlier. This is not a matter of working backward and saying that since we know there have been K-waves, leaders, and global wars that they must all somehow be related. Rather, the causal logic begins with the notion that new, rising leaders require an appropriate economic foundation to ascend in the political-economic hierarchy. However, the relatively abrupt development of the economic innovations that support this foundation is destabilizing for the system. The positions of some rivals may be advantaged by the new ways of doing things. Other rivals will perceive themselves as falling behind. The incumbent, declining (i.e., falling behind) leadership is particularly apt to feel threatened by changes it can no longer harness for its own purposes. Intensive conflict, thus, becomes more probable. The innovative edge of the rising leader makes its victory in this conflict more probable. Waging the conflict and winning it also improve the probability of another postwar growth surge. For these reasons, we expect the innovation surges to cluster in the temporal vicinity of the global wars.

It is also reasonable to expect some preliminary, start-up activities for each K-wave that precedes the peak in the growth of the new ways of doing things. More precisely, K-wave peaks are anticipated to fall within the phases of coalition building and execution—the two phases immediately adjacent to the macrodecision/global war phase. Start-up phases are linked to the phases that precede the peak growth phases—agenda-setting and macrodecision respectively. Since leadership long cycle theory links global wars to the attainment of military-political leadership, we should also expect that some appropriate leadership threshold is achieved, if at all, after the first and before the second K-wave peak.

Keep in mind that the temporal sequence displayed in table 5.1 was developed well in advance (by more than a decade) of the construction of the three hypotheses about the timing of K-wave peaks and the attainment of political-military leadership. This permits us to pinpoint the anticipated timing of these phenomena within two or three decades without biasing the test outcome with post hoc knowledge of how things worked out historically. The question then becomes just how accurate are the predictions in specifying the timing of long-term economic growth and systemic leadership.

In every case, the growth peak of the first K-wave in each pair is located within the coalition-building phase. Moreover, the peak is usually observed immediately prior to the decade in which the macrodecision/global war phase commenced. The second K-wave peak also follows the macrodecision phase. However, its precise timing—even though it does consistently fall within the predicted execution/world power window—varies. Sometimes, the peak occurs right at the end of the macrodecision conflict. Sometimes, it is located toward the end of the execution/world power phase. But since these phase windows are fairly short, we probably should not make too much of these slight variations in timing.

GLOBAL WAR

Within this macrocontext of fluctuations in long-term economic growth and political leadership, we see global war as an outcome of the combination of processes of concentration and deconcentration operating at both the global and key regional levels of analysis. That is, a focus solely fixed on global structures and processes will miss an important part of the puzzle in explaining why the world system periodically erupts into large-scale violence. Global and regional processes move in and out of synch with one another. But not all regions are equally important. For much of the past five hundred years, western Europe has been the key region of the world system. When the European and global processes became fused, the probability of a global war breaking out was greatly enhanced.

GLOBAL CONCENTRATION PROCESSES

From a systemic perspective, the global political economy is characterized by undulating patterns of capability concentration, followed by deconcentration, and then followed again by reconcentration. Figure 5.1 illustrates this pattern by showing the fluctuations in naval power concentration between the 1490s and 1990s. We attribute this pattern primarily to the emergence and relative decline of lead economies. The linkage to global war is straightforward. When the global political economy is highly concentrated, the outbreak of a global war is unlikely. After the global political economy has experienced considerable deconcentration, the outbreak of a global war becomes more probable because global wars, inherently, can be seen as succession struggles over which economy will replace the incumbent as the global system's political-military center. In fact, we designate as global wars only those intensive conflicts that lead to a new phase of significant reconcentration and global political-military and economic leadership. In this respect, we admit to being more interested in these wars' roles in the concentration-deconcentration process than we are in their identities as increasingly lethal wars among major powers. Put another way, we think

Figure 5.1   The Long Cycle of Global Leadership

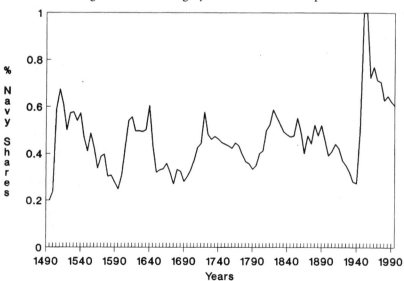

global wars merit special attention as a distinctive set of wars that are a critical part of the global political economy's functioning.

The tendency toward concentration, deconcentration, and reconcentration is much older than the "institution" of global war. We can find instances of concentration and deconcentration going back to at least 3500 B.C. and the Sumerians, but, in our perspective, the concentration-deconcentration-reconcentration sequence only emerged as a continuous process with the advent of Sung Chinese economic and maritime innovations a millennium ago. In the period roughly between 1000 and 1500, we can trace early, transitional versions of successive lead economies in the global, transcontinental sense (Northern and Southern Sung, Genoa, Venice, Portugal). Their fluctuations in relative prosperity appear to be associated with periods of intense conflict that intervene between the twin peaks of economic growth described earlier, but they do not take on the form of the global wars with which we have become more familiar in the twentieth century. After 1500 or, more precisely, 1494, the global war institution began to emerge in its modern form. In 1494, the French invaded Italy thereby inaugurating a recurring sequence of attempts to coercively unify the western European region that involved aspiring regional hegemons attempting to capture past and fading centers of economic innovation. The attempts to seize these centers at the regional level mobilized resistance coalitions organized increasingly by global powers. Presumably, this emergence reflects an evolving system experiencing environmental change. The global political economy (as did regional politics in Europe) evolved in such a way that it became increasingly susceptible to intermittent fusion with European regional politics. The global war institution is one of the consequences of that evolutionary change.

The regional dimensions of global war and the implications for system transformation will be returned to in later sections. For now, we need to focus on further elaborating the global processes that are most important. Five global wars are identified: the Italian and Indian Ocean Wars (1494–1516), the Dutch-Spanish Wars (1580–1608), the Wars of the Grand Alliance (1688–1713), the French Revolutionary and Napoleonic Wars (1792–1815), and World Wars I and II (1914–45). These wars are fought by coalitions of global and other types of powers, as identified in table 5.2.

The pattern is essentially one of the incumbent global system leader and its allies arrayed against a principal challenger and its allies. So far, the challenger has never won. On the other hand, the incumbent leader may also lose its status to one of its allies if the most active economic zone (i.e., where the new leading sectors are emerging) has shifted away from the old leader's control. In both the Dutch-British and British-American transitions, the political-military shifts in relative status took place during the respective global wars. The junior partner going into the war emerged as the senior partner and the new system leader.

In this respect, we should emphasize that the structure of conflict is more complicated than a simple challenger versus incumbent situation. Declining incumbents select, to some extent, which challengers they will fight and with whom they will

Table 5.2   Global Wars and Their Participants

Global Wars	Central Participants
Italian and Indian Ocean Wars 1494–1516	Spain vs. France, Portugal vs. Venice, Mamelukes, and Ottoman Empire
Dutch Independence 1580–1609	Netherlands, England, and France vs. Spain
Wars of the Grand Alliance 1688–1713	Britain and Netherlands vs. France (and Spain)
French Revolutionary and Napoleonic Wars 1792–1815	Britain and Russia vs. France
World War I and II 1914–1945	Britain, Russia/U.S.S.R., France, and the United States vs. Germany (and Japan)

Note: The list of war participants focuses on global powers and is hardly exhaustive of the full slate of coalition members.

ally to meet the intensive challenge. Thompson (1997) argues that this selection process is primarily a function of four variables: maritime-commercial orientation, proximity, similarity (regime type, culture, ideology, and race), and innovative nature. The threats that are seen as most dangerous are those associated with explicitly premeditated challenges that come from dissimilar types of states with fundamentally different strategic orientations. States with strategic foci on utilizing land forces to expand territorial control have found it difficult to compete with seapowers other than via attempts at direct conquest. Nearby challengers are less easy to ignore than those located farther away. The more "alien" the challenger, the greater is the likely level of suspicion and misperception in divining motivations and intentions. A challenger and incumbent leader are also more likely to fight if their economic competition is based on similar commercial-technological commodities. If the challenger perceives that the leader will thwart any peaceful positional encroachments, a nonpeaceful competition is more probable.

Similarly, potential challengers adopt different strategies of confrontation. The most traditional approach can be referred to as "capture the center," in which the challenger attempts to seize control of the lead economy and its commercial networks. An alternative approach is to avoid attacking the leader on its home ground and instead to focus on attacks on its far-flung commercial networks and the development of alternative networks, as demonstrated by warfare among Portugal, Spain, England, France, and the Netherlands in Asian and American waters. A third strategy involves creating a relatively autonomous subsystem within the world economy that excludes economic competition with the system leader. Napoleon's continental system, German *Mitteleuropa* aspirations, Japan's coprosperity sphere, and the communist international system of the second half of the twentieth century all illustrate this third strategy. How threatening this strategy appears will depend on how coercive the subsystem creation and maintenance processes are and who suffers most from the exclusionary policies. The capture-the-center strategy has gradually lost much of its appeal. The flanking, alternative network approach became increasingly popular in the period most focused on long-distance commerce while the exclusionary subsystemic approach has become more prevalent in the movement toward increased emphasis on industrial production.

Throughout the past five hundred years, the global power elite has remained a small group: Portugal (1494–1580), Spain (1494–1808), England/Britain (1494–1945), France (1494–1945), the Netherlands (1579–1810), Russia/the Soviet Union/Russia (1714 to the present), the United States (1816 to the present), Germany (1871–1945), and Japan (1875–1945). Of this group, only four global powers have qualified as world powers: Portugal, the Netherlands, Britain, and the United States. To qualify for these designations, world powers have to have exceeded control over 50 percent or more of the global reach capabilities, which, historically, we equate with sea power. Global powers need to demonstrate sea power activity in more than one regional sea and control over at least 10 percent of global reach capabilities.

However, the development of these global reach capabilities does not take place in a vacuum. The military ascent of the world powers is very much geared to their economic fortunes. The Chinese and Italian leads were prototypical predecessors confined to less geographically ambitious theaters. Their economic activities were certainly linked to naval superiority but not on the planetary scale achieved after the 1490s. Nevertheless, the observed pattern after the 1490s is remarkably consistent. In four cases, the Portuguese, Dutch, Britain II, and the United States, naval leadership was clearly attained prior to the second K-wave peak. Only in the case of Britain I did the naval threshold attainment occur at about the same time as the second K-wave peak. Hence, we conclude that naval leadership is a function in large part of sufficient resources to pay for the fleets (the first K-wave) and a strong security incentive to expand one's arsenal of coercive maritime resources (participation in the global economy and global war).

While emphasizing the significance of naval capability in the leadership long cycle, it is of course assumed that the naval capability leadership is based upon economic leadership. It has been demonstrated empirically that this is the case for nineteenth- and twentieth-century data centered on the British and American leadership eras. Figure 5.2 summarizes the "causal" relationships that were found in time series analyses (Rasler and Thompson 1994; Reuveny and Thompson 1997). Rapid leading sector growth leads to finite periods of economic leadership in those leading sectors and to somewhat longer-lasting naval power leads.

Figure 5.2 also connects innovation and global concentration processes to global warfare. Based on our theoretical arguments and empirical findings (on nineteenth- and twentieth-century data), we see systemic warfare as a product of economic innovation and leadership processes. In turn, systemic warfare influences innovation,

Figure 5.2   Innovation, Concentration, and Warfare

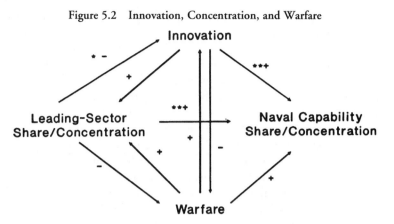

Note: *Britain only; **United States only.

economic concentration, and naval concentration. In this sense, long waves of economic and technological change, the political-military leadership long cycle, and warfare are all highly interdependent dynamics that lie at the heart of the global political economy's functioning.

THE INTERMITTENT FUSION OF GLOBAL AND REGIONAL PROCESSES

Nevertheless, an exclusive focus on global politics is inadequate, for the global system is not an autonomous sphere of activity. On occasion, global politics have become tightly fused with regional politics. These fusions can take many forms. Ambitious states in any region may make coercive bids for regional leadership. Vietnam in Southeast Asia or Iraq in the Middle East come to mind as recent examples. Just how dangerous these bids are depends in part on how salient is the region in which they occur. Regional concentration processes in more peripheral regions are apt to be less destabilizing than similar processes in more central regions. The appropriate comparison is between the Third Indochina war and the Gulf War versus, say, World War II. All three events were lethal, but the first two contests were unlikely to become "globalized." The third one spread throughout the planet relatively quickly. They all began as subregional or regional contests. The difference is that World War II emerged in part from a contest over the control of Europe—still one of the most salient or central regions of that time.

It also mattered that the European region was the home base for a number of global powers. It is easier to remain aloof from more distant contests than ones that take place in one's own backyard. Salience and proximity help explain why European regional international relations, on occasion, have been so explosive for the global political economy. This intermittent fusion of European regional and global politics is absolutely essential to an explanation of structural change and global war. At the same time, it is important to acknowledge the strong reliance on Dehio's (1962) interpretation of the history of European international relations that has been co-opted for the purposes of model construction and elaboration.

Unlike other regions of the world, especially eastern Asia, no single power ever established hegemony over Europe for very long. The basic Dehioan insight is that this outcome was due to what appears to be a relatively unique geopolitical pattern. Before a would-be regional hegemon could unify Europe coercively, counterweights emerged from areas immediately adjacent to the region. Introducing extraregional resources, they were repeatedly able to block the creation of European hegemony.

The eastern counterweight supplied brute land force. The western counterweight increasingly specialized in sea power, which was, in turn, predicated on the development of specializations in the role of commercial intermediary among Europe, Asia, and America. When both counterweights were operative, an aspiring regional hegemon was forced to fight a resource draining war on two fronts that it was likely to lose. The outcome was an intermittently renewed balance of European power that depended on the region remaining open to extraregional resources controlled by flanking states.

The regional motor of the balancing dynamic hinged on an intermittent rise of a hegemonic aspirant and the concentration of regional capabilities. France inaugurated this system in its 1494 attack on Italy. It was resisted by Spain primarily and then for a short time by a unified Hapsburg entity. A Franco-Ottoman coalition thwarted the second bid, this time on the part of the Hapsburgs. Both of these initial efforts preceded the emergence of a western maritime power capable of functioning as a counterweight. With some English assistance, the Netherlands provided the first maritime counterweight to Philip II's bid for supremacy. By the mid–seventeenth century, Spain had surrendered its regional lead to a restrengthened France. Louis XIV's late seventeenth-century activities came to be perceived as a direct threat to Europe and the global political economy. A second Anglo-Dutch coalition developed the first large-scale maritime blockade of the European continent and defeated the expanding navy of France in 1692. Between 1692 and the next destruction of the French fleet in 1805 at Trafalgar, the generally eroding, relative strength of the French kept the midcentury Anglo-French fighting from turning into a full-fledged struggle over either regional or global supremacy.

Unlike the earlier, more gradual bids for regional hegemony, the third French bid in 1792 emerged abruptly and was unusually successful for a few years before the Napoleonic variant was crushed in 1814 and 1815. After 1815, the main emphasis of global concern shifted away from the European region to the Russo-British sparring along their mutual Eurasian imperial boundaries. The British remained worried about the French potential for causing trouble in Europe for some time after

1815, but a fourth French bid, with hindsight, was increasingly unlikely. One reason was the emergence of a unified Germany.

Whether or not the ascending Germany of the late nineteenth century was merely seeking equality with other leading powers or European domination, a mixture of commercial and naval rivalries combined with geographical proximity increased the probability that Britain would identify Germany as its primary threat. In World War I, Germany then proceeded as if it were indeed seeking regional supremacy. By World War II, which can be seen as a continuation of World War I, both Germany and Japan had become more overt and ambitious about the extent of their regional aspirations. The end of that war led to the territorial dismemberment of the principal challenger (Germany). The division of the entire region into American and Soviet spheres completed the process of diminishing the regional autonomy of Europe and, presumably, some of its ability to generate local problems that could intrude into the functioning of the global political economy. While the significance of European economies for the global political economy remains high, a renewed, coercive bid for European regional domination seems unlikely.

From a regional perspective, the principal dynamic of this system has been the movement from a peak in the strength of the leading regional power through a long trough to next peak and so on. The long troughs were characterized by leveling process. The regional leader that had peaked earlier was in gradual decline, thereby encouraging and facilitating the emergence of new regional contenders. During the troughs, relative power relationships and alignments were unstable. The troughs not only provided windows of opportunity for the emergence of new land powers. They also encouraged the upwardly mobile to challenge the regional status quo. At the same time, the strength of the Western maritime powers should also be most concentrated during the troughs in regional concentration. The less the threat from adjacent land empires, the more the maritime powers could thrive.

The rhythms of two dissynchronized cycles or waves of power concentration, centered on two different types of major powers, are envisioned. On land, the leading regional power waxed and waned. At sea, the leading global power ascended and declined. For the most part, the one declined as the other peaked, but not in a completely dissynchronized fashion. Declining global leaders encouraged would-be regional hegemons. Suppressing would-be regional hegemons galvanized new global leaders to emerge or, in the case of Britain, to reemerge. It is not too much of an exaggeration to say that regional and global powers represent two very different "species" of power. To be sure, there was overlap. Some strong regional powers in Europe were also contenders in the global political economy. But they were never quite as successful as they might have been, given their roots in regional/territorial orientations as opposed to maritime orientations. The leading regional powers rose to primacy on basis of absolute autocracies, large armies and bureaucracies, and the success of expansionist foreign policies. Spain, France, and Germany were all created via coercive expansion within the region. Neighboring enemies could be beneficial in the sense that they provided rulers with incentives for developing military and economic strength. Global powers were more oriented toward long-distance trade than territorial expansion close to home. To varying degrees they were able to restrain their autocracies. Global powers had good reasons to favor navies over armies. They also led in the movement away from command bureaucracies toward more representative regimes for resource mobilization. Security depended on some type of geographical insularity, or at least the relative absence of proximate adversaries. Without some form of natural protection, they were likely to succumb to the superior military strength of adjacent, land-based empires.

But the iterative introduction of extraregional resources could not be repeated infinitely. Drawing in the flanking powers and their resources increasingly reduced the ratio of power that could be mustered within the European region relative to what could be mobilized away from Europe. Eventually, challengers from the European region could no longer expect to compete with stronger states outside Europe.

This finite durability of the classical European regional system may have depended on a unique constellation of geohistorical factors. Would-be regional conquerors found themselves caught between offshore rocks and eastern hard places. Despite repeated attempts, European hegemons could not overcome the western and eastern flanks with much more access to the resources useful for war and economic growth. Regionally biased strategies to overcome these barriers to supremacy proved to be largely self-defeating. The European subsystem retained its pluralistic structure but, ultimately, at the expense of its onetime autonomy and salience. The Soviet-American cold war subordinated the European region to a global contest after centuries of regional problems diverting global interests and resources.

WHITHER THE FUTURE?

It has been the interactions between global and regional structural changes that have generated the contexts for the world's most significant and serious wars. From a regional vantage point, the relationship between concentration and the probability of global war is positive. Greater concentration leads to the greater likelihood of intensive conflict. From the global perspective, though, the relationship is reversed. High concentration leads to a decreased likelihood of intensive conflict. The problem has been that these structural rhythms have been out of synch with one another. Global concentration levels tend to be low when regional concentration levels are on the rise.

Very much at the risk of oversimplification, possible future scenarios can be reduced to two: one is pessimistic and the other is optimistic. Both scenarios assume that the K-wave and leadership cycle dynamics will continue to characterize global politics and economics. Both scenarios also assume that the salience of the European region has been greatly diminished, thereby "neutralizing" the possibility of lethal regional-global structural interactions. The pessimistic scenario transfers the salient region focus to eastern Asia, which has some potential for reproducing some of the circumstances that characterized the period between 1494 and 1945. It is a multipolar region in which predominance is likely to be contested (just as it has been contested in the past). The most likely challenger is China, and, if an attempt at perceived regional hegemony were to be mounted in the twenty-first century, China could anticipate an opposing coalition of land and sea powers. The only difference might be that the directions would be reversed from the historical European pattern, with the land forces coming from the west (and northwest) and the sea powers from the east. While it seems difficult to imagine given the lethality of military capabilities, it is possible that some form of intensive conflict (global war) might reoccur around the middle of the twenty-first century. This pessimistic scenario is then premised on a "more-of-the-same" assumption.

The optimistic scenario is predicated on an extinction of the historical pattern of intermittent fusion with regional hegemonic schemes. It is conceivable that no single region will ever reproduce the interaction between western Europe and the global system between 1494 and 1945. And, just as it took five hundred years to bring together the elements that characterized global politics and economics for the next five hundred years, there may be grounds to anticipate other types of fundamental parameter shifts. One possibility is that the K-wave/leadership cycle dynamics will sputter out as technological change comes so quickly that it is difficult to monopolize for very long. Technology and highly technological production has become increasingly multinationalized; therefore, it may also be increasingly difficult for one world power to monopolize leading sectors. Another possibility is that trends toward increasing democratization will fundamentally alter international relations. This "democratic peace" process (i.e., democratic states do not tend to go to war with other democratic states) is not entirely independent of the leadership long cycle. World powers have also been leaders in democratization and defending the democratized community. One way to look at democratization is to view it as an expansion of the group that has governed global politics and the global economy for the past half-millennium. The expansion of that group should permit more complex processes of governance to emerge. More complex governance processes would include less reliance on a single leader and primitive trial-by-combat succession techniques as well as increased participation by (other) global institutions. It should also decrease the size of the pool from which militarized challengers issue.

That the world system could proceed down either fork in the road underlines the inherently nondeterministic nature of the explanatory apparatus. Global politics and economics have evolved over the past millennium. It may be possible to capture the basic parameters and their major transformations after the fact. However, it is not possible to predict what will happen in the future. It can only be presumed that the system will continue to function as before unless it undergoes significant transformation in its operating parameters. These transformations have occurred before—and they may again.

FURTHER READING ON THE LEADERSHIP LONG CYCLE PROJECT

The core analyses in the leadership long cycle project are Modelski (1978, 1981, 1982, 1987a, 1987b, 1996, 1999), Modelski and Modelski (1988), Modelski and Thompson (1988, 1996, 2000), Rasler and Thompson (1989, 1994), and Thompson (1983, 1988, 1997, 2000). Modelski (1978) is the first publication that outlines the leadership long cycle perspective. Modelski's views are elaborated considerably in the 1987 book, while the 1996 article outlines a shift toward a more explicitly evolutionary paradigm of world politics. Modelski and Modelski (1988) represent an effort to document instances of leadership over the past five hundred years, while Modelski and Thompson (1988) focus on developing data on global reach (naval capabilities) to test the idea of cyclical leadership patterns. Modelski and Thompson (1996) push the long cycle calendar back another five hundred years to Sung China and demonstrate how the Chinese behavior was transferred to the Mediterranean. This study also emphasizes the theoretical and empirical relationships between K-waves and leadership cycles with serial data on innovation and economic leadership extending back to the fourteenth century. Modelski and Thompson (2000) focus on theoretically projecting the evolutionary implications of leadership long cycle analyses into the twenty-first century. Rasler and Thompson (1989) highlight how global powers emerged and with what consequences for state and war making. Rasler and Thompson (1994) elaborate the links among regional and global processes of concentration and warfare and take a markedly Dehioan approach (Dehio 1962), as well as empirically modeling processes of political-economic decline in system leaders. Thompson's (1983) work is an edited volume that presents multiple ways of looking at world systems analysis and international relations. Thompson (1988) provides an early elaboration of how leadership long cycle analyses compare with other approaches in international relations, with a variety of empirical analyses of economic and political processes. Thompson (1997, 2000) develops a model to explain which states have been the most likely to become challengers over the past thousand years. Thompson's (1998) book is an edited volume on interstate rivalry processes in which some of the analyses are explicitly long cycle in form.

Among other topics explored in leadership long cycle terms include deterrence (Modelski and Morgan 1985; Thompson 1997/98), democratization (Modelski and Perry 1991; Modelski 1998), protectionism (Thompson and Vescera 1992; Reuveny and Thompson 1997; Thompson and Reuveny 1998), pre-1500 phenomena including older succession crises (Thompson 1995a) and an examination of Eurasian migration and incursions that pushes the long cycle calendar back in some respects to 4000 B.C. (Modelski and Thompson 1998), interstate rivalries (Thompson 1995b, 1999), naval history (Thompson 1995c), European expansion in the third world (Thompson 1999), and intelligence assessments (Alexseev 1997).

Not surprisingly, a number of criticisms of leadership long cycle analysis have been generated. Interested readers are invited to consider the perspectives advanced in Rapkin (1983, 1986, 1987), Zolberg (1983), Levy (1985, 1991), Rosecrance (1987), Nye (1990), Goldstein and Rapkin (1991), Arquilla (1992), Richards (1993), Vasquez (1993), Houweling and Siccama (1993, 1994), Williams and Huckfeldt (1996), Denemark (1997), Frank (1998), and Ingram (forthcoming). Specific responses to some of these complaints can be found in Modelski (1983), Thompson and Modelski (1994), and Thompson (forthcoming).

NOTES

1. In addition to the differences of emphasis on capital accumulation, capitalism, and core-periphery processes, leadership long cycle and world-systems analyses also have different historical scripts. Our modern world system began some five hundred years before the conventional world-system starting date (around 1500) and far from Europe (in Sung China). Leadership long cycle analysis also stresses the leadership qualifications not only of Sung China but also Genoa, Venice, Portugal, and two bouts of British leadership (whereas world-systems analysts see only one British phase). As a consequence, there is some disagreement about which wars were most significant. Leadership long cycle's global wars are the ones that usher in new phases of high economic/political/military capability concentration. These global wars only began to emerge in the 1490s and then reoccurred roughly every century or so (1494–1517, 1580–1608, 1688–1713, 1792–1815, 1914–1945). Fewer wars are accorded special status in world-system analysis that customarily begins with the Thirty Years War (1618–48), only a regional war by long cycle standards, but accepts the distinctive status of the French Revolutionary/Napoleonic Wars and World Wars I and II. Moreover, there are also different normative preferences. Conventional world-systems analysis is inclined to welcome the demise of capitalism and the movement toward socialism. In contrast, leadership long cycle's emphasis is placed instead on the primitive nature of a political system that depends on intensive bouts of coercion to decide issues of policy management and leadership succession. Survival depends on the development of new forms of, and institutions for, global policy management.

2. The term *cycles* does not imply a period of fixed duration that some analysts insist on as a minimal criterion. Leadership cycles or waves are irregular in duration but reoccurring in such a fashion that one can regard each iteration as part of the same process. There are differences of opinion on the methodological criteria associated with testing the presence of cycles in international relations (see, e.g., the debate between Beck 1991 and Goldstein 1991a). There are also differences of opinion on the extent to which economic and political cycles are tightly or loosely connected. Leadership long cycle analysis argues for a fairly tight connection, as does world-system analysis (see, e.g., Boswell and Sweat 1991). Much looser types of cyclical relationships are described in Goldstein (1991b), Pollins (1996), and Pollins and Murrin (1997). The interest in pushing back the leadership cycle to tenth-century China stems from, among other influences, McNeill's (1982) assertions about modern economic growth beginning with the Sung era in China.

REFERENCES

Alexseev, Mikhail. 1997. *Without Warning: Threat Assessment, Intelligence and Global Struggle.* New York: St. Martin's.

Arquilla, John. 1992. *Dubious Battles: Aggression, Defeat, and the International System.* Washington, D.C.: Crane Russak.

Beck, Nathaniel. 1991. "The Illusion of Cycles in International Relations." *International Studies Quarterly* 35:4(Dec.):455–476.

Boswell, Terry, and Mike Sweat. 1991. "Hegemony, Long Waves, and Major Wars: A Time Series Analysis of Systemic Dynamics, 1496–1967." *International Studies Quarterly* 35:2(June):123–149.

Dehio, Ludwig. 1962. *The Precarious Balance.* New York: Vintage.

Denemark, Robert. 1997. "Toward a Social Science of Long-Term Change." *Review of International Political Economy* 4:2(Summer):416–430.

Frank, Andre Gunder. 1998. *ReOrient: Global Economy in the Asian Age.* Berkeley: University of California Press.

Goldstein, Joshua. 1991a. "A War-Economy Theory of the Long Wave." In *Business Cycles: Theories, Evidence and Analysis,* edited by Niels Thygesen, Kumaraswamy Velupillai, and Stefano Zambelli. New York: New York University Press.

———. 1991b. "The Possibility of Cycles in International Relations." *International Studies Quarterly* 35:4(Dec.):477–480.

Goldstein, Joshua, and David P. Rapkin. 1991. "Hegemony and the Future of World Order." *Futures* 23:9(Nov.):935–959.

Hopkins, Terence K., and Immanuel Wallerstein, eds. 1996. *The Age of Transition: Trajectory of the World-System.* London: Zed.

Houweling, Henk W., and Jan G. Siccama. 1993. "A Neo-functionalist Theory of War." *International Interactions* 18:4:387–408.

———. 1994. "Long Cycle Theory: A Further Discussion." *International Interactions* 20:3:223–226.

Ingram, Edward. Forthcoming. "Hegemony, Global Power and World Power: Britain Two as World Leader." In *History and International Relations Theory: Respecting Differences and Crossing Boundaries,* edited by Colin Elman and Miriam F. Elman.

Levy, Jack S. 1985. "Theories of General War." *World Politics* 37:3(April):344–374.

———. 1991. "Long Cycles, Hegemonic Transitions and the Long Peace." In *The Long Postwar Peace,* edited by Charles W. Kegley, Jr. New York: HarperCollins.

McNeill, William H. 1982. *The Pursuit of Power: Technology, Armed Force, and Society since AD 1000.* Chicago: University of Chicago Press.

Modelski, George. 1978. "The Long Cycle of Global Politics and Nation State." *Comparative Studies in Society and History* 20:2(April):214–235.

———. 1981. "Long Cycles, Kondratieffs and Alternating Innovations: Implications for U.S. Foreign Policy." Pp. 63–83 in *The Political Economy of Foreign Policy Behavior,* edited by Charles W. Kegley, Jr., and Patrick J. McGowan. Beverly Hills, Calif.: Sage.

———. 1982. "Long Cycles and the Strategy of U.S. International Political Economy." Pp. 97–110 in *America in a Changing World Political Economy,* edited by William Avery and David P. Rapkin. New York: Longman.

———. 1983. "Of Global Politics, Portugal, and Kindred Issues: A Rejoinder." Pp. 115–139 in *Contending Approaches to World System Analysis,* edited by William R. Thompson. Beverly Hills, Calif.: Sage.

———. 1987a. *Long Cycles in World Politics.* London: Macmillan.

———. 1987b. *Exploring Long Cycles.* Boulder, Colo.: Reinner.

———. 1996. "Evolutionary Paradigm for Global Politics." *International Studies Quarterly* 40:3(Sept.):321–342.

———. 1998. "Enduring Rivalry in the Democratic Lineage: The Venice-Portugal Case." In *Great Power Rivalries,* edited by William R. Thompson. Columbia: University of South Carolina Press.

———. 1999. "From Leadership to Organization: The Evolution of Global Politics." Pp. 11–39 in *Future of Global Conflict,* edited by Volker Bornschier and Christopher Chase-Dunn. London: Sage.

Modelski, George, and Sylvia Modelski. 1988. *Documenting Global Leadership.* London: Macmillan.

Modelski, George, and Patrick Morgan. 1985. "Understanding Global War." *Journal of Conflict Resolution* 29:3(Sept.):371–419.

Modelski, George, and Gardner Perry III. 1991. "Democratization in Long Perspective." *Technological Forecasting and Social Change* 39:1(March):22–34.

Modelski, George, and William R. Thompson. 1988. *Seapower and Global Politics, 1494–1993.* London: Macmillan.

———. 1996. *Leading Sectors and World Politics: The Coevolution of Global Economics and Politics.* Columbia: University of South Carolina Press.

———. 1999. "The Evolutionary Pulse of the World System: Hinterland Incursions and Migrations, 4000 B.C. to A.D. 1500." Pp. 241–274 in *World-Systems Theory in Practice: Leadership, Production, and Exchange,* edited by P. Nick Kardulias. Lanham, Md.: Rowman & Littlefield.

———. 2000. "The Short and Long of Global Politics in the Twenty-first Century: An Evolutionary Approach." *International Studies Review* (forthcoming).

Nye, Joseph S., Jr. 1990. "The Changing Nature of World Powers." *Political Science Quarterly* 105:2(Summer):177–192.

Pollins, Brian M. 1996. "Global Political Order, Economic Change, and Armed Conflict: Coevolving Systems and the Use of Force." *American Political Science Review* 90:1(March):103–117.

Pollins, Brian M., and Kevin P. Murrin. 1997. "Where Hobbes Meets Hobson: Core Conflict and Colonialism, 1495–1995." Paper delivered to the annual meeting of the International Studies Association, Toronto, Canada.

Rapkin, David A. 1983. "The Inadequacy of a Single Logic: Integrating Political and Material Approaches to the World System." Pp. 241–268 in *Contending Approaches to World System Analysis,* edited by William R. Thompson. Beverly Hills, Calif.: Sage.

———. 1986. "World Leadership." Pp. 129–157 in *Exploring Long Cycles,* edited by George Modelski. Boulder, Colo.: Reinner.

———. 1987. "The Contested Concept of Hegemonic Leadership." In *World Leadership and Hegemony,* edited by David P. Rapkin. Boulder, Colo.: Reinner.

Rasler, Karen, and William R. Thompson. 1989. *War and State Making: The Shaping of the Global Powers.* Boston: Unwin Hyman.

————. 1994. *The Great Powers and Global Struggle, 1490–1990*. Lexington: University Press of Kentucky.

Reuveny, Rafael, and William R. Thompson. 1997. "The Timing of Protectionism." *Review of International Political Economy* 4:1(Spring):179–213.

Richards, Diana. 1993. "A Chaotic Model of Power Concentration in the International System." *International Studies Quarterly* 37:1(Jan.):55–72.

Rosecrance, Richard. 1987. "Long Cycle Theory and International Relations." *International Organization* 41:2(Spring):283–301.

Thompson, William R., ed. 1983. *Contending Approaches to World System Analysis*. Beverly Hills, Calif.: Sage.

————. 1988. *On Global War: Historical-Structural Approaches to World Politics*. Columbia: University of South Carolina Press.

————. 1995a. "Comparing World Systems: Systemic Leadership Succession and the Peloponnesian War Case." In *The Historical Evolution of International Political Economy*, vol. 1, edited by Christopher Chase-Dunn. London: Elgar.

————. 1995b. "Principal Rivalries." *Journal of Conflict Resolution* 39:2(June):195–223.

————. 1995c. "Some Moderate and Radical Observations on Desiderata in Comparative Naval History." Pp. 93–114 in *Doing Naval History: Essays toward Improvement*, edited by John B. Hattendorf. Annapolis, Md.: Naval Institute Press.

————. 1997. "The Evolution of Political-Commercial Challenges in the Active Zone." *Review of International Political Economy* 4:2(Summer):286–318.

————. 1997/98. "The Anglo-German Rivalry and the 1939 Failure of Deterrence." *Security Studies* 7:2(Winter):58–89.

————, ed. 1998. *Great Power Rivalries*. Columbia: University of South Carolina Press.

————. 1999. "The Military Superiority Thesis and the Ascendancy of Western Eurasia in the World System." *Journal of World History* 10:1(Spring):143–178.

————. Forthcoming. "Martian and Venusian Perspectives on International Relations: Britain as System Leader in the Nineteenth and Twentieth Centuries." In *History and International Relations Theory: Respecting Differences and Crossing Boundaries*, edited by Colin Elman and Miriam F. Elman.

————. 2000. *The Emergence of the Global Political Economy*. London: University College of London Press (forthcoming).

Thompson, William R., and George Modelski. 1994. "Long Cycle Critiques and Deja Vu All Over Again: A Rejoinder to Houweling and Siccama." *International Interactions* 20:3:209–222.

Thompson, William R., and Rafael Reuveny. 1998. "Tariffs and Trade Fluctuations: Does Protectionism Matter as Much as We Think?" *International Organization* 52:2(Spring):421–440.

Thompson, William R., and Lawrence Vescera. 1992. "Growth Waves, Systemic Openness and Protectionism." *International Organization* 46:2(Spring):493–532.

Vasquez, John A. 1993. *The War Puzzle*. Cambridge: Cambridge University Press.

Williams, John, and Robert Huckfeldt. 1996. "Empirically Discriminating between Chaotic and Stochastic Time Series." *Political Analysis* 6:125–149.

Zolberg, Aristide R. 1983. "'World' and 'System': A Misalliance." Pp. 269–290 in *Contending Approaches to World System Analysis*, edited by William R. Thompson. Beverly Hills, Calif.: Sage.

6

Gender and the World-System: Engaging the Feminist Literature on Development

Joya Misra

In short, "gender" is not just another variable to be thrown into analyses, but is an integral component of the world-system evolution. Focusing on gender points to new theoretical insights into the factors that shape group consciousness, into subtle forms of resistance to oppression, into the ways capital exploits extant cultural values, and into ways micro and macro social processes are linked. Thus, the study of gender is not only politically "in," but leads to improved sociological theorizing. (Day and Hall 1991, 4)

One of the most potent critiques that the world-systems perspective faces comes from researchers concerned about the lack of attention within world-systems analyses to the way the world-system shapes and is shaped by gender relations. In the last three decades, a vast literature focused on gender and development has emerged. In fact, a number of scholars use a world-systems perspective to address gender relations (Fernandez-Kelly 1983, 1994; Moghadam 1996; Nash 1995; Pyle 1990; Seidman 1993; Smith, Wallerstein, and Evers 1984; Smith et al. 1988; Truelove 1990; Ward 1984, 1985, 1988, 1990, 1993). Yet, this literature and the broader literature on gender and development have been typically disregarded by many world-systems scholars, who generally do not address the varied and critical roles that women play in the world-system (see Ward 1993 for a detailed critique of world-systems scholarship).

In this chapter, I provide an overview of the progression of scholarship focused on gender and development and summarize some of the most important contributions this literature has made. Finally, I discuss some basic differences between the feminist scholarship on development and world-systems scholarship and emphasize the importance of creating greater dialogue between world-systems and feminist scholars, while also arguing for maintaining their distinct perspectives.

HISTORY OF FEMINIST SCHOLARSHIP ON DEVELOPMENT

Feminist scholarship on development follows many of the basic contours of the larger development literature, as well as more general feminist theory.[1] Table 6.1 summarizes the progression of development theory in sociology. Just as development scholarship in sociology has shifted from modernization to dependency and world-systems theory, gender scholarship has shifted from a modernization-oriented approach (women in development, or WID) to more critical approaches (gender and development, or GAD). Additionally, just as postcolonial critiques have been raised against mainstream development theory, postcolonial feminist critiques have emerged.

Shifts in feminist development theory also relate to perspectives within more general feminist theory. While liberal feminist theorists focus on creating equality between men and women in the political arena, socialist and Marxist feminists have focused attention on creating economic equality between men and women (Eisenstein 1981; Barrett 1988; Hartmann 1976). Other strands, particularly black and postcolonial feminist theory, focus on relaying the diversity of women's experience (Collins 1990; Mohanty 1991; Trinh 1989; Kristeva 1981). The feminist literature on development has drawn primarily from these strands of feminist theorizing.

During the 1950s and 1960s, the modernization approach was the predominant perspective used to explain development. According to modernization theorists, development is long-term, irreversible, and socially progressive change, occurring in all societies, leading eventually to a convergence to a model along the lines of Western Europe and North America (Rostow 1964; Inkeles 1964; Huntington 1968). Modernization theorists argue that with the appropriate internal characteristics and the support of industrialized countries, developing countries will follow a sequence of stages and become modernized (Rostow 1964).

Neo-Marxist theorists, noting the inadequacy of modernization theory to explain Latin American development, suggest an alternative argument in the form of dependency theory. Rather than focusing on internal characteristics, such as traditional culture, dependency theorists stress the roles played by colonialism and neocolonialism in shaping the underdevelopment of third world countries (Dos Santos 1970;

Table 6.1 Development Theory in Sociology

Traditional Development Theory	Feminist Development Theory	General Feminist Theory
Modernization Theory	Women in Development (WID)	Liberal Feminist Theory
Dependency/World-systems	Gender and Development (GAD)	Marxist/Socialist Feminist Theory
Postcolonial	Postcolonial Feminist	Postcolonial Feminist Theory

Frank 1967, 1969; Amin 1976; Baran 1957). Drawing on the work of Fernand Braudel (1972) and the French *Annales* school, Wallerstein further refined the dependency perspective and posited the world-system perspective (Wallerstein 1974, 1976, 1984). Agreeing that colonialism and neocolonialism are critical to understanding development, Wallerstein shifts the unit of analysis from the nation-state to the world-system, arguing that understanding a nation's position in the world-economy requires understanding its relationship to all of the other nations in the world-economy. World-systems scholarship suggests a dynamic model of nations stratified into the core, semiperiphery, and periphery, with countries able to move both upward and downward in this stratification system

Postcolonial theorists have provided a critique of traditional development theory. First popularized with the publication of Edward Said's (1978) *Orientalism,* this theory shifts the critique of colonialism from strictly political and economic to cultural analyses. Julia Emberley (1993, 7) argues, "[I]deology functions in support of economic and political institutions to maintain the relations of domination and exploitation between those subjects positioned as 'colonizer' and 'colonized.'" Postcolonial theorists argue that the third world needs to identify and respond to these sorts of ideological assaults, as well as economic and political exploitation.

Tracing the pattern of mainstream development scholarship helps clarify how feminist scholarship on development has progressed, since feminist approaches connect with mainstream theorizing. Within the modernization literature, women were rarely discussed in debates about how to stimulate development. For example, Inkeles (1964; Inkeles and Smith 1974) suggests that development will occur as *men* are "made modern." This prevailing gender blindness in development theorizing was challenged by Ester Boserup's (1970) groundbreaking *Women's Roles in Economic Development.* Boserup inspired a burgeoning "women in development" (WID) literature that focuses on how development processes need to incorporate women fully, particularly in terms of formal waged work.

While mainstream modernization theorists assumed that modernization benefited women, the WID literature showed that development processes could actually disadvantage women. WID began by pointing out that colonialism had disempowered women (Boserup 1970; Tinker 1976; Charlton 1984). By making women peripheral to production and favoring men in land tenure, education, and production, colonialism radically changed power relations between men and women. In the postcolonial period, development programs continued to rest on Eurocentric assumptions about women's roles that disempowered women in these regions. For example, even in regions where women were traditionally responsible for agriculture, agricultural development programs were directed at men, while educational programs for girls were focused on creating Western-style homemakers (Boserup 1970).

Boserup and other WID theorists suggest that women should be incorporated as active participants in development. Moser (1995, 224) recounts, "this approach argues that women must be 'brought in' to the development process." As a first step,

WID theorists advance the idea that women must be educated in the skills critical for employment in the modernizing sectors of the economy. Yet, as neo-Marxist dependency and world-systems critiques of modernization theory grew, the WID approach also lost ground.

Drawing from these neo-Marxist critiques, gender and development (GAD) researchers question whether economic development is, in fact, an appropriate aim. Working from a dependency/world-systems perspective, GAD scholars explore the ways that integration into the world economy leads to increasing capitalist exploitation. These scholars also argue that research must analyze the interlinkages among local, regional, national, and international processes, rather than focusing on only one level (Mies 1982a, 1982b, 1986, 1988).

Yet, the feminist critique extends beyond those levied by the world-systems perspective, to also reconceptualize development from the standpoint of women. GAD scholars suggest that incorporating women into development processes can actually lead to a strengthening of not only capitalist but also patriarchal exploitation (Fernandez-Kelly 1994; Moon 1997). GAD theorists argue that development programs addressed toward women have a "tendency to *intensify* the existing forms of gender subordination; a tendency to *decompose* existing forms of gender subordination; and a tendency to *recompose* new forms of gender subordination" (Elson and Pearson 1981, 31).

In GAD, scholars critique the idea that educating women, or giving women access to jobs, will ameliorate gender inequality. Instead, GAD makes a structural critique about the nature of production and reproduction, by examining how processes of capital accumulation impact women's and men's lives (Fernandez-Kelly 1994). While WID scholars focus their attention on bringing women into the institutions dominated by men, GAD scholars use their gender lens to reconsider whether these institutions are the only ones of importance, while also showing how processes of capital accumulation affect men and women differently (Fernandez-Kelly 1994). For example, GAD scholars focus on women's roles in informal, subsistence, and domestic labor, while WID scholars focus on bringing women into formal labor processes.

Most recently, postcolonial feminist critiques of development have emerged. Chandra Mohanty's (1991) "Under Western Eyes," first published in 1984, is the most influential of these critiques. Mohanty agrees that Western feminist scholarship has added to what we know about women in the third world and that some of this research has been excellent at generating a deeper understanding of the interconnections between first and third world economies and these effects on women. Yet, Mohanty (1991, 64) also critiques Western feminists for creating a colonial discourse that homogenizes the lives of third world women, implying that third world women belong to a single group with common interests, regardless of class, ethnic, or racial location, and suggesting that gender and patriarchy are universal and nonvarying:

What is problematic about this kind of use of "women" as a group, as a stable category of analysis, is that it assumes an ahistorical, universal unity between women based on a generalized notion of their subordination. Instead of analytically *demonstrating* the production of women as socioeconomic political groups within particularly local contexts, this analytical move limits the definition of the female subject to gender identity, completely bypassing social class and ethnic identities. . . . Because women are thus constituted as a coherent group, sexual differences become coterminous with female subordination, and power is automatically defined in binary terms: people who have it (read: men), and people who do not (read: women). Men exploit, women are exploited. Such simplistic formulations are historically reductive; they are also ineffectual in designing strategies to combat oppressions.

Mohanty's critiques are echoed by other postcolonial theorists, including Minh ha Trinh (1989), who points out that feminist researchers have established an image of the third world woman that resembles the neocolonialist researchers' image of "natives."

These critiques have had an important impact on the field. Third world feminist scholars have become more influential in defining the problems and issues facing third world women (Beneria and Sen 1981; Sen and Grown 1987; Pongchaipat 1988; Trinh 1989; Rios 1990; Mohanty 1991; Acevedo 1995; Haj 1995; Sternbach et al. 1995; Hsiung 1996). Additionally, Western scholars have been more careful to address class and racial/ethnic differences among third world women and to specify the historical and political-economic circumstances related to women's varied roles (Mies 1982a, 1982b, 1986, 1988; Werlhof 1988; Hossfeld 1990, 1996; Pyle 1990; Tinker 1990; Truelove 1990; Ward 1990, 1993; Wolf 1990, 1992; Karpadia 1995; Ward and Pyle 1995).

Three decades ago, women were almost invisible in the scholarship on development (Baran 1957; Rostow 1964; Inkeles 1964; Huntington 1968; Dos Santos 1970). Yet today, feminist scholarship on development is vital and rapidly expanding. Current feminist scholarship critiques the concept of development, while also recasting basic understandings of the development process by bringing women to the center of the analysis. WID strategies remain the most influential in governmental and nongovernmental agencies, just as modernization approaches remain powerful (Moon 1997). Yet, in recent years, GAD and postcolonial critiques have gained prominence, as they reconceptualize development by incorporating a dynamic model that looks for differences in the ways that women interact with the economy between countries, over time, and by class, status, region, and racial and ethnic group.

In the following two sections, I review some of the major contributions that feminist researchers have made to knowledge about the third world and summarize the work of Maria Mies, as an example of this research. I focus on two components of this literature: women's roles in the economy and women's resistance to economic exploitation.

FEMINIST CONTRIBUTIONS TO THE DEVELOPMENT LITERATURE

Work is integral to understanding development, particularly since much development scholarship has stressed *economic* development. With development, the economic systems of third world countries have been markedly changed, impacting the types of work done by both men and women. Feminist scholarship on women's work includes discussions of work in the export industry, work in nonexport industry, work in the informal sector, and domestic work.

Initially, WID scholarship focused on the ways that integrating women into the economic system empowered them (Boserup 1970). GAD scholarship suggested that this integration merely led to further marginalization and the exploitation of women (Ward 1988; Acevedo 1995). Recent scholarship, influenced by postcolonial theorizing, reflects the contradictions inherent within women's employment, seeing women as *both* empowered and exploited by it, with effects that change over time, between countries, and relate to factors such as race/ethnicity, age, marital status, education, family structure, and family composition (Ward and Pyle 1995). For example, Wolf (1990) shows that although single women in Javanese export factories are paid such low wages that parents must continue to financially support their daughters, workers do benefit from wages by being able to meet consumer needs or invest in savings associations.

Although jobs in transnational corporations (TNCs) compose less than a quarter of the positions held by women in developing areas, they are a critical sector. Export-oriented production is growing in many regions of the world and has important effects on the lives of men and women. Where in some regions, women were primarily responsible for subsistence production and domestic work, export-oriented production has drawn women into waged work, reshaped men's work, and changed gender relations within the home (Truelove 1990). Export-oriented firms are focused on maintaining profits and remaining competitive on the world market (Fernandez-Kelly 1983; Safa 1986; Acevedo 1995). Women have been favored workers for many export firms, because they are almost always cheaper workers, and are seen as a more docile and pliant workforce. Because women are not perceived as "breadwinners," factory owners use this traditional gender ideology to justify paying women lower wages (Ward 1988; Hsiung 1996). Faced with few job opportunities, women accept positions that are frequently gender segregated and low paying and offer little chance for promotion (Fernandez-Kelly 1983; Beneria and Roldan 1987; Rios 1990; Tiano 1990; Truelove 1990). In addition, owing to racial discrimination, in some regions ethnic and racial minority women are often paid even lower wages (e.g., aborigines in Taiwan [Hsiung 1996]). Mohanty (1991b, 28) comments:

> World market factories relocate in search of cheap labor, and find a home in countries with unstable (or dependent) political regimes, low levels of unionization, and high unemployment. What is significant about this particular situation is that it is young third world women who overwhelmingly constitute the labor force. And it is these

women who embody and personify the intersection of sexual, class, and racial ideologies.

However, with rising unemployment, men have been more likely to take factory jobs once considered too low paying to be men's work. In addition, when factories increase job benefits in order to attract workers, men become a larger proportion of their workforce (Cantanzarite and Strober 1993).

Lim (1990) argues that workers in export firms are not necessarily as exploited as some of the literature reports. Instead, she suggests that there is a great deal of variation in wages, job security, and labor conditions, based on the age of workers and their organization in labor movements. Lim (1990, 113) argues that the stereotypical portrayal of workers as merely exploited is related to the "tendency to employ a static, ahistorical approach to the subject despite the dynamism of developing economies" that assumes that the labor conditions as export firms are first established do not change. In addition, scholars note that this literature is "disproportionate to the relative importance of such employment for Third World women, the vast majority of whom are employed in agriculture, services, and non-export, nonmultinational activities" (Lim 1990, 101). Acevedo (1993, 72) remarks that women were working long before TNCs arose: "the new research has dispelled what I call the *export processing fallacy,* that is, the notion popularized by some dependency and world-systems theories that the massive integration of women into the labor force is directly caused by export processing industrialization strategies promoted by multinationals."

Feminist scholarship has also attended to the role the state plays in maintaining profitable environments for export firms. For example, the state can establish export-processing zones, where these firms do not have to pay tariffs or duties on the goods they import or export. By subsidizing the costs of industry, the state helps the firms realize greater profits. The state also plays regulatory functions. To maintain an attractive environment for industry, the state can attempt to create a pliant workforce, by making unions or strikes illegal, or ignoring labor code violations. For example, in Taiwan, labor law limits unionization to firms with more than thirty workers. This has created the growth of numerous small firms, many of which employ women who work in their own homes (Hsiung 1996). The state's role also reflects gender ideologies, as illustrated by Jean Pyle's (1990) study of Ireland. Because the Irish government miscalculated in its benefits to export firms, TNCs flocked to Ireland, but Ireland did not profit from their presence (O'Hearn 1989). In addition, the state acted to keep women out of the workforce to support traditional gender norms. Pyle (1990, 111–112) discovers, counter to the expectations found in the development literature, that "[t]hrough explicit and informal policies, the Irish state was able to curtail the use of female labor by multinational firms, contributing to unexpectedly lackluster measures of women's economic activity throughout the period." Pongpaichit (1988) similarly finds an explicit gender ideology in state policy making in Singapore. The state plays a key role in structuring women's role in work, as well as development.

Women's work in export firms can serve to subsidize capital accumulation. True-love (1990) shows that a committee of coffee growers established mini-*macquilas* in rural Colombia focused on shoe and garment production in 1974. In these *macquilas,* women were employed at below-subsistence wages as informal workers in a "coop-erative." It was in the interests of the coffee industry to provide work for women, since the seasonal coffee industry could not provide year-round employment for men. Rather than increasing wages to men to keep them from migrating to urban areas, the *macquilas* were established. In this way, men and women must pool their wages but survive, while the coffee industry maintains its profits.

In recent years, women have increasingly been recruited to work as subcontractors, producing goods for export firms but working within in their homes (Beneria and Roldan 1987; Standing 1989; Wilson 1991). Although publicized as permitting women to combine work and domestic duties, these jobs actually provide women with lower wages and fewer benefits, which along with lower infrastructural costs, allows the companies to profit more. For example, while women workers have composed much of the workforce in labor-intensive industries in the newly industrializing countries of East Asia (Gallin 1990; Wolf 1990), the recent economic slowdown has led to the proliferation of "living rooms as factories" (the title of one Taiwanese community development program) (Hsiung 1996). Both factory and homework are premised on the notion that women's work is unimportant and unnecessary (Wolf 1990, 1996; Acevedo 1995). Even though women often serve as the only income producer for their families, the shift to "living rooms as factories" includes a lowering of wages, based on the assumption that women workers are economically dependent on their husbands (Hsiung 1996).

The homework that women do for export firms points to the gray area between work and home for many women. Acevedo (1995, 71) remarks, "[W]omen's industrial homework is at the intersection of the formal and informal sectors of the economy, making the household an intermediate point where labor, capital, and gender relations are negotiated to accommodate changing production requirements." Much of the homework done for firms is actually work done in the informal sector.

Despite expectations by both modernization and dependency theorists that the informal sector would eventually disappear (Higgins 1959; see Portes et al. 1989 for a discussion of these expectations), this sector has been growing in both developing and industrialized countries (Portes et al. 1989; Portes and Sassen-Koob 1987; Sassen 1988; Karides 1997). Informal activities run a wide gamut—from garment manufacturing to telemarketing to selling home-brewed beer (Portes et al. 1989). Informal work is often gender segregated, with women involved in production and men serving as the subcontractors, directing women's labor (Mies 1982; Hsiung 1996). As Ward and Pyle (1995, 48) comment, "Women's participation in the informal sector globally is higher than their formal participation rates and is expanding throughout the world in both rural and urban areas." Racial and ethnic minority women and men are also more frequently employed in the informal sector. In regions that require agricultural labor, during off-seasons, men are drawn into the informal work that women usually dominate (Enloe 1989; Truelove 1990).

Since informal work is unregulated and "underground," it is usually not counted in statistics on economic production (Beneria 1995). Yet informal work does not necessarily differ a great deal from formal work. For example, home workers making textile goods may be formal or informal workers—depending on whether their bosses report their labor force participation. Much informal work has nevertheless been invisible until recently (Beneria 1995; Ward and Pyle 1993). In 1976, Boulding et al. (1976, 6) remarked:

> There is one major distorting device operating on all data collection concerning women, above and beyond interpretation differences and collection facilities. This is a set of cultural assumptions about the secondary importance of anything women do; it produces underregistration of women from birth to death, and the underenumeration of women in employment.

Because informal work is missing from national employment statistics by definition, it has often been left out of development scholarship, creating significant holes in theorizing about development processes. Although also addressed in some mainstream research (Portes et al. 1989; Portes and Sassen-Koob 1987; Sassen 1988), feminist research has particularly documented the importance of women's informal labor, as well as domestic labor and subsistence activities, to the overall economy.

While men are more usually drawn into the formal workforce, women frequently are involved with several different forms of work (Beneria 1995; Ward and Pyle 1995). Informal work can be combined with formal wage work or subsistence activities, and almost always is combined with domestic labor (Beneria and Roldan 1987). Informal sector work can take place in the home, so women may work in the formal workforce during the day and work to produce items for the informal market at night. In rural areas of developing countries, women are often still involved in subsistence activities, growing food, gathering water and fuel, but they intersperse these activities with informal market production. Despite being unpaid, domestic labor (maintaining the household, child care, food preparation, making clothing, etc.) and subsistence labor are critical to the maintenance of the household, as women provide food, clothing, and services. In addition, a number of studies have noted that while men often contribute only a portion of their earnings to meet family needs, women contribute all or almost all of their income (Beneria and Roldan 1987; Blumberg 1989; Safilios-Rothschild 1984). Safilios-Rothschild (1984, 50) notes:

> Studies have documented, for example, that overall household income is not as significant a factor in the status of child nutrition as the income of the mother. Because women's income is most often used to buy food, whether or not women are responsible for feeding the family, increases in this income tend to improve the quantity and quality of food, but increases in men's income do not.

Therefore, women's work is critical to the continued maintenance of the capitalist world-economy, through a variety of means.

Feminist scholarship has demonstrated that women's work is *not* of secondary importance but actually critical to an understanding of local, regional, national, and international economies. By acknowledging the importance of what women do, scholars have begun to recognize how flexible capitalism actually is, and how the accumulation crisis has been ameliorated by the exploitation of nonwaged as well as waged workers. This work has also contributed to a better understanding of social change and resistance. While world-systems scholarship has generally focused on the idea of proletarian resistance through labor movements organized around particular industries, feminist scholarship points out that resistance must occur among all workers—women and men, formal and informal, waged and nonwaged. Feminist scholarship has also pointed to new forms of resistance, solidarity, and collective action.

In traditional terms, women have always been active members of labor movements. Although traditional scholarship often missed women's union membership, feminist accounts have helped point out the existence of women's labor organizing (Ward 1990; Ward and Pyle 1995). Work itself can provide women with independence and power, and worker resistance and organization have helped improve wages and working conditions in some areas (Lim 1990). For example, despite depictions of Asian workers and women as "passive," women in the South Korean labor movements have been among the most militant strikers in recent decades (Park 1998).

Workers also resist in ways other than through large-scale labor movements. Ong's (1987) work in Malaysia points out the various "spirits of resistance" factory women use, including being possessed by spirits. Hossfeld's (1990, 1996) study of third world women workers in the United States shows how these women use the racist and sexist biases of the managers against them. For example, Hossfeld (1990, 171) recounts:

> A Salvadoran woman, fed up with her supervisor for referring to his Hispanic workers as "mamacitas" and "little mothers" and with admonishing them to "work faster if you want your children to eat," had her husband bring both her own children and several nieces and nephews to pick her up one day. She lined all the children up in front of the supervisor and asked him how fast she would have to work to feed all those mouths. One of the children had been coached, and he told the supervisor that his mother was so tired from working that she did not have time to play with them anymore. The guilt-ridden supervisor, astonished by the large number of children and the responsibility they entailed, eased up on his admonishments and speed-up efforts and started treating the women with more consideration.

For example, workers *invent* cultural traditions banning women from working at night or requiring young women to work with women of the same ethnic background. In Taiwan, Hsiung (1996) shows how women workers "wrangle" with their managers through verbal jousting and receive higher wages when they win. While these examples of individual resistance may not change the system and ultimately affect less change than large-scale labor movements, these cases help call attention to the potential for women's resistance. In addition, connections between women in the labor movements in different countries, like those fostered through the UN Beijing conference, have helped spread strategies among workers.

Feminist scholars have also pointed out that women (and men) in the third world are involved with activism outside the workplace but related to working issues. For example, in Latin America, women have formed social change groups that organize collective meals, health cooperatives, water rights groups, and textile and craft collectivities as a response to the breakdown in the subsistence economy (Nash 1995; Sternbach et al. 1995). Rather than organizing in the workplace, these women organize in the neighborhood, bringing class- and gender-related concerns together in their attempts at change. Chinchilla (1993) traces how women in rural areas have organized around agricultural issues, with struggles that relate not only to working conditions or land tenure, but also focus on ethnic identity and women's roles. Haj (1995) shows how women in Palestine have been politically active in the nationalist movement and how women's rights are integrally related to this movement. Feminist scholars have also shown that women in the informal sector organize, as in India (Sen and Grown 1987; Shiva 1989). In some areas these movements have allied with worker's movements, to create a more powerful resistance movement.

Although feminist research has added much more to development scholarship, the major contributions I have highlighted are new approaches to the meaning of work in the global economy, and how redefining work has led to redefinitions of resistance to economic exploitation (Ward and Pyle 1995). In the next section, I discuss Maria Mies's work to provide a concrete example of how feminist research provides an alternate view of the workings of the world economy.

MARIA MIES: WOMEN AS THE LAST COLONY

Maria Mies's (1982a, 1982b, 1986) research serves as an excellent example of feminist scholarship on third world women (Mohanty 1991; Acosta-Belen and Bose 1995). Mies's (1982a, 1982b, 1986, 1988) classic work on the lace makers of Narpasur, India, and her recent work with colleagues Veronika Bennholdt-Thomsen and Claudia von Werlhof (1988) shows how integration into the world market has led to changes in the sexual division of labor, but that these changes have led to *greater* class and gender polarization than before.

Mies (1982a, 1982b, 1986, 1988) analyzes the household industry of lace making in Narsapur by focusing on the "housewives" who actually produce the lace. The crocheting of lace is done by women and girls in the home, who make lace to supplement the insufficient wages of their agricultural laborer husbands and fathers. First developed in the late nineteenth century, this industry produced nine million rupees (more than a million U.S. dollars) worth of lace in 1978, mostly exported to Australia, West Germany, Italy, Denmark, Sweden, Britain, and the United States.

Mies provides a detailed study of the structure of the lace-making industry over time, showing how first world consumers of lace are linked to the third world producers. In addition, she explores the division of labor within the lace-making industry in India and the different roles played by men and women at different times. For example, initially lace making centered around women—women in India produced the lace, and the lace was sold by women volunteers in Scotland and England to support missionary work. At the turn of the century, two Indian brothers began exporting lace as a business, completely restructuring the industry. The lace exporters reorganized production into a division of labor, with each worker making a part of the final product, while "finishers" sewed the different parts together. In addition, lace exporters provide thread to the workers and pay them wages. Since this restructuring, about sixty other "manufacturers" have been set up to export lace, each organized along these lines. In each, the exporters are men, and men have also displaced women as the agents (between the producers and the firm), and as the finishers of lace production.

Mies analyzes how considering these women to be housewives supports their exploitation. Since women's work in the home is not considered valuable, the lace making is also not seen as work, and the producers of lace become invisible. Mies (1982,14) notes that, in Narsapur, the 150,000 to 200,000 women working in the lace-making industry are not included in the nation's census statistics of workers:

> Thus, officially, these women do not exist as workers. No wonder, therefore, that the nonproducing "manufacturers" appear to be the initiators and active agents of the lace industry. Their capital accumulation seems to be a miraculous process, since they do not have to invest in factories or equipment.

Mies (1982, 21) argues that "the lacemakers' work, and the time spent on it, is 'invisible' not only to the exporters or the outside world, but also to the husbands under whose noses this work takes place. Even they define it as nonwork." Mies (1982, 21) recounts the argument of one the Agnikulakshatriya women of Narsapur:

> Our men feel that we just sit in the house and eat, doing nothing. They think that we are investing *their* money [for the thread] and then show Rs. 10 [the profit earned] as *our* earnings. As if we had won it by playing cards. We say that we are also working along with them.

Yet, although many men are unemployed, they will not take part in this "nonwork," since the pay is so low. The class polarization affecting many of the men, as the Green Revolution and other agricultural changes have taken away their livelihood, is related to the polarization between men and women in the lace-making industry. Some husbands of lace makers have entered the lace trading industry, working as small merchants. Mies (1982, 22) discusses Mariamma, who feels dependent on her retired husband who sells her lace in Calcutta, rather than realizing that her husband is dependent on her for her lace production: "She has already internalized the mystification that he who controls the marketing or the product and the capital 'gives' work to the actual producers. She has no control over the profit he makes or over the share he sends back to her."

In *Women: The Last Colony,* coauthored with Bennholdt-Thomsen and von Werlhof, Mies (1988) continues to explore the links between patriarchy and capital accumulation. The accumulation crisis in capitalism in the 1970s and 1980s (which produced global restructuring) has led to increasing exploitation of peripheral countries and the labor in those countries, including women's informal sector and subsistence work. Drawing on Marxist theorizing, Mies et al. (1988) argue that capitalist exploitation does not simply occur as the owners of capital exploit wage workers (the traditional proletariat). "Housewives" and subsistence producers in colonial countries are another key component of exploitation, critical to maintaining profits. Nonwage labor allows for an even greater extraction of surplus value. Werlhof (1988) suggests that men and women workers in the third world and housewives have replaced traditional proletarian wage workers as new "pillars of accumulation."

By concentrating on nonwaged production and informal sector work, Mies (1988) shows how resistance strategies must be broadened to include all of these workers. Wage workers may focus only on advancing their own standing, without incorporating the demands of the more marginalized workers. Mies (1988) argues that the labor movement must emphasize showing classical proletariats that they are themselves threatened by marginalization. In addition, the fight against sexism and racism *must* go hand in hand with strategies focused on eliminating class rule. As long as exploitation of women (and racial and ethnic minorities) can be used to subsidize capitalist accumulation and ameliorate the concerns of wage workers, resistance movements will not succeed.

Clearly, Mies's work addresses a host of issues that should be of concern to world-systems theorists. In this brief summary of her scholarship, I have attempted to highlight some of the more provocative arguments Mies makes. However, it is important to note that her work also addresses the way the organization of labor in Narsapur and elsewhere is related to the larger world-economy. Mohanty (1991, 65) comments:

> Mies' analysis shows the effect of a certain historically and culturally specific mode of patriarchal organization, an organization constructed on the basis of the definition of the lace makers as "non-working" housewives, at familial, local, regional, statewide, and international levels. The intricacies and the effects of particular power networks not only are emphasized, but they form the basis of Mies' analysis of how this particular group of women is situated at the center of a hegemonic, exploitative world market.

Through her example of the lace makers of Narsapur, Mies shows how the global accumulation crisis has been pacified by the exploitation of informal and nonwaged workers, and how first world and third world markets are linked. Yet, Mies's insights have yet to be incorporated into mainstream world-systems approaches to development. In the concluding section, I attempt to address the disjuncture between feminist and world-systems scholarship.

CONCLUSIONS: THE MICRO-MACRO DEBATE

World-systems scholarship and feminist scholarship share a number of key characteristics. Both feminist and neo-Marxist inspired world-systems scholarship are critiques of standard understandings of political economy and suggest the possibility of positive social transformation. Wallerstein (1997,1254) discusses exploring possible futures in social science, arguing "Why are we so afraid of discussing the possible, of analyzing the possible, or exploring the possible?" Feminist researchers similarly argue that feminist scholarship "must be designed to provide a vision of the future as well as a structural picture of the present" (Cook and Fonow 1990, 80). With their strong focus on social change, feminist and world-systems research both suggest that scholarship can and should lead to greater empowerment for the people who make up society.

Similarly, both feminist and world-systems theorists critique definitions of capitalism that focus solely on wage labor. Although traditional Marxist analyses presume that capitalism requires the expansion and dominance of wage labor, Wallerstein (1984) instead suggests that capitalism is premised on the notion of many different forms of production, including but not limited to wage labor. In fact, world-systems theorists argue that the increasing exploitation of wage laborers will *require* the occurrence of other forms of production, particularly in the periphery. Fernandez-Kelly (1994, 162), a feminist world-systems theorist, notes:

> [O]ne of the lasting contributions of world-system analysis has been the elaboration of the notion of articulation, which contradicted conceptualizations of capitalism as a system whose expansion had destroyed all preexisting modalities of labor organization. The modern world-system may be best understood as a composite formed by overlapping modes of production—some characterized by subsistence activities—and one predominant realm where advanced capitalist exchanges occur.

Feminist scholarship similarly points out the ways that women's subsistence agricultural production, domestic labor, and informal labor subsidize the formal wage labor that some workers take part in, since the exploitation of wage workers means that wages frequently do not provide enough for survival (Mies 1986, 1988). By redefining the way that productive work is viewed, to include domestic, informal, and subsistence work, feminists, like world-systems scholars, have challenged basic conceptualizations about the nature of production.

With these shared understandings, what creates the schism between feminist and world-systems research on development? Both world-systems and feminist perspectives argue that research must be grounded in the detailed context and should pursue interdisciplinary modes of research that create a more fluid system of knowledge and research (Wallerstein 1976, 1995). The world-system perspective requires scholars to create an understanding of historical patterns from exploring historical specificities. According to Wallerstein (1976, 1989, 1996, 1997), social scientists have focused on ahistorical and abstract theorizing in their quest to create general laws and to model the approaches of the natural sciences, while historians have focused on discussing specific historical moments with little interest in using these moments to delineate larger patterns. Yet to fully understand the social world, Wallerstein argues that social researchers must both study particular events *and* explore general patterns. The world-systems perspective requires scholars to bring together historical specificity with an understanding of larger patterns and trends.

Similarly, feminist researchers argue that social scientific theories must be grounded in empirical specificities, rather than being deductively created abstractions (Cook and Fonow 1990; Reinharz 1992). For feminist researchers, this critique means that social researchers should create knowledge based on their own lived experience. Smith (1987, 85) suggests that social research should be reorganized, by "first placing the sociologist where she is actually situated . . . and second, making her direct experience of the everyday world the primary ground of her knowledge." From both perspectives, the social scientist must explore *both* particular events *and* general patterns. But at the most basic level, the divergences between the two perspectives rely on differences in the unit of the analysis.

Feminist and world-systems scholars critique the nation-state as the appropriate unit of analysis for exploring development. World-systems researchers argue that by focusing almost exclusively at the level of the nation-state, scholars have been blind to the systemwide dynamics that affect nation-states. By adopting the world-system as the unit of analysis, the world-systems perspective attempts to make sense of the larger structures and trends that make up the world-system, affecting the interrelations and positioning of specific nation-states (Wallerstein 1974). Feminist researchers argue that by focusing almost exclusively at the level of the nation-state, scholars have been blind to the microdynamics that affect nation-states. By adopting individuals as the unit of analysis, feminist scholars attempt to make sense of the microprocesses and relationships that both compose and create the world-system.

This focus on women's experience and knowledge, while providing feminist scholars with key insights into the workings of the world-economy by making sense of women's (and racial and ethnic minorities') contributions, also sets their models further from the more transcendent approaches taken by many world-systems scholars. Fernandez-Kelly (1994, 61) supports this assertion by arguing, "The world-system perspective steered analysis to high levels of abstraction where it is difficult to unfold the statistic self or to examine the particulars of actual human interaction."

By focusing on the macroworkings of the world-economy, the world-systems perspective has in some ways been disconnected from the human experiences that *make up* the world-economy. But feminist research, as Smith (1987) suggests, uses those experiences to radically critique social theorizing. For many feminist scholars, theory must be grounded in *lived experience*. Much of the feminist research on developing contexts is ethnographic and focused on microprocesses, even as it relates back to the larger world-system. Harcourt (1994, 5) notes in the introduction of her book:

> [I]n order to take into account women's diverse experiences across age, class, race and geographic boundaries, we need more than just a change of policy—we need a substantial rethinking and recasting of the development enterprise. Because women's experience and knowledge have been obscured in the male bias of Western academe, including economic development theory and practice, the task is not simply to add women into the known equation but to work with new epistemologies and methodologies.

Just as the world-systems perspective created a new understanding of development by focusing attention on the world-system as the unit of analysis, feminist scholarship has created a new understanding of development by focusing attention on individual women's lives as the unit of analysis. Both perspectives make what *was* invisible (systemic processes, gender relations) visible, by shifting the frame of reference.

What is the solution to this problem? How can world-systems theorists integrate feminist research on development when these scholars have a basic disagreement about the unit of analysis needed to study development? Ward (1993, 60) has argued "that when theories continually fail to respond to feminist critiques, and thus to incorporate gender, race, and class at their centers, this omission results in theories that fail to fully capture the experiences of diverse groups of women and men." Similarly, how can feminist theorists integrate world-systems research on development? Surely, feminist research would also be strengthened by attending to the macroprocesses of the world-economy. Indeed, the work done by feminist world-systems scholars such as Kathryn Ward, Patricia Fernandez-Kelly, and others incorporates critical insights from both strains of theorizing.

Yet, both feminist and world-systems researchers have more often than not ignored the scholarship done by the other group of researchers (Ward 1993). Rather than arguing for an integration of these perspectives, I would like to suggest the importance of creating dialogue between world-systems and feminist researchers. Korzeniewicz (1996, 2) suggests that the world-systems perspective will become even more relevant if it is used to "push the boundaries of knowledge in specific areas of inquiry . . . such a strategy has the potential not only of facilitating the pursuit of new ties to other theoretical approaches or interests, but of further developing the specific content that serves to distinguish a world-systems approach from other perspectives." Boswell (1996, 2) supports this argument by stating, "To advance the theory, we need to encourage people to disagree with it. We need to engage alternative perspectives, beyond just modernization, and force into new fields of inquiry."

Similarly, in examining the position of feminist theorizing in sociology, Stacey and Thorne (1996) argue, "As feminist ideas have been absorbed into the academy, they have inevitably lost much of their political force." The revolutionary potential of feminist theorizing (and the revolutionary potential of world-systems theorizing) requires interdisciplinary debate but distinct identities. As Baron (1998, 28) argues:

> [Feminist] scholars have been successful precisely because they have successfully traversed disciplinary boundaries and have tended to develop interdisciplinary scholarly networks. Instead of being "disciplined" by the disciplines, they challenge the disciplinary boundaries. Rather than becoming "integrated" into the academy as a narrow academic enterprise, they continue the feminist tradition of political engagement and concern for contemporary social change.

Both feminist micro-insights and the world-systems macro-insights are important to understanding development. I argue that the debate between feminist and world-systems scholarship should continue, with clearer communication between the two sides, but without either side losing its distinctive approach or line of vision.

With the insights brought through feminist research, world-systems scholarship could be improved, particularly in terms of theorizing the role of *local* sites within the development process. Similarly, with the insights brought through world-systems research, feminist scholarship could be improved, particularly in terms of theorizing the role of the *global* context. By engaging one another, both theoretical approaches can and should be strengthened, yet this does not require either being folded into the other.

SUGGESTED READINGS

Beneria and Sen (1981)
Bose and Acosta-Belen (1995)
Boserup (1970)
Enloe (1989)
Fernandez-Kelly (1983)
Hossfeld (1996)
Hsiung (1996)
Mies (1986)
Mohanty, Russo, and Torres (1991)
Ward (1990, 1993)
Wolf (1992)

NOTES

The author acknowledges the extremely insightful and helpful comments of Tom Hall, Marina Karides, David Mednicoff, and Stephanie Moller, and most particularly Terry Boswell. Please address all correspondence to the author, at University of Massachusetts, Amherst, Massachusetts 01003, or via E-mail at misra@soc.umass.edu.

1. By "feminist" scholarship, I refer to research that is grounded in the experience of women and that seeks in some way to advance the position of women in society (Ostrander 1989).

2. Harcourt (1994, 3) frames this shift as one from WID to WED (women and environment and alternatives to development): "WED . . . mounts a profound critique on the whole development process. Proponents of the WED position . . . argue that development theory and practice based on Western biases and assumptions excludes both women and nature from its understanding of development and, in so doing, has contributed to the current economic and ecological crisis."

3. In this context, "homework" refers to either waged or piece-rate work produced in the home.

4. For example, Maria Mies (1988, 44) states:

> Following this analysis it may be said that capitalist penetration does not mean the transformation of all non-capitalist subsistence production into capitalist production units and of the use-value producers into "free" wage labourers. . . . This preservation of strata and regions of subsistence production does not mean, however, that they remain unaffected or that they can retain control of their product and means of production once capitalist penetration has started.

Although Mies takes Rosa Luxumberg as her intellectual predecessor, it seems clear that she shares world system's conceptualization of capitalism.

5. Wallerstein (1997, 1246) remarks:

> [Historians] wished to stick extremely close to their data and to restrict causal statements to statements of immediate sequences—immediate particular sequences. They balked at "generalizations". . . they were haunted by the fear that to generalize was to philosophize, that is, to be unscientific. . . . The emerging disciplines of economics, sociology, and political science by and large wrapped themselves in the mantle and mantra of "social science," appropriating the methods and honors of triumphant science (often, to be noted, to the scorn or despair of the natural scientists). These social science disciplines considered themselves "nomothetic," in search of universal laws, consciously modeling themselves on the good example of physics (as nearly as they could).

REFERENCES

Acevedo, Luz Del Alba. 1995. "Feminist Inroads in the Study of Women's Work and Development." Pp. 65–98 in *Women in the Latin American Development Process,* edited by Christine E. Bose and Edna Acosta-Belen. Philadelphia: Temple University Press.

Acosta-Belen, Edna, and Christine E. Bose. 1995. "Colonialism, Structural Subordination, and Empowerment: Women in the Development Process in Latin America and the Caribbean." Pp. 15–36 in *Women in the Latin American Development Process,* edited by Christine E. Bose and Edna Acosta-Belen. Philadelphia: Temple University Press.

Amin, Samir. 1976. *Unequal Development: An Essay on the Social Formation of Peripheral Capitalism.* New York: Monthly Review Press.

Baran, Paul. 1957. *The Political Economy of Growth.* New York: Monthly Review Press.

Baron, Ava. 1998. "Romancing the Field: The Marriage of Feminism and Historical Sociology." *Social Politics* 5:17–37.

Barrett, Michele. 1988. *Women's Oppression Today: The Marxist/Feminist Encounter.* London: Verso.

Beneria, Lourdes. 1995. "Toward a Greater Integration of Gender in Economics." *World Development* 23:11(Nov.):1839–1850.

Beneria, Lourdes, and Martha Roldan. 1987. *The Crossroads of Class and Gender: Industrial Homework, Subcontracting, and Household Dynamics in Mexico City.* Chicago: University of Chicago Press.

Beneria, Lourdes, and Gita Sen. 1981. "Accumulation, Reproduction, and Women's Role in Economic Development: Boserup Revisited." *Signs* 7:2(Winter):279–298.

Blumberg, Rae Lessor. 1989. "Toward a Feminist Theory of Development." Pp. 161–199 in *Feminism and Sociological Theory,* edited by Ruth Wallace. Beverly Hills: Sage.

Bose, Christine E., and Edna Acosta-Belen, eds. 1995. *Women in the Latin American Development Process.* Philadelphia: Temple University Press.

Boserup, Ester. 1970. *Woman's Role in Economic Development.* New York: St. Martin's.

Boswell, Terry. 1996. "World-System Theory or World-System Analysis?" *PEWS News* Fall:1–3.

Boulding, Elise, et al. 1976. *Handbook of International Data on Women.* New York: Sage.

Braudel, Fernand. 1972. *The Mediterranean and the Mediterranean World in the Age of Philip II.* New York: Harper & Row.

Cantanzarite, Lisa, and Myra Strober. 1993. "Gender Recomposition of the Macquiladora Workforce." *Industrial Relations* 32:1(Winter):133–147.

Charlton, Sue Ellen M. 1984. "Development as History and Process." Pp. 7–55 in *Women in Third World Development.* Boulder, Colo.: Westview.

Chinchilla, Norma Stoltz. 1993. "Gender and National Politics: Issues and Trends in Women's Participation in Latin American Movements." Pp. 37–54 in *Researching Women in Latin America and the Caribbean,* edited by Edna Acosta-Belen and Christine E. Bose. Boulder, Colo.: Westview.

Collins, Patricia Hill. 1990. *Black Feminist Thought.* New York: Routledge.

Cook, Judith A., and Mary Margaret Fonow. 1990. "Knowledge and Women's Interest: Issues of Epistemology and Methodology in Feminist Sociological Research." Pp. 69–93 in *Feminist Research Methods,* edited by Joyce McCarl Nielsen. Boulder, Colo.: Westview.

Day, Catherine, and Thomas D. Hall. 1991. "Are There Women in the World Economy?" *PEWS News* Winter:3–5.

Dos Santos, Theotonio. 1970. "The Structure of Dependence." *American Economic Review* 60(May):231–236.

Eisenstein, Zillah. 1981. *The Radical Future of Liberal Feminism.* New York: Longman.

Elson, Diane, and Ruth Pearson. 1981. "The Subordination of Women and the Internationalization of Factory Production." In *Of Marriage and the Market,* edited by Kate Young et al. London: CSE Books.

Emberley, Julia. 1993. *Thresholds of Difference: Feminist Critique, Native Women's Writings, Postcolonial Theory.* Toronto: University of Toronto Press.

Enloe, Cynthia. 1989. *Bananas, Beaches, and Bases: Making Feminist Sense of International Politics.* Berkeley: University of California Press.

Fernandez-Kelly, M. Patricia. 1983. *For We Are Sold, I and My People: Women and Industry in Mexico's Frontier.* Albany: State University of New York Press.

———. 1994. "Broadening the Scope: Gender and the Study of International Development." Pp. 143–168 in *Comparative National Development,* edited by Douglas Kincaid and Alejandro Portes. Chapel Hill: University of North Carolina Press.

Frank, Andre Gunder. 1967. *Capitalism and Underdevelopment in Latin America.* New York: Monthly Review Press.

———. 1969. *Latin America: Underdevelopment or Revolution.* New York: Monthly Review Press.

Gallin, Rita. 1990. "Women and the Export Industry in Taiwan: the Muting of Class Consciousness." Pp. 179–192 in *Women Workers and Global Restructuring,* edited by Kathryn B. Ward. Ithaca, N.Y.: ILR.

Haj, Samira. 1995. "Palestinian Women and Patriarchal Relations." Pp. 167–184 in *Rethinking the Political: Gender, Resistance, and the State,* edited by Barbara Laslett, Johanna Brenner, and Yesim Arat. Chicago: University of Chicago Press.

Harcourt, Wendy, ed. 1994. *Feminist Perspectives on Sustainable Development.* London: Zed.

Hartmann, Heidi. 1976. "Capitalism, Patriarchy, and Job Segregation by Sex." *Signs* 1:167–169.

Higgins, Benjamin Howard. 1959. *Economic Development.* New York: Norton.

Hossfeld, Karen. 1990. "'Their Logic against Them': Contradictions in Sex, Race, and Class in Silicon Valley." Pp. 149–178 in *Women Workers and Global Restructuring,* edited by Kathryn Ward. Ithaca, N.Y.: ILR.

———. 1996. *Small, Foreign, and Female: Immigrant Women Workers in Silicon Valley.* Berkeley: University of California Press.

Hsiung, Ping-Chun. 1996. *Living Rooms as Factories: Class, Gender, and the Satellite Factory System in Taiwan.* Philadelphia: Temple University Press.

Huntington, Samuel P. 1968. *Political Order in Changing Society.* New Haven, Conn.: Yale University Press.

Inkeles, Alex. 1964. "Making Men Modern." Pp. 342–361 in *Social Change,* edited by Amitai Etzioni and Eva Etzioni. New York: Basic Books.

Inkeles, Alex, and David Smith. 1974. *Becoming Modern: Individual Change in Six Developing Countries.* Cambridge, Mass.: Harvard University Press.

Karides, Marina. 1997. "Working Off the Books: Women, Race, and the Informal Economy in New Orleans." Paper presented at the Annual Meetings of the Southern Sociological Society, New Orleans, April.

Karpadia, Karin. 1995. "Where Angels Fear to Trade? 'Third World Women' and 'Development.'" *Journal of Peasant Studies* 22:3(April):356–368.

Korzeniewicz, Roberto Patricio. 1996. "The Future of PEWS." *PEWS News* (Winter):1–2.

Kristeva, Julia. 1981. "Women's Time." *Signs* 7:13–35.

Lim, Linda. 1990. "Women's Work in Export Factories: The Politics of a Cause." Pp. 109–119 in *Persistent Inequalities,* edited by Irene Tinker. New York: Oxford University Press.

Mies, Maria. 1982a. *The Lace Makers of Narsapur: Indian Housewives Produce for the World Market.* London: Zed.

———. 1982b. "The Dynamics of the Sexual Division of Labor and Integration of Rural Women into the World Market." Pp. 1–28 in *Women and Development,* edited by Lourdes Beneria. New York: Praeger.

———. 1986. *Patriarchy and Accumulation on a World Scale: Women in the International Division of Labor.* London: Zed.

———. 1988. "Capitalist Development and Subsistence Production: Rural Women in India." Pp. 27–50 in *Women: The Last Colony,* edited by Maria Mies, Veronika Bennholdt-Thomsen, and Claudia van Werlhof. London: Zed.

Mies, Maria, Veronika Bennholdt-Thomsen, and Claudia van Werlhof. 1988. *Women: The Last Colony.* London: Zed.

Moghadam, Valentine. 1996. *Patriarchy and Economic Development.* Oxford: Clarendon.

Mohanty, Chandra. 1991. "Under Western Eyes: Feminist Scholarship and Colonial Discourses." Pp. 51–79 in *Third World Women and the Politics of Feminism,* edited by Chandra Mohanty, Ann Russo, and Lourdes Torres. Bloomington: Indiana University Press.

Mohanty, Chandra, Ann Russo, and Lourdes Torres, eds. 1991. *Third World Women and the Politics of Feminism.* Bloomington: Indiana University Press.

Moon, Paul. 1997. "The Cross-Cultural Compatibility of Western Feminist Development Theory." *Journal of World-Systems Research* 3:241–249 (electronic journal: http://csf.colorado.edu/wsystems/jwsr.html; ftp and gopher: csf.colorado.edu/wsystems/journals/).

Moser, Caroline O. N. 1995. "Women, Gender, and Urban Development Policy." *Third World Planning Review* 17:2(May):223–235.

Nash, June. 1995. "Latin American Women in the World Capitalist Crisis." Pp. 151–166 in *Women in the Latin American Development Process,* edited by Christine E. Bose and Edna Acosta-Belen. Philadelphia: Temple University Press.

O'Hearn, Denis. 1989. "The Irish Case of Dependency: An Exception to the Exception?" *American Sociological Review* 54:4(Aug.):578–596.

Ong, Aihwa. 1987. *Spirits of Resistance and Capitalist Discipline: Factory Women in Malaysia.* Albany: State University of New York Press.

Ostrander, Susan A. 1989. "Feminism, Voluntarism, and the Welfare State: Toward a Feminist Sociological Theory of Social Welfare." *American Sociologist* 20:1(Spring):29–41.

Park, MiKyoung. 1999. "Lived Experiences and Collective Action: An Analysis of South Korean Women Textile Workers' Labor Protests in the 1970s." Unpublished Ph.D. dissertation, Department of Sociology, University of Georgia, Athens.

Pongchaipat, Pusat. 1988. "Two Roads to the Factory: Industrialization Strategies and Womens' Employment in South East Asia." Pp. 151–163 in *Structures of Patriarchy: The State, the Community, and the Household,* edited by Bina Agarwal. London: Zed.

Portes, Alejandro, Manuel Castells, and Lauren A. Benton, eds. 1989. *The Informal Economy: Studies in Advanced and Less Developed Countries.* Baltimore, Md.: Johns Hopkins University Press.

Portes, Alejandro, and Saskia Sassen-Koob. 1987. "Making It Underground: Comparative Material on the Informal Sector in Western Market Economies." *American Journal of Sociology* 93:1(July):30–61.

Pyle, Jean. 1990. "Export-Led Development and the Underemployment of Women: the Impact of Discriminatory Development Policy in the Republic of Ireland." Pp. 85–112 in *Women Workers and Global Restructuring,* edited by Kathryn B. Ward. Ithaca, N.Y.: ILR.

Reinharz, Shulamit. 1992. *Feminist Methods in Social Research.* New York: Oxford University Press.

Rios, Palmira N. 1990. "Export-Oriented Industrialization and the Demand for Female Labor: Puerto Rican Women in the Manufacturing Sector, 1952–1980." *Gender & Society* 4:3(Sept.):321–337.

Rostow, W. W. 1964. "The Takeoff into Self-Sustained Growth." Pp. 285–300 in *Social Change,* edited by Amitai Etzioni and Eva Etzioni. New York: Basic Books.

Safa, Helen. 1986. "Runaway Shops and Female Employment: the Search for Cheap Labor." Pp. 84–106 in *Women and Change,* edited by June Nash and Helen I. Safa. South Hadley, Mass.: Bergin & Garvey.

Safilios-Rothschild, Constantina. 1984. "The Role of the Family in Development." Pp. 45–51 in *Women in Third World Development,* edited by Sue Ellen Charlton. Boulder, Colo.: Westview.

Said, Edward. 1978. *Orientalism.* New York: Pantheon.

Sassen, Saskia. 1988. *Mobility of Labour and Capital.* Cambridge: Cambridge University Press.

Seidman, Gay. 1993. "No Freedom without the Women: Mobilization and Gender in South Africa, 1970–1992." *Signs* 18:2(Winter):291–320.

Sen, Gita, and Caren Grown. 1987. *Development, Crises, and Alternative Visions.* New York: New Feminist Library.

Shiva, Vandana. 1989. *Staying Alive.* London: Zed.

Smith, Dorothy. 1987. "Women's Perspective as Radical Critique of Sociology." Pp. 84–96 in *Feminism and Methodology: Social Science Issues,* edited by Sandra Harding. Bloomington: Indiana University Press.

Smith, Joan, Immanuel Wallerstein, and Hans-Dieter Evers, eds. 1984. *Households and the World Economy.* Beverly Hills: Sage.

Smith, Joan, J. Collins, Terence Hopkins, and A. Muhammed, eds. 1988. *Racism, Sexism, and the World-System.* New York: Greenwood.

Stacey, Judith, and Barrie Thorne. 1996. "The Missing Feminist Revolution: Ten Years Later." *Perspectives: The ASA Theory Section Newsletter* 18:1–3.

Standing, Guy. 1989. "Global Feminization through Flexible Labor." *World Development* 17:7(July):1077–1095.

Sternbach, Nancy Saporta, Marysa Navarro-Aranguren, Patricia Chuchryk, and Sonia E. Alvarez. 1995. "Feminisms in Latin America: From Bogota to San Bernardo." Pp. 240–281 in *Rethinking the Political: Gender, Resistance, and the State,* edited by Barbara Laslett, Johanna Brenner, and Yesim Arat. Chicago: University of Chicago Press.

Tiano, Susan. 1990. "Macquiladora Women: A New Category of Workers?" Pp. 193–223 in *Women Workers and Global Restructuring,* edited by Kathryn B. Ward. Ithaca, N.Y.: ILR.

Tinker, Irene. 1976. "The Adverse Effect of Development on Women." Pp. 22–34 in *Women and World Development,* edited by I. Tinker and M. B. Bramsen. Washington, D.C.: Overseas Development Council.

———, ed. 1990. *Persistent Inequalities.* New York: Oxford University Press.

Trinh, T. Minh-ha. 1989. *Women, Native, Other.* Bloomington: Indiana University Press.

Truelove, Cynthia. 1990. "Disguised Industrial Proletariats in Rural Latin America." Pp. 48–63 in *Women Workers and Global Restructuring,* edited by Kathryn B. Ward. Ithaca, N.Y.: ILR.

Wallerstein, Immanuel. 1974. *The Modern World-System: Capitalist Agriculture and the Origins of European World-Economy in the Sixteenth Century.* New York: Academic Press.

———. 1976. "A World-System Perspective on the Social Sciences." *British Journal of Sociology* 27:3(Sept.):343–354.

———. 1984. "Patterns and Perspectives of the Capitalist World-Economy." Pp. 13–26 in *The Politics of the World-Economy: The States, the Movements, and the Civilizations,* edited by Immanuel Wallerstein. Cambridge: Cambridge University Press.

———. 1995. "What Are We Bounding, and Whom, When We Bound Social Research?" *Social Research* 62:4(Winter):839–856.

———. 1996. "Social Science and Contemporary Society—The Vanishing Guarantees of Rationality." *International Sociology* 11:1(March):7–25.

———. 1997. "Social Science and the Quest for a Just Society." *American Journal of Sociology* 102:5(March):1241–1257.

Ward, Kathryn. 1984. *Women in the World System: Its Impact on Status and Fertility.* New York: Praeger.

———. 1985. "The Social Consequences of the World Economic System: The Economic Status of Women and Fertility." *Review* 8:4(Spring): 561–593.

———. 1988. "Women in the Global Economy." Pp. 17–48 in *Women and Work #3,* edited by Barbara Gutek, Laurie Larwood, and Ann Stromberg. Beverly Hills: Sage.

———, ed. 1990. *Women Workers and Global Restructuring.* Ithaca, N.Y.: ILR.

———. 1993. "Reconceptualizing World System Theory to Include Women." Pp. 43–68 in *Theory on Gender/Feminism on Theory,* edited by Paula England. New York: Aldine de Gruyter.

Ward, Kathryn, and Jean Pyle. 1995. "Gender, Industrialization, Transnational Corporations, and Development: An Overview of Trends and Patterns." Pp. 37–63 in *Researching Women in Latin America and the Caribbean,* edited by Edna Acosta-Belen and Christine E. Bose. Philadelphia: Temple University Press.

Werlhof, Claudia von. 1988. "The Proletarian is Dead: Long Live the Housewife!" Pp. 168–181 in *Women: The Last Colony,* edited by Maria Mies, Veronika Bennholdt-Thomsen, and Claudia von Werlhof. London: Zed.

Wilson, Fiona. 1991. *Sweaters: Gender, Class, and Workshop-Based Industry in Mexico.* New York: St. Martin's.

Wolf, Diane. 1990. "Linking Women's Labor with the Global Economy: Factory Workers and Their Wives in Rural Java." Pp. 25–47 in *Women Workers and Global Restructuring,* edited by Kathryn B. Ward. Ithaca, N.Y.: ILR.

———. 1992. *Factory Daughters: Gender, Household Dynamics, and Rural Industrialization in Java.* Berkeley: University of California Press.

Part III

World-Systems Overviews

7

Canada's Linguistic and Ethnic Dynamics in an Evolving World-System

Leslie S. Laczko

A basic message of the world-systems perspective is that processes occurring within nation-states in any given period cannot be fully understood without proper consideration of the larger world context. The goal of this chapter is to illustrate this basic idea by focussing on the issue of the changing internal ethnic and linguistic diversity of states in the contemporary world-system, with special reference to the case of Canada.

ETHNIC PLURALISM IN THE WORLD-SYSTEM

A useful starting point is to look at how the world's total volume of ethnic and linguistic diversity or pluralism is distributed. The world-system contains many more languages and ethnic groups than it does independent states, as the data from the 1970s presented in table 7.1 clearly illustrate. Since the ratio of large language groups to states is over 4 to 1, and the ratio of distinct ethnic groups to states is over 40 to 1, it is clear that most of the world's states are necessarily multiethnic and multilingual. This particular empirical configuration can be put into perspective if we consider that it lies somewhere between the following two hypothetical contexts. Suppose the entire world-system were governed by a single state, a world government of some sort. In this situation, the world's total diversity or pluralism would be of the "intrastate" variety. If, on the other hand, each of the nearly seven thousand ethnic groups in table 7.1 had its own state (homogeneous by definition), then the world's total volume of ethnic pluralism would be of the "interstate" variety.

This suggests that the world total of linguistic and ethnic pluralism can be broken down into the fraction within states and the fraction between states. The latter

Table 7.1   Number of States, Languages, and
Ethnic Groups in the World-System, c. 1970

1	World-system
5–6	Continents
159	Independent states
656	Major languages
6,876	Ethnic groups

Source: Constructed in part from Boulding (1979, 271).

term is simply proportional to the number of states in the system. This analysis-of-variance-inspired decomposition of a total yields the following two terms:

P total = P within states + P between states (i.e., number of states)

If the world-system were divided into only one state, the value of the second term would be zero, and the world's total diversity would be represented by the first term. If, on the other hand, each of the world-system's distinct ethnic groups had its own state, the value of the first term would be very low or zero, and all of the world's diversity would be represented by the second term.

With this framework defined, we can ask, what world-systemic factors determine the relative importance of these two terms—namely, the volume of pluralism found within states and the number of states? Chase-Dunn's (1998) structural analysis of world-system processes divides characteristics of the world-system into three categories: structural constants, systemic cycles (such as Kondratieff waves), and (usually monotonic) increasing long-term trends. In this framework, the interstate system is listed as a constant, but little attention is given to the properties under discussion here, especially the total number of states in the system. Similarly, although Wallerstein (1991), Chase-Dunn, and many others characterize world culture as multiple and diverse, the overall packaging of ethnic groups and language groups into states has not been systematically studied.

Applying this tripartite categorization to the problem at hand, we can say that our second term, the number of nominally independent states, is probably best classified as an (increasing) trend. The number of states has gone from forty-five to nearly two hundred since midcentury, and this trend shows no signs of imminent reversal. What determines the slope of this apparently monotonic increase? The number of new states coming into official existence seems to follow a wave pattern, with strong waves occurring in the 1960s and since 1989. As the number of states increases, it is likely that the average state size and average population size decreases and also that the variance of many state attributes increases. While the popularity of the territorial nation-state as an organizational form has spread, the range and variability of social forms found within states has increased as well, with micro- and "quasi-states" (Elkins 1995, 73) making up a nonnegligible proportion of the total. What about the first term, the volume of pluralism found within states? Are there any systemic

Table 7.2 Distribution of States within Which at Least 90 Percent of the Population Can Speak the Main Language, 1960s

Region	Index	Number of States
Europe*	63	27
The Americas	58	26
Asia and Oceania	30	23
Middle East and Africa	20	55
Total World-System	38	131

Source: Reconstructed from Rustow (1967), table 3.

*In 63 percent of the twenty-seven states in Europe, at least 90 percent of the population could speak the main language.

factors shaping the ratio of language groups to states, ethnic groups to states, and ethnic groups to language groups? How do these ratios change over time?

To return now to the empirical distribution of ethnic and linguistic groups in the world-system, we can ask, if some states obviously have more internal pluralism than others, which are they? Tables 7.2 and 7.3 provide some evidence on the ethnic structure of states over the past few decades. From table 7.2, based on data from the 1960s, it is clear that the world's linguistically homogeneous states (defined here as those where over 90 percent of the population can speak the language most spoken) were disproportionately located in Europe and the Americas, with the rest of the world's states being much less homogeneous.

Table 7.3 presents data from the mid-1980s, gathered before the major changes that have occurred in eastern and central Europe since 1989. Here, we see that although the number of independent states increased over these two decades, the overall pattern remained very much the same. Only a small minority of the world's states are highly homogeneous, and these states are disproportionately concentrated in Europe and the rest of the developed world. More generally, it has been well estab-

Table 7.3 Ethnic Composition of States in the World-System, Late 1980s

Ethnic Composition	Number of States
Largest group over 95% of population	30*
Largest group between 85% and 95% of population	36
Largest group between 50% and 85% of population	64
Largest group less than 50% of population	34
Total number of states 1980s	164

Source: Reconstructed from Wilmer's summary of United Nations data (1993, 56–57, note 12).

*Of which fifteen are in Europe.

lished that the world's rich and developed countries are much more homogeneous than the majority of less developed countries. In fact, there is an overall inverse relationship between a state's level of socioeconomic development and its level of internal ethnic and linguistic pluralism. Broadly speaking, the more developed the country, the lower its level of internal ethnic and linguistic pluralism.

What is responsible for this inverse relationship? From a world-systems perspective, it reveals the net result of centuries of state and nation building in core areas and the relative recency of such activities in third world areas. The forces of modernization, mass schooling, mass communications, and state-sponsored cultural homogenization have all combined to reduce the volume of ethnic and linguistic pluralism within core states (see Meyer and Hannan 1979). When viewed in world historical perspective, the world's relatively few homogeneous states confirm William McNeill's argument (1986) that throughout human history, polyethnicity has been the normal state of affairs, and that the (homogeneous) nation-state has been an aberrant and temporary exceptional development concentrated in the world's core regions over the past two centuries. The influence of the ideal of the homogeneous nation-state, despite its empirical rarity, is one particularly modern characteristic of the ethnic structure of the world-system (see Hall 1998).

CANADA'S PLURALISM IN COMPARATIVE PERSPECTIVE

Let us now turn to Canada.[1] As I have shown elsewhere (Laczko 1994), Canada is a partial exception to the overall trend shown in figure 7.1. In technical terms, Canada is an outlier, or exceptional case, in the overall inverse relationship between a state's level of socioeconomic development and its degree of internal pluralism. Canada, Belgium, Switzerland, and the United States all display higher degrees of pluralism than would be expected given their high level of development. Even compared to these other exceptional cases, Canada's level of pluralism is high. Because these four countries display pluralism levels that are quite a bit above the general trend line, they are referred to as positive outliers. Canada shares with the United States its "aboriginals and immigrants" pattern that is typical of new world settler societies. Canada also shares with Belgium and Switzerland the existence of parallel institutions[2] organized along linguistic lines, as well as a consociational[3] way of organizing linguistic and territorial differences. Canada's linguistic structure, for example, is such that it would not even come close to being defined as homogeneous in table 7.2, seen earlier. Because it combines these two types of patterns, Canada comes out as the most extreme outlier of them all in quantitative terms.

Although the data used to produce figure 7.1 are now decades old, Canada's distinct position is maintained in Gurr's recent (1993) classification of "minorities at risk" around the world. His data from the 1980s reveal that in the Western democracies and Japan, the minorities at risk average 5.5 percent of each state's population, whereas in Canada the total is 25 percent, including French speakers in Quebec

Figure 7.1 Index of Pluralism by GNP/Capita

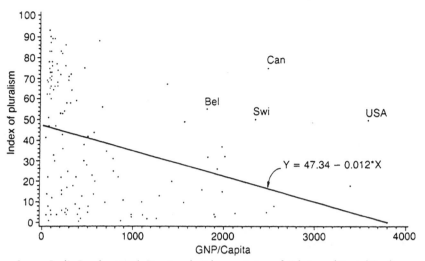

Source: Leslie Laczko, 1994. Reprinted with permission of *Ethnic and Racial Studies.*

and in other provinces, as well as aboriginal peoples. Canada's exceptional level of pluralism given its high level of development can be traced to geopolitical factors. Canada's relatively decentralized federal system and its historical linguistic dualism, along with its higher proportion of aboriginals and immigrants compared to the United States, are all features whose origins are linked to the way the Canadian state has developed alongside its much larger and more powerful neighbor and competitor. I have argued in more detail that this historical pattern of development in the shadow of a much larger neighbor has made the Canadian state more accommodating with internal corporate actors than would otherwise have been the case (Laczko 1994).

Depending on the criteria used, of course, many other subsets of the Canadian population could be classified as minorities. The historic religious cleavage between Protestants and Catholics has faded in importance. The historic French-English conflict, once played out between ascriptive and putatively homogeneous "ethnic" blocs, has been transformed by the rise of modern Québécois nationalism since the 1960s and the new centrality of language as a real and symbolic marker.

As a result, while French Quebecers have developed more of a "majority outlook" in recent decades, the historical minority reflexes are still present in many ways. One characteristic distinctive feature of the political culture of modern Quebec is that Quebecers are more supportive of collective action and state intervention to reduce inequalities than other Canadians (Laczko 1996), in much the same way that Canadians are more in favor of collective action and state intervention than

Americans (Laczko 1998). At the same time, Francophones outside Quebec have developed their own distinct identities in recent decades, and Anglophones in Quebec have, in turn, begun to display many typical minority reflexes in response to Québécois nationalism (Laczko 1995). The most recent development on this scene is the rise of a "partitionist" movement among non-Francophones in Quebec since the close results of the October 1995 referendum on sovereignty. It is a reflection of the closeness of the vote that various English-speaking groups are now advocating the right of certain regions of Quebec to remain part of Canada should a majority of Quebecers decide in favor of sovereignty.

Perhaps more significantly, aboriginal politics in Quebec and the rest of Canada has been reshaped in recent decades, with the growth of a sophisticated aboriginal or "First Nations" leadership that has recast its historical claims using the language and logic of nationhood. The federal government's need to accommodate Quebec has provided aboriginal leaders across Canada with a window of opportunity to advance their own parallel claims at the same time (see Laczko 1997). In the past decade, two attempts at constitutional reform have proved unsuccessful. The Meech Lake Accord, a constitutional amendment that would have recognized Quebec as a "distinct society within Canada," failed to gain the approval of all the provincial legislatures by the agreed-on deadline in 1990. The more complex Charlottetown Accord was defeated by referendum in a majority of provinces in 1992. Canada's two failed attempts at constitutional reform in the past decade have revealed the difficulties involved in simultaneously recognizing Quebec's distinctiveness and the place of First Nations, and at the same time accommodating the view of Canada, popular in the western provinces, as a country of "equal individuals and equal provinces." All of these developments reveal that Quebec is the key, directly or indirectly, to Canada's exceptionally high level of pluralism given its level of economic development.

Canada in mid-1998 was still in a constitutional impasse. The recent failures have hardened attitudes in both Quebec and English Canada. What is minimally acceptable in Quebec is likely to be unpopular in the other provinces, and no evident solutions are in sight, even as the next Quebec referendum may be less than two years away.[4] So Canada is facing the real possibility of scission or division into two successor states, an independent Quebec and a rump rest of Canada (see Freeman and Grady 1995; Lemco 1994; McRoberts 1995; Young 1995).

DISCUSSION

This situation is not surprising if the data we have shown are viewed from a world-systems perspective. In one sense, the statistics on the changing intra- and interstate distribution of the total volume of the world's pluralism can be interpreted as evidence of a tug-of-war between simultaneous world-systemic pressures toward both greater homogeneity and greater diversity within states. Traditional nation-building strategies were often aimed, and often quite explicitly, at producing more linguistically and ethnically homogeneous societies as part of the development process, and the net result is the negative trend line in figure 7.1. The diffusion of nationalist ideologies outward from the core areas over the past century fueled the hope that the number of such states would continue to expand, leading to a world-system composed of a growing number of (relatively) homogeneous states. From a structural perspective, the states that are significant positive outliers are more at risk than others as candidates for fission and dissolution. It is worth noting that, in the data from the 1960s and 1970s, South Africa, the Soviet Union, Yugoslavia, and Czechoslovakia all display levels of internal pluralism that are higher than what would be expected given their level of economic development.

In chapter 1 of this volume, it is mentioned that state formation, capital intensive production, commodification, and the size of economic enterprises are all continuously increasing and that these are characteristic trends of the modern world-system. We might add that long-term trends toward state formation do not necessarily mean the expansion in the size of existing states. State formation also takes the form of an increase in the number of states, either by the creation of new entities or by the scission or division of existing states. The exceptional positive outliers in figure 7.1 are prime candidates for fission and division.

Is there a cyclical pattern to state fissions and divisions? Friedman (1994, 1998) argues that periods of declining hegemony of a core power are marked by an increased incidence of ethnic mobilization and identity politics. The dissolution of the Soviet Union, Yugoslavia, and Czechoslovakia in the last decade has created a "climate of opportunity" for regional autonomist secessionist movements within the world's core states, as various recent developments in Canada, Belgium, Spain, the United Kingdom, France, and Italy readily attest. That Canada is at risk of being a candidate for dissolution is of course a statement of structural probabilities, rather than an iron law that such a course of events is predetermined. While it is impossible to predict the course of future events, we can note that in the current context Canada is buffeted between sets of contradictory tendencies. In a sense, Canada is at the cusp, balancing between opposing world-systemic trends.

Except for its use of a different language, Quebec is culturally less different from the rest of North America than ever before. At the same time, Quebec is the Western world's most powerful subnational government, and is closer to political sovereignty than ever before. The two broad tendencies at work here, in opposite directions, are the following: On the one hand, as Dion (1996) has argued, there has never been a case of fission/secession/separation of an advanced industrial democracy.[5] It is significant that the closest candidate, the former Czechoslovakia, was economically developed but not a democracy, and that no referendum was held on the elite-organized division of the single state into two. The historical momentum of a century and a half of shared institutions, as well as the complexity of the economic and political links between Quebec and the rest of Canada, is so extensive that in many ways Canada is more integrated than ever before. In this view, Canada's status as a long-standing democracy (instances of which are disproportionately located in the world's core countries; see Rueschemeyer, Stephens, and Stephens 1992) give it a stockpile of resources operating in favor of continued compromise and accommodation that were not found in the recently dissolved federations in the Balkans and in Central and Eastern Europe. Dion's argument is that support for secession among a region's voters is a function of two factors: the gains and confidence inspired by the idea of secession, and the fears for survival associated with continued union, and that high levels on both factors are likely to be rare in functioning democracies. Building on Dion's argument in world-systems terms, we can say that one reason secessionist movements are less likely to be successful within core states is that these states are likely to be well-established democracies with a full range of mechanisms at hand for accommodating regional minority grievances.

On the other hand, however, as Chase-Dunn has pointed out, there has never been any concerted move toward a pan-core political unit. Why? Because elites of core political units have always been in competition with each other, and because core economic elites are often well served by the multiplicity of states to play off against each other. Indeed, throughout the 1990s, the juxtaposition of news items about business enterprises getting larger while independent states get smaller has become commonplace. The upshot of this is that the world's core political and economic elites, faced with the real possibility of a Canadian divorce, might not be overly threatened by it, provided that a stable transition was in the offing. This is, of course, a big unknown. Indeed, there are many reasons to believe that Canada's core economic rivals, specifically those based in the United States, would not mind taking advantage of the division. Indeed, one effect of free trade and the North American Free Trade Agreement (NAFTA), both strongly supported in Quebec, is that it makes the Canadian state and the Canada-U.S. border less important. Similarly, the political elites of Canada's middle power rivals would probably not object to Canada being weakened and replaced as a member of the G-8, for example.

The century-long transition from a world of great power dominance to a world of independent nation-states has involved a continual monotonic increase in the number of states since the turn of the century. The first sharp increase was in the wave of decolonization in the 1960s. Most of this increase naturally occurred in peripheral areas of the world. More recently, a second wave of increases has pushed the total number of states to nearly two hundred, and the exact total is changing yearly. The recent wave has been brought about by the breakdown of several large federal systems in eastern and central Europe as a result of the end of the cold war (Soviet Union, Yugoslavia, Czechoslovakia). Unlike the earlier wave of new states centered in the periphery, these recent scissions have created new states in the developed semiperiphery. Canada, as a relatively small country at the perimeter of the core (much like Belgium), is naturally facing the same pressures, all the more so in light of the demonstration effect of the more recent state fissions in Europe. The high level of pluralism within Canada (and Belgium) is due to a series of historical compromises linked to the necessity of maintaining peace between competing communal groups and regions in the geopolitical context of living with much larger core power neighbors, and in both states the internal compromise is being renegotiated as the external environment is modified.

Does Canada consist of one nation or two? This century-old existential debate has in recent decades become more complex and more acute than ever before. Aboriginal issues are now central to "national unity" discussions, leading many observers to increasingly speak of three nations, Quebec, English Canada, and aboriginal peoples. At the same time, the meaning of many traditional Canadian symbols is being transformed by immigration, by the new emphasis on market solutions represented by NAFTA, and by the global economy. Throughout Canadian history both English-Canadian and French-Canadian nationalists have proposed solutions inspired by the homogeneous nation-state ideal, a recent version of which holds that Canada ought to properly be English-speaking and that Quebec ought to be an independent French-language state. In world-historical terms, of course, as McNeill has argued, bilingual and polyethnic societies such as Canada are much closer to the normal state of affairs than is the unilingual homogeneous nation-state.

To conclude, the contemporary world-system displays simultaneous tendencies toward larger state units as well as secessionist movements pushing for smaller states. It is useful to view both as the product of nonrandom forces, with world-systemic trends and cycles interacting with national and local factors. It is worth noting that the ongoing creation of larger state units such as the European Union and the more modest step toward trade unification represented by NAFTA disproportionately involve states that are in core areas of the world. The ongoing movements in favor of secession and division of existing states are no longer restricted to peripheral and semiperipheral states as in earlier decades but are in full bloom in the perimeter of the core, where they have higher visibility than ever before. At the same time, the core status and democratic history of these states make secession less likely in these core states than it has been in other parts of the world over the past decades.

In this context, the next rounds of the ongoing debate about Quebec's place in Canada are likely to be of special importance in understanding the dynamics of core states, of polyethnic federations, and of the world-system as a whole.

NOTES

This chapter is a revised version of a paper prepared for a roundtable presentation at the Annual Meeting of the American Sociological Association, Toronto, August 1997. My thanks to the editor, Thomas Hall, for many helpful comments and suggestions.

1. Very readable sociological introductions to Canada are provided by Hiller (1996) and Goyder (1990). The earlier book by Marsden and Harvey (1979), although now dated, is still valuable for its attempt to explicitly interpret social change in Canada using a world-systems framework. Two classic works of Canadian sociology are Hughes (1963) and Porter (1965). Good overviews of contemporary Quebec society and politics are given in McRoberts (1988) and Gagnon (1993). Canada's French- and English-language sociological communities

have followed separate parallel paths of development with contact and collaboration being the exception rather than the rule. For an overview of English-Canadian sociology, see Brym and Fox (1989), and for French-language Quebec sociology, see Juteau and Maheu (1989). Russell (1994) compares the development of racial and ethnic relations in Canada, the United States, and Mexico.

2. Parallel institutions refer to the top-to-bottom institutional duplication between communal groups. In Quebec, for example, there have long been separate French (historically Catholic) and English (most often Protestant) schools, hospitals, voluntary organizations, and even cemeteries. This meant that, until recent decades, most members of each language community could live most of their lives without more than passing contact with members of the other group. Societies with such a high degree of institutional parallelism are often referred to as segmented societies. The institutions that bring different communities together in such contexts are usually the political system, and, especially in Quebec and in Belgium, the labor market and the work world. Hughes (1963) gives a detailed description of the French-English segmentation that prevailed in one typical Quebec community in the 1930s.

3. *Consociationalism* is the term applied to the pattern of extensive accommodation by elites of segmented societies, often to consciously counterbalance the low level of intercommunal contacts of the mass of the population. The term was popularized by the work of the political scientist Arend Lijphart on the historic divisions between Protestants and Catholics in pre-1960s Netherlands (Lijphart 1970). McRae (1973) provides applications of the concept to society and politics in Canada, Belgium, Switzerland, and the Netherlands.

4. Within Quebec the terms *sovereignist* and *independentist* are now used along with the term *separatist* to describe the political orientation of the pro-independence Parti Québécois (PQ). The first term is often preferred by many who are sympathetic to the movement, and the latter term is often preferred by many who are opposed to the movement. The Parti Québécois was first elected in 1976 and held the first referendum on Quebec sovereignty in 1980. Over 60 percent of the Quebec population voted against the proposal, and the PQ was subsequently defeated in a general election. The PQ regained power in Quebec in the 1994 elections. A second referendum on sovereignty, held in October 1995, was narrowly defeated by a margin of less than 1 percent. Party rules stipulate that the PQ cannot hold more than one referendum on sovereignty per mandate. The leader, Lucien Bouchard, has promised that if he is reelected in the next provincial election, which was to be held in 1998 or in 1999 at the latest, his next mandate will include another (third) referendum on Quebec's political future.

5. The creation of two German states, East and West, out of the defeated Germany after World War II was a result of the war and the ascent of the Soviet Union to superpower status, not of internal German regional tensions. The rapid move toward German reunification since 1989 was, of course, successful because of the weakening and eventual dissolution of the USSR.

REFERENCES

Boulding, Elise. 1979. "Ethnic Separatism and World Development." Pp. 259–281 in *Research in Social Movements, Conflicts, and Change*, edited by Louis Kriesberg, vol. 2. Greenwich, Conn.: JAI.

Brym, Robert, and Bonnie Fox. 1989. *From Culture to Power: The Sociology of English Canada.* Toronto: Oxford University Press.

Chase-Dunn, Christopher. 1998. *Global Formation: Structures of the World Economy.* Lanham, Md.: Rowman & Littlefield. (Originally published 1989, Cambridge, Mass.: Blackwell.)

Dion, Stéphane. 1996. "Why Is Secession Difficult in Well-Established Democracies? Lessons from Quebec." *British Journal of Political Science* 26:269–283.

Elkins, David. 1995. *Beyond Sovereignty: Territory and Political Economy in the Twenty-first Century.* Toronto: University of Toronto Press.

Freeman, Alan, and Patrick Grady. 1995. *Dividing the House: Planning for a Canada Without Quebec.* Toronto: HarperCollins.

Friedman, Jonathan. 1994. *Cultural Identity and Global Process.* Thousand Oaks, Calif.: Sage.

———. 1998. "Transnationalization, Socio-political Disorder and Ethnification as Expressions of Declining Global Hegemony." *International Political Science Review* 19:3(July):233–250.

Gagnon, Alain-G., ed. 1993. *Québec: State and Society.* 2d ed. Scarborough, Ontario: Nelson Canada.

Goyder, John. 1990. *Essentials of Canadian Society.* Toronto: McClelland & Stewart.

Gurr, Ted Robert. 1993. *Minorities at Risk: A Global View of Ethnopolitical Conflicts.* Washington, D.C.: United States Institute of Peace Press.

Hall, Thomas D. 1998. "The Effects of Incorporation into World-Systems on Ethnic Processes: Lessons from the Ancient World for the Modern World." *International Political Science Review* 19:3(July):251–267.

Hiller, Harry H. 1996. *Canadian Society: A Macro Analysis.* 3d ed. Scarborough, Ontario: Prentice Hall.

Hughes, Everett C. 1963. *French Canada in Transition.* Chicago: University of Chicago Press.

Juteau, Danielle, and Louis Maheu, eds. 1989. "State of the Art Issue: Francophone Québécois Sociology." *Canadian Review of Sociology and Anthropology* 26:3.

Laczko, Leslie S. 1994. "Canada's Pluralism in Comparative Perspective." *Ethnic and Racial Studies* 17:1(Jan.):20–41.

———. 1995. *Pluralism and Inequality in Quebec.* New York: St. Martin's, and Toronto: University of Toronto Press.

———. 1996. "Language, Region, Race, Gender, and Income: Perceptions of Inequalities in Quebec and English Canada." Pp. 107–126 in *Social Inequality in Canada,* edited by Alan Frizzell and Jon H. Pammett. Ottawa: Carleton University Press.

———. 1997. "Attitudes toward Aboriginal Issues in Canada: The Changing Role of the Language Cleavage." *Quebec Studies* 23(Spring):3–12.

———. 1998. "Inégalités et État-providence: le Québec, le Canada, et le monde." *Recherches sociographiques* 39:2–3:317–340.

Lemco, Jonathan. 1994. *Turmoil in the Peaceable Kingdom: The Quebec Sovereignty Movement and Its Implications for Canada and the United States.* Toronto: University of Toronto Press.

Lijphart, Arend. 1970. *The Politics of Accommodation: Pluralism and Democracy in the Netherlands.* Berkeley: University of California Press.

Marsden, Lorna, and Edward B. Harvey. 1979. *Fragile Federation: Social Change in Canada.* Toronto: McGraw-Hill Ryerson.

McNeill, William H. 1986. *Polyethnicity and National Unity in World History.* Toronto: University of Toronto Press.

McRae, Kenneth D. 1973. *Consociational Democracy: Political Accommodation in Segmented Societies.* Toronto: McClelland & Stewart.

McRoberts, Kenneth. 1988. *Quebec: Social Change and Political Crisis.* 3d ed. Toronto: McClelland & Stewart.

———, ed. 1995. *Beyond Quebec: Taking Stock of Canada.* Montreal: McGill-Queen's University Press.

Meyer, John W., and Michael T. Hannan, eds. 1979. *National Development and the World System: Educational, Economic, and Political Change, 1950–1970.* Chicago: University of Chicago Press.

Porter, John. 1965. *The Vertical Mosaic: An Analysis of Social Class and Power in Canada.* Toronto: University of Toronto Press.

Rueschemeyer, Dietrich, Evelyn Huber Stephens, and John Stephens. 1992. *Capitalist Development and Democracy.* Chicago: University of Chicago Press.

Russell, James. 1994. *After the Fifth Sun: Class and Race in North America.* Upper Saddle River, N.J.: Prentice Hall.

Rustow, Dankwart. 1967. *A World of Nations.* Washington, D.C.: Brookings.

Wallerstein, Immanuel. 1991. *Geopolitics and Geoculture: Essays on the Changing World-System.* New York: Cambridge University Press.

Wilmer, Franke. 1993. *The Indigenous Voice in World Politics: Since Time Immemorial.* Newbury Park, Calif.: Sage.

Young, Robert A. 1995. *The Secession of Quebec and the Future of Canada.* Montreal: McGill-Queen's University Press.

8

Urbanization in the World-System: A Retrospective and Prospective

David A. Smith

It may only be a slight hyperbole to claim that efforts to understand cities and urbanization were a critical impetus to the rise of the discipline of sociology in the late nineteenth century. The profound changes that accompanied migration from the countryside to towns and great metropolises of Europe animated the central questions probed by both Emile Durkheim and Karl Marx. In his extended essay *The City,* Max Weber systematically argued that the emergence of western European towns was deeply implicated in the development of industrialization and capitalism. Similarly, the institutionalization of sociology in the United States as a distinct, legitimate academic discipline was also inextricably linked to fascination with cities and urban life. The empirical engine driving "the Chicago School" was the explosive growth of that midwestern city with its migrants, distinctive neighborhoods, wealth and poverty, influential business district, and infamous slums (e.g., Park and Burgess 1925; on the early days of "urban ecology," see Smith 1995).

Now, in the closing decades of the twentieth century, urbanization has, in effect, "gone global." The population of the world's cities has grown very rapidly: in 1900 about 3 percent of the world's people lived in cities, by 2000 close to half will be urban dwellers (Hay 1977; United Nations 1995).[1] There are now over two billion urban residents, making the growth of cities across the globe in this century the most massive shift of population in human history. Significantly, in the post–World War II era the locus of the most explosive urban growth shifted from the economically more advanced countries to the less developed countries that came to be known as "the South" or "the third world" (Davis 1972; Roberts 1978).[2] Today the largest fastest-growing "megacities" are clustered in these countries (Dogan and Kasarda 1988). The unbridled growth of cities in the relatively poor countries of Africa, Latin America, and Southeast Asia presents both pressing policy dilemmas and compelling cases for comparative sociological research; the undeniable empirical *diversity* in the patterns of "third world urbanization" makes this a particularly intriguing area for investigation.

While the sheer scale and shifting geography of urban demography (and all its consequences) is breathtaking, it may not be the most significant aspect of urbanization "going global." Even before "globalization" became the intellectual and media buzzword of the 1990s, it was becoming clear that relationships between cities around the world were changing. Major changes in the organization of production and finance—variously labeled as the end of "global Fordism" (Lipietz 1987), the rise of "flexible production" (Scott 1988) or the emergence of a "New Leviathan" (Ross and Tracte 1990)—made cities, particularly the most prominent "world cities" (Friedmann and Wolff 1982) or "global cities" (Sassen 1991), especially important nodes in economic webs defining the world-economy. It was becoming increasingly clear that we live in a "smaller world" with thoroughly globalized production of goods and services in which flows of capital, people, goods, and information are moving—and with increasing velocity (Sassen 1996). As cities become crucial nodes in these networks, an increasingly integrated, hierarchical world city system solidifies and becomes increasingly salient (Knox and Taylor 1995; Smith and Timberlake 1995).

These epochal transformations in global urbanization in the late twentieth century provide a stiff challenge for conventional urban sociology. Dominant mainstream paradigms such as urban ecology, developed to explain city dynamics in North America and Western Europe under very different historical circumstances, are poorly equipped to deal with the dizzying array of changes cities and their residents confront in a today's globalizing world. The times require visions that will put cities in global perspective. The world-systems perspective represents a promising conceptual framework for this task. Indeed, a group of comparative urban scholars recognized this almost two decades ago and began to develop a world-systems perspective on cities that has been incorporated into the emerging "urban political economy" paradigm (see Walton 1979; Timberlake 1985). This chapter will summarize and evaluate some key contributions to that work, attempt to synthesize current conceptualizations of cities in the world-system, and offer some suggestions for priorities for future research.

AN EMERGING PARADIGM: URBANIZATION IN THE WORLD-ECONOMY

Like the broader world-systems approach, the international political economy perspective on urbanization initially emerged in the 1970s. Interest in "dependent urbanization" occurred in a context in which there was a great revival of interest in the effects of colonialism and imperialism on global development/inequality, epitomized in the writings of Latin American *dependistas* and others such as Andre Gunder

Frank, Samir Amin, and Immanuel Wallerstein. Theoretical analogies between urban systems and the emerging concept of a world-system were explicit. In an early polemic book, Frank (1969) described Latin American underdevelopment as a result of "a chain of constellations of metropolis and satellites"; urban theorist David Harvey evoked similar imagery when he claimed, "Within countries functioning hierarchies of city types provide channels for the circulation of surplus value . . . [there is] a massive global circulation of surplus value in which contemporary metropolitanism is embedded" (Harvey 1973, 232). Perhaps more significantly, at about the same time scholars who were doing comparative urban research in different parts of the world were simultaneously developing international political economy notions. Manuel Castells (1977) and Bryan Roberts (1978) used the idea of "dependent urbanization" to comprehend cities and underdevelopment in Latin America; even earlier Terence G. (Terry) McGee (1976, 9) concluded that urban poverty and "dualism" in Southeast Asian cities were the result of "the dependency of the peripheral economies upon the metropolis within the international system." John Walton (1977), comparing Latin American and African urban systems, cogently argued that patterns of city growth are shaped by the historical context of a region's initial incorporation into, and changing role within, the capitalist world-system.

Needless to say, all of this work challenged the older "developmentalist" assumptions of "modernization theory" and human ecology, with their emphasis on "generative" cities, intersocietal convergence, and national urban trajectories. But, while this research linked cities and underdevelopment in the South to global processes, paid much more attention to economic classes and political elites than ecological "variables" such as technological change, and even offered hints about hierarchic world city systems, it did not clearly formulate a general new urban paradigm. But by the 1980s U.S.-based sociologists, inspired by the aforementioned scholarship as well as Wallerstein's version of a "political economy of the world-system" (see Wallerstein 1974, 1979), were eager to systematically develop an "urbanization in the world-economy" approach. Useful essays attempted to programmatically outline the key elements of a "new urban sociology" based on international political economy, and pressing research priorities followed (Walton 1981, 1982; Chase-Dunn 1984).

Michael Timberlake's edited volume in 1985 represented the most thorough-going effort to synthesize a comprehensive "urbanization in the world-economy" perspective. The editor's introduction succinctly summarized the shared premise of a coalescing body of research:

> Urbanization must be studied holistically—part of the logic of a larger process of socioeconomic development that encompasses it, and that entails systematic unevenness across regions of the world. The dependence relation is an important theoretical concept used to pry into ways in which the processes embodied in the world-system produce various manifestations of this unevenness, including divergent patterns of urbanization. (Timberlake 1985, 10)

The volume's remaining chapters brought together some extremely promising examples of "state of the art" work reflecting this international political economy perspectives on cities. There were several theoretical essays. The most important of these was a concise version of Alejandro Portes's view of "the informal sector and the world-economy" in which he argues that third world informal workers are actually providing a low wage subsidy to the rest of the world economy—which explains away the "paradox" of the "survival" of what modernization theorists had pegged as the "traditional" part of "dual economies."[3]

Another section of the book examined regional dynamics. Some chapters focused on urban "unevenness" in the third world, examining primacy in Guatemala (Smith 1985) and Thailand (London 1985), and presenting a comparative analysis of urban hierarchies in South Korea and the Philippines (Nemeth and Smith). Sassen-Koob's essay in this section examines Los Angeles and New York as emerging centers with "global control capability" in the reconfiguring world-economy but notes that this economic restructuring has led to increased numbers of labor migrants, class polarization, and poverty within these leading cities. In retrospect, this brief chapter contains most of the seeds that would later blossom into her fully developed image of "the global city" (Sassen 1991).

The final portion of the book examined "global patterns." In addition to interesting chapters about "women and urbanization in the world-system" (Ward) and systematic patterns of labor force structures in the core, semiperiphery, and periphery (Timberlake and Lunday), there was an important essay by Christopher Chase-Dunn reporting the results of a major project to look at the world city system hierarchy over the past 1,200 years. While the statistical results based on estimated urban population sizes from 800 A.D. until 1975 are interesting, the lasting contribution of this essay may be its conceptual discussions of a world city system and its links to the capitalist world-system, and the author's determination to find ways to map out this world urban system empirically.

In capsule form, the Timberlake (1985) book included seminal work in each of the three areas that will be highlighted in the rest of this essay. First, there was the theme of rapid but extremely "uneven" urbanization in the less developed noncore countries. The third world case studies documented how global processes were leading to urban outcomes like extreme "primacy" where giant cities become disproportionately large or powerful, growing material disparities between urban and rural areas, and burgeoning inequality between groups within these cities. Portes's essay highlighted a key mechanism lying behind urban poverty in the periphery and exposes it as a key mechanism of global "unequal exchange." Second, there were the beginnings of Sassen's conception of the "global city" in her essay. While she later moved on to study other crucial cases (London and Tokyo), this essay included both the idea of these centers as loci for global "command and control" functions, and the seemingly paradoxical idea that it will be precisely in these core metropolises at the lofty heights of world-system where we are most likely to see the class and race polarization and poverty characteristic of "the reperipheralization at the core" (Timberlake and Lunday's analysis also supported the latter view). Third, Chase-

Dunn clearly explicated the idea of a hierarchically organized world-city system, linked to the processes of the generic capitalist world-economy, and amenable to measurement and empirical description.

Research contextualizing cities and urbanization in the world-system has continued into the 1990s. Many of the same scholars who did the pioneering work in this area have pushed ahead with ambitious research agendas. But there is much less coherence to this body of work, as individual studies and their authors go off in distinct directions, leading to findings that sometimes seem fragmented and poorly integrated with other "strands" of research. As a result, there is much less sense of some synthetic understanding of cities in the world-economy than there was in 1985. There may be several reasons. Like many other areas of sociology, there has been a tendency toward methodological division in comparative sociology (Smith 1991) that has obscured common themes and results. And urban studies is multidisciplinary, sometimes done by "area specialists." Finally, there is a sense of being "done in" by sheer success: today we are *all* keenly aware of global processes and world-system dynamics—the paradigm shift away from the old nation-centered developmentalism is dramatic.

So to some extent some scholars may find it unnecessary to reassert their grounding in the world-systems perspective, since they more or less take it for granted as a starting point, and have moved onto the details of "normal (social) science." On the other hand, it may also be true that some scholars using the global political economy approach to cities may have neglected a basic insight of the world-systems approach: the importance of accounting for historical variation and changing economic phases—if the idea of "peripheral" (or, for that matter, "core") urbanization has any analytical utility it *must* be sensitive to relevant transformations in the global economy (including those bundled up in today's notions of globalization).

In fact, the most strident counterattacks on the assumptions of a world-systems perspective on cities from the "old urban sociology" (see, e.g., the discussion below of Kasarda and Crenshaw 1991 on "third world urbanization") have targeted the most static and oversimplified notions of "dependent urbanization." However caricatured and unfair, dismissive critiques of urban political economy (like those of world-systems analysis more generally) resonate with both traditional mainstream sociologists and the self-styled intellectual avant garde who identify with "postmodernism." And, surprising as it may seem, even today's urban sociology textbooks remain dominated by the traditional ecological perspective.[4] So an essay like this one that attempts to "pull together" three related but somewhat distinct areas of urbanization in the world-systems research—on peripheral urbanization, global cities, and world city networks—seems particularly timely.

DEPENDENT URBANIZATION: CITIES OUTSIDE THE CORE

An international political economy perspective on urbanization emphasizes the extent to which cities are located and articulated in a hierarchical global system dominated by the logic of competitive capitalism. World-systems analysis commonly refers to "zones" of the global economy and elaborates on the exploitative relationships that develop between the dominant "core" and subordinate "periphery" and "semiperiphery" areas. But in fact this "structural" language should not obscure the reality that it is groups of people (political and business elites, capitalist class) located in particular places (increasingly, the major "world cities" discussed later) who exercise power over decisions that affect the geography of space that produces cities (through the creation/destruction of land, resources, and the "built environment") and determine the material fate of the people who live in them (for a more systematic elaboration, see Smith and Timberlake 1997, 116–117).

In its simplest form, the implication of this view for the less developed or "third world" countries casts the spotlight on the ways in which global "dependency" creates various forms of uneven growth and inequality (Castells 1977, chap. 3; Roberts 1978; Armstrong and McGee 1986). This "urban unevenness" is manifest at multiple levels: very rapid city growth that drains people and resources from rural areas (referred to as "urban bias" or "overurbanization"); concentration of people and resources in the very largest cities (leading to debates over "optimal city sizes" or "urban primacy") and inequalities and polarization within cities (often manifest in "dualistic" patterns of housing, consumption or employment, urban poverty, and the expansion of the "urban informal sector") (Walton 1982; Timberlake 1985; Smith 1996a). Demonstrating a generic relationship between "dependency" and "urban unevenness" may not seem like a compelling research problem to readers today— Abu-Lughod (1996, 206) characterizes this type of research on cities as "a rather mechanical application of dependency theory." But world-systems analysis reminds us to put everything in historical context: in the 1970s and 1980s identifying the level of economic involvement between countries in "the South" and core countries with "dependency" and further linking this to these patterns of "uneven urbanization" presented a frontal assault on the "modernization/ecology" approach to cities and development. Analysis that seems self-evident or oversimplified now was at that time debunking the entrenched mainstream assumptions of urban sociology.

A series of studies did exactly that. Research on "dependent" or "peripheral" urbanization proceeded along two different methodological tracks.[5] First, there were a number of in-depth case studies of particular cities, national urban systems, or regional patterns of urbanization. Particularly useful work in this genera included McGee's (1969, 1976) pioneering case studies on Southeast Asian cities, Roberts's (1978) masterful overview of dependent urbanization in Latin America, London's (1980) political-economic analysis of Bangkok's extreme urban primacy, and work by Walton (1977), Gugler and Flanagan (1977), and Salau (1978) placing African city growth in an international political economy perspective. Armstrong and McGee (1986) compare urban growth and inequality in Asia and Latin America using dependency, unequal exchange, and capital accumulation as organizing concepts. While these studies did not gloss over national and regional differences they all highlighted patterns of urban unevenness, the salience of the global political-economic context, and the inadequacy of modernization/ecology/developmentalism arguments.

The second major methodological approach used quantitative cross-national analysis. This strategy, which attempted to "prove" various causal propositions about urbanization in the world-system, was more ambitious but also severely constrained by the available statistical data. While the results were not always consistent, this research adduced evidence that relates subordinate world-systems position to "bloated tertiary sectors" (Evans and Timberlake 1980), "overurbanization" (Timberlake and Kentor 1983), "overurbanization" and "urban bias" in Africa (Bradshaw 1987), and "urban bias" and economic stagnation (London and Smith 1988).

One measure of how influential this research was can be gauged by its critics. In a 1991 article in the *Annual Review of Sociology,* Kasarda and Crenshaw proclaimed that the modernization/ecology paradigm remained dominant in the study of third world urbanization and launched a counterattack on the political-economy of the world-system approach. In addition to rather audacious suggestions that concepts such as "class relations" are irrelevant to urban studies and the idea of "uneven development" is unintelligible, they also express nervousness about words like "exploitation" and "capitalism" (especially if they are used together!), preferring instead to use allegedly value-neutral concepts such as "cultural diffusion" and "industrialism." In their eagerness to dismiss the urbanization in the world-economy approach, these authors blatantly distort (or totally misunderstand) world-system views on the pervasiveness of capitalist inequality and feasibility of national autarky in the modern world-economy. Despite obvious bias and inaccuracies this high-profile review was doubly damaging: first, it provided a warped view of the world-systems perspective's accomplishments to those unfamiliar with comparative urban research—and offered justification to ignore this work for mainstream scholars already predisposed to be unsympathetic. Perhaps more consequentially, this dismissive yet superficial critique failed to expose genuine weaknesses and lacunae in this perspective, at a time when some constructive criticism probably was needed.

Owing to a confluence of factors, research grounded in the urbanization in the world-economy perspective has waned. Quantitative studies on urbanization in the world-economy in the 1990s are sparse—the paucity of truly comparable data on key urban concepts along with models that were either contradictory and/or low on predictive power probably helps to explain why leading analysts' attention shifted to other variables. And most compelling comparative case studies undertaken in "the urban South" today focus on changes in specific institutional areas (gender roles, industrialization, political dynamics) (see Tiano 1994; Walton and Seddon 1994; Davis 1994) rather than holistic understandings of urban transformation and how cities are articulated with wider global economic forces.

Against this backdrop of drift away from the central themes of the urbanization in the world-economy perspective, a few recent anthologies exemplify contrasting responses to the current conceptual malaise in comparative studies of cities in the

South/third world. The first are edited by Josef Gugler (1996, 1997), whose research on African cities in the 1970s contributed to the emergence of the comparative political economy approach. Despite a presumptively optimistic title (referring to urbanization in "the developing world"), this book makes a good faith attempt to provide original up-to-date broad-stroke images of regional and national trajectories of urban change in Africa, the Middle East, Latin America and three major Asian countries (China, India, and Indonesia). Individually some of the chapters are interesting, informative, even insightful. Some include detailed historical and contemporary data on urban trends; others, trenchant observations and synthetic arguments about migration patterns, city growth, intraurban inequality, and so forth. But there is no underlying or unifying conceptual framework—indeed, the editor's preface claims that contributions span "the entire range of social science approaches to urbanization" (Gugler 1997, viii). While this may appeal to a reader's sense of fairness, after finishing the volume we have little sense that we know more, in any cumulative or comprehensive sense, about "the urban transformation of the developing world" than when we opened the book. No doubt this partially reflects the editor's preference for highlighting divergent regional trajectories and capturing nuances and complexities. But it also is indicative of contemporary conceptual confusion in comparative urban research: in the wake of the demolition of the old modernization/ecological approach, and the rejection of "mechanical" versions of "dependent urbanization," currently there is a theoretical vacuum that makes eclecticism attractive.

The other volume, however, *The Urban Caribbean: Transitions to a New Global Economy*, edited by Alejandro Portes, Carlos Dore-Cabral, and Patricia Landolt (1997), provides useful signposts toward rebuilding a revitalized global political economy approach to urbanization in less developed areas. This book is more modest in scope—it focuses on changes in urban patterns and process in five small Caribbean Basin countries (Costa Rica, the Dominican Republic, Guatemala, Haiti, and Jamaica). Most of the chapter authors are indigenous social scientists from the region who collected survey data and did ethnographic research on very poor people in poverty-stricken areas of politically volatile (if not openly authoritarian) societies. Under these difficult circumstances, one would expect uneven results—this book has its weak spots. Nevertheless, this is a more satisfying volume. Why? Because it reports on a collaborative research project, coordinated by Portes et al., that attempts to systematically examine the way the regional economic crisis and restructuring of the 1980s affected urban outcomes. Each author begins with the premise that there was a major shift in the role these societies played in the global economy, as they moved from "import substitution industrialization" to "export oriented development"[6] and reduced government intervention; they then link the way this economic restructuring impacts lead city primacy, urban inequality, poverty, spatial/residential polarization, the extent and character of "informalization," and patterns of political participation. The book includes an elegant introduction that provides an overview of urban patterns in the five countries and a preview of the project results

(Portes et al. 1997). This chapter outlines a testable theoretical approach that "combines global trends with specific national processes" (48). Specifically it argues that "prior level of development" and the character of the nation-state interact with each country's role in the global economy to determine urban outcomes. While this is not a simple model, and the results reported in the chapters are sometimes a bit "messy," this strategy has great potential. It combines an appreciation of national difference and regional distinctiveness in the context of a "new global economy," with a resolute insistence that sociological generalization remains the goal.

What lessons can we glean from this discussion that might contribute to revitalizing the global political economy perspective on cities in "the South"/"third world"? Portes et al. suggest some specific ones: the need to incorporate national and regional variations into explanations, the role of states and political struggles in shaping these cities, how urban masses' links to transnational migrant networks affect "on the ground" urban patterns, the advantages of multimethodological research, and so forth. The more generic lesson, though, is that comparative research can (and should!) examine how common experiences of global-economic transformation impact urbanization across societies. But this research must go beyond assuming uniform "dependency" relations and get at the more specific nature and character of the roles that subordinate cities/countries pay in particular historical phases of global capitalism. Core-periphery relations that impact cities are not static—they undergo various transformations. This insight is not new. Indeed, Bryan Robert's (1978) argument clearly explains how changing periods of core-periphery economic links *differentially* effected patterns of urban growth in Latin America. For instance, he claimed that "import substitution industrialization" (ISI), which puts a premium on the development of a adequate market of middle- and high-income consumers, tended to reinforce colonial primate urban systems characterized by high levels of income concentration, whereas countries that did not follow the ISI path exhibited less uneven urban growth (Roberts 1978, chap. 3).

The more recent (post-1970s) trend toward market liberalization and "export-oriented industrialization" (EOI) in Latin America (and elsewhere in the less developed world) represents a third stage in which the dispersion of manufacturing and the deemphasis on internal consumer markets may promote the growth of intermediate-sized cities while intensifying the "informalization" of work for the urban poor (Oliveira and Roberts 1996). This is borne out in the Portes et al. volume, which suggests that while the current EOI phase is complicated, the most "open" economies do see some lessening of urban primacy in combination with increasing poverty and informal employment in their cities. This is a step in the right direction. But I would argue that even this project/collection does not completely come to grips with the latest incarnation of economic globalization and the recent scholarship on the topic. The loose definition of "informal sector" throughout the book elides the fact that several empirical case studies are examples of subcontracted export manufacturing work, rather than the types of urban service provision normally associated with informalization. Perhaps a more useful conceptualization of these activities would center on their position as subordinate "links" in "global commodity chains" (for a discussion, see Gereffi and Korzeniewicz 1994).

One necessary task, to revitalize the urbanization in the world-economy approach to contemporary cities in the less developed world, would be to develop systematically the urban implications of the current global restructuring that is variously described as "the rise of flexible production" (Scott 1988) or the emergence of "post-Fordist production" (Storper and Walker 1989).[7] That means systematically exploring the spatiodemographic implications of subcontracted export manufacturing in dispersed locations, as well as the degree to which neoliberal "reforms" (which often erode the provision of social welfare and basic urban services) lead to economic polarization within third world cities. For example, the spread of global garment production to the South vis-à-vis various subcontracting arrangements exposes the exploitative "dark side" of "flexible production" (Smith 1996b) and has implications for urbanization in places like Guatemala (Perez-Sainz 1997) and the Dominican Republic (Lozano 1997), as well as cities in other regions of the global periphery. But, as yet, the conceptual linkages between that regime of capitalist accumulation and uneven urbanization have not been systematically explored. This is not the place to elaborate on this, but clearly the transformed dynamics of the current era of "globalization" will affect uneven urbanization in the South in historically distinctive ways.[8] The suggestive research of Portes et al. provides hints about what elements need to be part of such a theoretical synthesis: the role of states and popular politics, changing gender divisions of labor, transnational links that encompass friends and family who have migrated to the core—all these factors and their articulation with "the new global economy" need to be part of a renovated and reenergized perspective on peripheral cities in the world-economy in the twenty-first century.

WORLD CITY/GLOBAL CITY

The basic idea of systems of interconnected cities is quite old. And even human ecological theorists writing in the early twentieth century wrote some suggestive sketches of worldwide hierarchies of urban dominance (Gras 1922; McKenzie 1927). But it was only recently, in the current era of "globalization" and with explicit acknowledgments of debt to world-system analysis, that fully developed conceptualizations of a global city system emerged (Freidmann and Wolff 1982; Sassen 1991; Knox and Taylor 1995).

The pioneering statement on "world cities" came from urban planners John Friedmann and Goetz Wolff in 1982. They claimed that world cities are "control centers of the global economy" that share a distinct set of structural characteristics. In a later article Friedmann (1986) presents a qualitative hierarchical ranking of these cities and presents a figure that "maps" the "hierarchy of world cities" (his fig. A.1). The basic premise of a "world city" or "global city," in my view, implicitly incorporates this *relational* aspect in which these places are assume to be at or near the "top" of the global city system. So it is a bit artificial to discuss "world cities" abstracted from the transnational networks in which they are deeply implicated. However, for analytical purposes, in this section I will focus on scholarly work that attempts to understand the structure and dynamics with*in* particular global cities. All these analyses, of course, rely on putting the specific global city under examination in global economy perspective. But, since the bulk of this literature is *not* attempting to look at the more encompassing issue of world-city networks and hierarchies in the more generic and complete sense, I will leave that for the next section.

In the past decade, excellent extensive case studies were done on London (King 1990) and New York (Ross and Trachte 1990) as examples of contemporary global cities. But clearly the exemplary work is Sassen's 1991 book, *The Global City: New York, London, Tokyo*. This ambitious treatise combines a global overview, detailed case studies of the three urban centers, and a synthetic conceptual argument, which includes an overview of contemporary globalization and the dynamic role world cities play in it. These central nodes both "attract" and "repel": various functions relating to the "command and control" of the global economy concentrate in global cities even as manufacturing is increasingly dispersed to ever more far-flung regions of the world. Sassen specifies the key roles of these places:

> [T]hese cities now function in four new ways: first, as highly concentrated command posts in the organization of the world economy; second, as key locations for finance and specialized service firms, which have replaced manufacturing as the leading economic sectors; third, as sites of production, including the production of innovations, in these leading industries; and fourth, as markets for the products and innovations produced. (Sassen 1991, 3–4)

One of the most interesting claims to emerge from "the world city hypothesis" (Friedmann and Wolff 1982) developed in various case studies (e.g., King 1990 or Ross and Trachte 1990) is that global city status lead to heightened levels of intracity social inequality. So, as she documents the power and assets of firms in New York, London, and Tokyo, Sassen also describes how these global cities are sites of increasing social and economic polarization. They are home to large numbers of highly remunerated salaried professionals who staff "postindustrial production" (managers, designers, engineers, and other "symbolic analysts"; see Reich 1991) but also are places coping with fiscal crisis and the breakdown of basic services, rampant poverty and unemployment, sharp levels of residential segregation, and widening material gaps between haves and have nots. At the bottom socioeconomic layers of all three global cities (including Tokyo) is a growing low-wage labor force engaged in sweatshops and homework manufacturing, composed of isolated and politically marginalized immigrants and racial and ethnic minorities (Sassen 1991, chaps. 8 and 9; see also King's [1990] descriptions of London).

While others have noted this paradoxical pattern, juxtaposing global city status with social polarization, Sassen is among the few who try to explain it. At the most abstract level, she claims that there is a new logic of world-wide capital accumulation linked to economic globalization (see especially Sassen 1991, chaps. 2 and 10). The transition to this new global economy leads to the growth of business service complexes that include managerial, design, marketing, and financial expertise in the global city but also underlie the loss of blue-collar jobs and the expansion of very low-wage "casual" or "informal" employment in these places. Sassen argues that there are direct and visible links between the "postindustrial complex" and its upper-echelon professionals and the marginalization of the urban poor. For instance, the "gentrification" of global cities by high-income "symbolic analysts" creates pressure for low-income people to vacate established housing, while simultaneously generating a surge in demand for low-wage, often "casualized" service jobs attending to the needs of the affluent: "residential building attendants, dog walkers, housekeepers for the two-career family, workers in the gourmet restaurants and food shops, French hand launderers" (Sassen-Koob 1985, 262) are just some of the jobs reflecting the underside of the return of wealthy young professionals to newly fashionable urban neighborhoods.

The evidence of socioeconomic polarization in global cities such as New York and London is quite convincing (for additional corroborating research, see Mollenkopf and Castells 1991 or Fainstein, Gordon, and Harloe 1992). It would also seem to be the trend in the other large U.S. cities that are taking on global city functions—for instance, Miami (Portes and Stepick 1993) and Los Angeles (Soja 1991). The case of Los Angeles is interesting. In a highly readable, if rather impressionistic book, Mike Davis (1990) calls it a "city of quartz," implying that it is a place that is a hard-edged amalgamation of people and groups of various ethnic, racial, and class status. His rather apocryphal vision is of a city that is rapidly changing into a multicultural mosaic of political actors, neighborhoods, and interest groups dominated by steep power differentials and yawning material inequalities. Davis depicts a palpable tension pervading the city, which made his work look prophetic when the Los Angeles riot/rebellion of 1992 ignited two years after the book appeared. Urban geographer Edward Soja (1989, 193), in a chapter appropriately titled "It All Comes Together in Los Angeles," describes the city as a key challenge to contemporary urban theory with "[s]eemingly paradoxical but functionally interdependent juxtapositions that are the epitomizing features. . . . There may be no other comparable region which presents so vividly such a composite assemblage and articulation of urban restructuring processes."

While Los Angeles seems to share certain characteristics of a "global city" with a postindustrial complex that includes the "movers and shakers" in the managerial, finance and design worlds (not to mention "Hollywood's" lofty perch in production of world culture) and it seems to be a highly polarized "dual city" like New York, an empirical overview of economic restructuring in Los Angeles suggests a rather different dynamic (Wolff 1992). Consistent with the pattern in other world cities, service employment outstrips manufacturing, poverty rose steadily during the 1980s (and remained relatively high in the early 1990s), and the area's largest minority,

Latinos, make up an increasing proportion of all workers, especially in low-wage jobs. But Wolff notes that Los Angeles city and county now contain "the largest regional concentration of manufacturing in the nation" (Wolff 1992, 2). Indeed, despite growing poverty and unemployment during the same period, Los Angeles shows a job gain between 1983 and 1990. Part of the paradox of growth with polarization is attributable to a dramatic shift within the manufacturing sector: high-tech employment (such as aerospace) is declining, while nondurable goods production is rising. As of 1990, garment production was the leading employer in the manufacturing sector (Wolff 1992). The picture here is of a city that fancies itself as "the gateway to the Pacific Rim" and a leading center of design-related production but also enmeshed in "flexible production" (including grim aspects of its "darker side" usually associated with third world cities).

The L.A. case raises some key issues that set the agenda for further research linking global cities to a wider urbanization in the world-economy paradigm. First, there is the issue of disentangling the *distinctive* world city effects from the *generalized* impacts of worldwide (or at least corewide) economic restructuring.[9] Simply put, is there really a greater tendency toward inequality/polarization in global cities than in other cities in affluent core nations? Are New York, London, and Los Angeles following a distinctive dynamic from, say, Philadelphia, Manchester, and Phoenix? Or are the pervasive effects of economic globalization creating trends toward new types of inequality up and down the urban hierarchy? Of course, if there are general characteristics of the current "post-Fordist" organization of the world-economy that promote polarization throughout the core nations, that has implications that are quite different from those of "the world city hypothesis." At this point, it's an open (but very researchable) question.

A closely related second issue is specifying the causal mechanism in operation that promotes greater polarization in world cities. As noted earlier, Sassen (1991) points to an almost symbiotic relationship between the "producer service" providing well-off professionals and the swelling ranks of low-wage informalized jobs in the personal services sector. But, obviously this can only partially account for growing inequality, and Sassen (1991) also suggests that the rapid decline in manufacturing in global cities such as New York, London, and Tokyo is implicated in rising polarization. While she does show that the employment declines in industry are greater in the world cities than in the rest of their nations (Sassen 1991, chap. 6), it is not entirely clear what causal mechanisms are at work that would promote "deindustrialization" from New York City, for example, as opposed to more generic pressures reducing manufacturing production throughout the United States. The rather different situation in Los Angeles (where manufacturing employment seems to be rising but is mainly generating poorly paid jobs) would seem to require a different "model." The point is *not* that this part of "the world city hypothesis" is wrong or misguided;[10] to the contrary, it is very appealing since it has a certain "face validity." But more research needs to be done in this area that addresses the precise links between the growing socioeconomic gaps within global cities and the role these places play in the world-economy.

A final issue (also directly relevant to the L.A. case) is measuring world city status itself. Which cities do we include as "global cities" and which do we leave out? How do we know a "world city" when we see one? At the "top end" this is not much of a problem: New York, London, and Tokyo seem pretty clearly to "fit"; places such as Hong Kong or Paris are pretty likely candidates for inclusion, too. But when we start considering other "lesser" urban areas things get much fuzzier. Freidmann's (1986) list, however suggestive, is the subjective ranking of an expert, with no real operationalizable definition of a "world city." From his vantage point (at UCLA), Los Angeles ranked as one of the top five world cities. But Sassen (1991, 1994) clearly shows that L.A. ranks far below New York on various indices of "producer service" concentration. Indeed, Los Angeles scores low on measures of corporate headquarters, particularly in the financial sector. Sassen (1994, 62–63) presents tables that show that L.A. just makes the top ten among U.S. cities for commercial banking assets and rates twenty-fourth among U.S. cities as a headquarters for "fifty largest diversified financial companies." On the other hand, when we use multiple indicators (including population size, foreign trade/shipping/air freight, corporate headquarters, existence of "international markets," and the proliferation of transnational investment), both Los Angeles and Chicago seem to rate world city status (even if they do lag behind New York City) (Abu-Lughod 1994). One solution to this problem is to "unpack" the world city construct and analyze the various unique roles great cities play in the global economy or world city hierarchy. In actual historically grounded case studies, this makes good sense, since it should lead to a fuller, more nuanced understanding of a particular urban area and its dynamics. But if we hope to retain any generalizability for the world city hypothesis, we need to develop a way to measure the concept of "global city" with some precision. Both Sassen's "producer services" indexes and suggestions for more multiplex attributional measures are defensible. But perhaps the best way to make progress here is to pursue *relational* measures, which potentially should have the additional virtue of allowing us to rank the entire world city hierarchy.

WORLD CITY NETWORKS AND HIERARCHIES

Simply discussing the notion of world or global cities presupposes some sense of the global city hierarchy. In fact, in his early attempt to describe the world city hypothesis, John Friedmann (1986) provides a table listing world cities (ranked as "primary" or "secondary" and according to the location in the world-system core or semi-periphery) and figure showing "the hierarchy of world cities" (including the key "linkages" between them). This ranking is designed to capture the functional importance of cities as "basing points" for global capital. It has a certain intuitive appeal (e.g., New York, London, Tokyo, and Los Angeles head up the hierarchy as "core primary cities") and is sufficiently suggestive to become an oft-cited starting point for subsequent studies. But this Friedmann (1986) classification is based on the informed opinion of one comparative urban expert rather than any empirical data.

It is rather telling that very few studies have even attempted to empirically measure the world urban hierarchy. Two efforts are worth noting—and illustrate some of the challenges to this sort of research. First, Chase-Dunn's 1985 chapter attempted to use historical city size data (painstakingly compiled from various archival sources) to capture the dynamics of the changing global urban rankings since 800 C.E. Given very limited data from the distant past, ranking cities by estimated populations may be the best measurement of urban system hierarchy. But population rank does not provide any direct gauge of where a city is situated in an urban system and its patterns of linkages and relationships with other places. Research on the contemporary world city system needs to capture these relational elements. David Meyer's work tries to do this. In a 1986 article he examines and maps a portion of today's global city system using network imagery. In particular, he emphasizes the key role of banking and financial linkages, examining the distribution of transnational bank headquarters and branches. His sophisticated quantitative analysis reveals a nested hierarchical global city system with New York, London, and Tokyo controlling Latin American financial institutions (with other core cities lagging far behind). While promising, this analysis was geographically constrained and does not employ formal network analysis. Subsequently, Meyer (1991) attempted to develop a more globally generalizable model of the role of "business intermediaries" in world metropolises and their changing networks and hierarchies over time. This article is conceptually rich and includes interesting historical illustrations—but no systematic analysis of relational data.

Social network analysis offers a formal set of quantitative techniques to examine the ties and links between world cities and map the global urban hierarchy.[11] With appropriate data, this methodology permits researchers to analyze multiplex patterns of exchanges, linkages, or flows between cities (or other nodes) in a way that reveals the complex patterning of connections between them, as well as the structure of the entire network. It is a logic tool to measure the global city systems and hierarchies; yet, paradoxically, very few urban scholars even seem aware of this methodology, much less interested in applying it. However, it seems obvious that capturing the connections and dominance relations between cities—whether these be based on links on "global commodity chains" (Gereffi and Korzeniewicz 1990), international business, financial and monetary transactions (Meyer 1991; Sassen 1991), or critical flows of information (Castells 1994)—should begin to provide approximations of the global urban hierarchy.

Formal network analytical research requires matrix style data that provide information on all the connections between the nodes that make up the network. Using this methodology to gauge world city system structure requires information on the flows between world cities. Unfortunately, this type of data is difficult to obtain. Most statistics on trade, investment, migration, and other worldwide exchanges are aggregated to the *national* level. Not surprisingly, earlier network studies of the structure of the world-system tended to focus on nations (Snyder and Kick 1979; Smith and White 1992). But a theme of recent literature on globalization and global cities is the extent to which cities subsume nation-states as key units of analysis. A network analysis that measures some substantively important flow(s) between major global cities would offer an "initial cut" at providing a network measure of global urban hierarchy.

Smith and Timberlake (1995) provide that preliminary attempt. In the absence of more conceptually appealing data on intercity connections or flows, this research focuses on the volume of air passenger travel between major world cities. These data come from the International Civil Aviation Organization (ICAO 1993) and provide information on the actual number of travelers flying between each city. While it might be preferable to get data on things such as capital or commodity flows between urban areas, air passengers form strands in the web linking the world's cities. Corporate executives and government officials, international financiers and independent entrepreneurs, are among those flying between cities, greasing the wheels of production, finance or commerce through face-to-face contact. This air travel data provide an image of one way that global cities are linked.

Initial results, examining 1991 data on twenty-three of the world cities on Friedmann's (1986) list, provide an empirically grounded "ranking" of cities in the global city hierarchy. The results, which can not be fully summarized here, are quite interesting (Smith and Timberlake 1995). For instance, London, New York, and Tokyo all cluster near the top as the most "central" of world cities, confirming the qualitative assessments of Friedmann and Sassen. But Paris, somewhat surprisingly, scores just below London and above New York. Los Angeles places at the lower end of the next group of five cities—quite a bit below the top four and just above Chicago, which scores just above Mexico City (Smith and Timberlake 1995). While this initial study was limited by the small number of cities included and its one-year "snapshot" dimension, subsequent analysis is underway that incorporates many more cities and multiple time points—a recent version now examines 1986, 1991, and 1994 data on over ninety cities (Timberlake and Smith 1998). Algebraically, the addition of many additional nodes could significantly change the quantitative rankings of various cities. But the new results depict a global urban hierarchy that is relatively stable over time, but with a few interpretable changes (e.g., Tokyo and Singapore moved up a bit in the last decade, while secondary U.S. cities such as Chicago and Boston declined in rank) (Timberlake and Smith 1998).

There is little doubt that formal network analysis of flows between cities is well equipped to measure the structure of the world city system and relational dimensions of global urban hierarchy. While the initial results using air passenger flows are substantively interesting, they need to be cautiously interpreted: like other quantitative methodologies, the patterns that emerge from social network analysis are dependent on the specific nature of the data that are used. There is little doubt that air travel links between major cities really *does* help define the global web connecting world cities. But it is hard to argue that air travel data would be the preferred information to use to measure world city connections—at best, this would be one of several types of crucial flows that researchers would like to measure. The problem here is mainly about data: there is very little inflow/outflow data on *any* links or connections using cities as the key nodal units.

Conceptually, the literature on globalization/global cities suggests any number of crucial links, flows, and connections to examine. For instance, data on where global commodity chains "touch down" and how they link urban places together would be of enormous theoretical value. Alternatively, information that systematically revealed the flow of investments or other types of financial linkages measured in a comprehensive city-to-city fashion would make the Sassen/Meyer arguments amenable to rigorous network verification. Unfortunately, these type of connections are usually internalized within firms, often guarded as proprietary information, and notoriously difficult to gauge. Trying to compile even approximate matrix style data would be extremely difficult for nation-to-nation flows—and close to impossible for exchanges between cities.

Recent suggestions that *information* may be a critical (if intangible and invisible) sinew binding urban systems (if not entire economies) together may provide a more promising research strategy. For instance, prominent urban scholar Manuel Castells (1994, 29) recently claimed "the Information City is at the same time the Global City, as it articulates the directional functions of the global economy in a network of decision-making and information processing centres." Telecommunications flows and/or the infrastructural capacity of cables or channels between cities are critically important links between world cities—and may be amenable to measurement (see Garcia 1998; Graham 1998). The monthly telephone bill most of us receive shows that service providers keep data that are in a form suitable for network analysis (calls are listed city to city). In principle (and with corporate or government cooperation), it should be possible to obtain data matrices on telephone calls, faxes, or other electronic links by origin and destination. Both government and industry have an interest in recording the maximum information volume that can flow through communications satellites or fiber optic cables—these data would offer a fertile ground for network analytical research. So despite today's dearth of data, there may be prospects for formal network approaches to global urban hierarchy in the future.

Even in the absence of matrix-style data that can be "run" through formal algorithms, the images and analogies of social network can and should be used to help us understand world city systems and hierarchies. The language used to describe world cities in the world-system is relational. But, like it or not, formal quantitative methodologists do not "own" any franchise over the language of social network analysis. Both urban case studies and statistical studies based on variables that measure the attributes of particular cities will be more useful to the extent that they explicitly incorporate network analogies, since our theories are suffused with that imagery.

In this regard, schematic representations (like those found in Friedmann 1986 or Meyer 1991) can be very useful heuristics for understanding urbanization in the world-economy.

CONCLUSION

A guiding premise of this chapter is that it makes sense to study various dimensions of global urbanization beginning with the basic assumptions of world-systems analysis. Research on urbanization in the world-economy blossomed in the 1980s but appears to have faded in the last decade. After reviewing the early days and discussing the apparent fallow period, the focus shifted to three central, related, but distinct areas of past research: peripheral/dependent urbanization in the putative third world, the world city/global city hypothesis, and world city networks and hierarchies. In each section, I have tried to highlight the strengths and weaknesses of past research and indicate fruitful directions for future research. Although the three areas highlighted are important for piecing together a comprehensive vision of urbanization in the world-system, this essay made no attempt to be comprehensive.

My suggestions for revitalizing this perspective are also not intended to be exhaustive.[12] One involves recapturing a sensitivity to historical phases and transformations leading to more nuanced conceptions of city growth under different capitalist regimes—in the contemporary era of globalization this would mean, for example, rethinking the idea of "peripheral urbanization" (or "world city formation") under circumstances of "global post-Fordism" or "flexible production." Another is to contextualize urban patterns in terms of national distinctiveness and regional differences in the context of changes in the global economy, without abandoning the goal of sociological generalization. A third important task is to move beyond the observations that peripheral cities or world cities exhibit uneven growth and polarization. It is far more interesting to trace the causal mechanisms that run from the global-economic context (such as late twentieth-century "economic restructuring"), through local and national social structures, classes, and states, to the "on the ground" urban inequality. This may require detailed comparative-historical case analysis. Finally, with its thoroughgoing imagery of networks and hierarchies, the global urban political economy perspective also needs formal and quantitative analysis that attempts to rigorously examine the world urban hierarchy. Efforts to structurally map the world city system require methodological sophistication but also depend on the generation and compilation of better relational data on global city links and flows. In an era when urbanization has "gone global," revitalizing the urbanization in the world-economy approach makes eminent sense. To carry this off successfully, scholars will need a combination of qualitative and quantitative methodologies, rigorous cross-national analyses and nuanced case studies, historical imagination, and thoroughly up-to-date understanding of the present era of "globalization."

SUGGESTIONS FOR FURTHER READING

Weber's (1958) work is indispensable for a comparative historical perspective on cities and social change from a "classical" sociological theorist. For intelligent broadly drawn but readable American urban ecological view of the same topic, see Hawley (1968, 1971). Students interested in the origins of urban political economy should look at Harvey (1973) and Castells (1977); for foundational world-system notions, see Frank (1969) and Wallerstein (1974); and see McGee (1969 and 1976) and Roberts (1978) for the path-breaking work on third world cities. As argued in the text, Timberlake (1985) is a key book that attempted to establish an agenda for research on urbanization in the world-economy and covers a range of topics. For more recent overviews on cities in "the South," see Armstrong and McGee (1986) and Smith (1996b). While some students may want to read the formative article by Friedmann and Wolff (1982), the most accessible book on the world/global city is Sassen (1994). The detailed case studies of Los Angeles (Davis 1990) and Miami (Portes and Stepick 1993) are also interesting reads.

NOTES

1. These figures are based on country-specific classifications that vary in their definitions of what constitutes a city, including what population threshold a place must meet. While this may be sensitive to prevailing cultural interpretations of "urban," it makes cross-national comparisons a bit tenuous—in some nations the "percent urban" may include places with ten thousand people or fewer, while in others the minimum population is significantly higher. Most researchers prefer to use measures that are more standardized, such as the percentage of the population living in cities with at least a hundred thousand or a million residents. But even these data depend on somewhat arbitrary factors such as where various cities' political boundaries are drawn. Perhaps the most crucial point, for present purposes, is that *all* available indicators show an irrefutable trend toward explosive urban growth, particularly in world regions that were overwhelmingly rural at the beginning of the twentieth century.

2. There is currently some controversy over using these terms. Uncritical use could be taken to imply that there is a uniform pattern of development (urban or otherwise) throughout the less developed world. Early world-system arguments that insisted on differentiating peripheral and semiperipheral dynamics were pioneering in showing the divergent trajectories in the third world (Wallerstein 1976); my own empirical examination of distinct patterns of urbanization by world-system "zone" and geographic region reinforces that theme (Smith and London 1990; Smith 1996a, chap. 2). In this chapter I will occasionally use the expressions "less developed countries," "the South," or "the Third World"—this will not imply uniformity. I will *not* use the misleading blanket term "developing countries" that has slipped back into comparative urban studies recently (Gugler 1996, 1997) because of its implicit "neo-modernization" bias (it is foolish to assume that all less developed countries are now "making progress"!).

3. The "urban informal sector" refers to a variety of activities involving "casual work" (e.g., unregulated street vending) and "self-help" activities (e.g., construction of shantytowns dwellings on occupied land) designed to provide subsistence for "marginal" urban residents. These "underground" activities are distinct from formal wage labor in that they often rely on family and friendship networks, are extremely labor-intensive, and avoid formal state supervision and regulation (Portes 1985).

4. While many texts implicitly take an ecological approach and provide only brief (sometimes rather dismissive) summaries of other perspectives, John Palen's *The Urban World* (1997) includes a chapter on "urban political economy" written by Michael Timberlake and me. While this is better than the alternative, even in this case the consignment of the dominant paradigm in contemporary comparative urban research to a single chapter less than twenty pages in length attests to the staying power of the old "modernization/ecology" theory.

5. The following discussion is very sketchy. For much more detail, see Smith (1996a); for an earlier analytic discussion of the divergent methodological approaches and an attempt to synthesize their results, see Smith (1991).

6. "Import substitution industrialization" (ISI) refers to an economic strategy in which poor countries work to develop the manufacturing capability to make products for their own consumers that were formally imported from already industrialized nations. The ISI approach was very popular in Latin America in the 1950s and 1960s and increased production of a variety of items (from processed food or mass-produced clothing to automobiles). As a growth strategy, ISI is constrained by the size and wealth of the potential consumer market of each country (a particularly serious problem in small Caribbean countries). "Export oriented industrialization" (EOI), on the other hand, involves the development of factory production that is targeted on overseas customers and markets. This strategy was the engine driving the "economic miracle" in the newly industrialized nations of East Asia during the 1970s and 1980s. With its global focus EOI avoids the limits of small domestic markets, but it also can create an economy that is very vulnerable to fluctuating world economic conditions and highly dependent on foreign markets and capital (as illustrated by the recent East Asian economic meltdown of late 1997/early 1998).

7. Ironically, the exact meaning of "flexibility" is often rather pliable! The general idea is that today production needs to respond quickly to changes in styles and markets, so there is a premium on being able to quickly alter product lines, shift production to different sites (or even countries), maintain "lean" inventories and organizational infrastructure, and so forth. Global subcontracting arrangements in many contemporary industries reflect the intense competitive pressure to maximize this sort of flexibility. This "post-Fordist" mode contrasts with the more stable, hierarchical structure of leading sectors of older mass production manufacturing in the mid–twentieth century (epitomized by domestically located, corporate owned, capital intensive factories of the U.S. auto industry).

8. The tone of this section may suggest that a revitalized global political economy approach must be case study oriented. I do not accept this. Quantitative studies can also be designed that capture the unique "dependency"/world-system effects of the contemporary era. For example, York Bradshaw's statistical operationalization of "debt dependence" offers a promising start (Bradshaw and Wahl 1991; Bradshaw, Noonan, Gash, and Sershen 1993). Global commodity chain analysis suggests that fine-grained measures of commodity export profiles might also be a useful way to measure a country's or city's relative position in the global production system.

9. Abu-Lughod's (1994) chapter outlining her plan to do comparative case studies of Chicago, New York, and Los Angeles provocatively raises similar questions about whether increased polarization is really greater in world cities—and why (see my next research issue). She does not provide clear-cut answers but argues that these researchable questions "are a matter for empirical investigation" (184).

10. Sassen's version of the "global city hypothesis" is attacked from two very distinct conceptual angles in a very recent symposium in *Urban Affairs Review*. In my opinion, neither fusillade strikes more than a glancing blow to her basic claims—the targeting itself suggests how influential the global city hypothesis has become. James White's (1998) critique, while raising some legitimate questions about "empirical verification," inaccurately accuses Sassen of economic reductionism and seems to confuse generic "globalization" with specific arguments about global cities. The postmodernist critique offered by Michael Smith (1998), which problematizes "the global city as an objective reality" and reminds urban scholars to continually examine "the ideology of globalization," makes some interesting points, but it slides too far down the slippery slope of antipositivism (and, ironically, ignores the attention that Sassen herself gives to "the social constructivist" perspective on global cities—see especially Sassen 1996).

11. There is no space in this review article to describe the methodology of formal network analysis or the variety of structural properties that it can measure. For a fuller discussion and an inventory of relevant network measures, see Smith and Timberlake (1995).

12. For example, despite some efforts in the 1980s (see Ward 1985), integrating gender issues into the "urbanization in the world-economy" perspective remains a crucial but unfulfilled task. Similarly, the role of race and ethnicity in urban political economy needs better specification. In the last decade of the twentieth century, the effects of various "transitions from socialism" on cities and city systems in places like the former Soviet Europe, east-central Europe, China, or Vietnam remain poorly theorized. All these important conceptual problems are beyond the purview of this chapter.

REFERENCES

Abu-Lughod, Janet. 1994. "Comparing Chicago, New York, and Los Angeles: Testing Some World City Hypotheses." Pp. 171–191 in *World Cities in a World-System,* edited by Paul Knox and Peter Taylor. New York: Cambridge University Press.

———. 1996. "Urbanization in the Arab World and the International System." Pp. 185–208 in *The Urban Transformation in the Developing World,* edited by Josef Gugler. New York: Oxford University Press.

Armstrong, Warwick, and Terence G. McGee. 1986. *Theatres of Accumulation: Studies in Asian and Latin American Urbanization.* New York: Methuen.

Bradshaw, York. 1987. "Urbanization and Underdevelopment: A Global Study of Modernization, Urban Bias, and Economic Dependency." *American Sociological Review* 52:2(April):224–239.

Bradshaw, York, Rita Noonan, Laura Gash, and Claudia Sershen. 1993. "Borrowing against the Future: Children and Third World Indebtedness." *Social Forces* 71:3(March):629–656.

Bradshaw, York, and Ana-Maria Wahl. 1991. "Foreign Debt Expansion, the International

Monetary Fund, and Regional Variation in Third World Poverty." *International Studies Quarterly* 35:3(Sept.):251–272.

Castells, Manuel. 1977. *The Urban Question: A Marxist Approach.* Cambridge, Mass.: MIT Press.

———. 1994. "European Cities, the Information Society, and the Global Economy." *New Left Review* 204(March):18–32.

Chase-Dunn, Christopher. 1984. "Urbanization in the World-System: New Directions for Research." Pp. 111–120 in *Cities in Transformation,* edited by Michael P. Smith. Beverly Hills: Sage.

———. 1985. "The System of World Cities, A.D. 800–1975." Pp. 269–292 *in Urbanization in the World-Economy,* edited by Michael Timberlake. New York: Academic Press.

Davis, Diane. 1994. *Urban Leviathan: Mexico City in the Twentieth Century.* Philadelphia: Temple University Press.

Davis, Kingley. 1972. *World Urbanization: 1950–1970.* Berkeley: University of California Press.

Davis, Mike. 1990. *City of Quartz: Excavating the Future in Los Angeles.* London: Verso.

Dogan, Matthei, and John Kasarda. 1988. *The Metropolis Era: A World of Giant Cities,* vol. 1. Newbury Park, Calif.: Sage.

Evans, Peter, and Michael Timberlake. 1980. "Dependence, Inequality, and the Growth of the Tertiary: A Comparative Analysis of Less Developed Countries." *American Sociological Review* 45:4(Aug.):531–555.

Fainstein, Susan S., Ian Gordon, and Michael Harloe, eds. 1992. *Divided Cities: New York and London in the Contemporary World.* Cambridge, Mass.: Blackwell.

Frank, Andre Gunder. 1969. *Latin America: Underdevelopment or Revolution?* New York: Monthly Review Press.

Friedmann, John. 1986. "The World City Hypothesis." *Development and Change* 17:1(Jan.):69–83.

Friedmann, John, and Goetz Wolff. 1982. "World City Formation: An Agenda for Research and Action." *International Journal of Urban and Regional Research* 6:3:309–344.

Garcia, Linda. 1998. "Electronic Commerce and the Future of Cities." Paper presented at a conference on "Global Cities and their Cross Border Networks: The Impact of Trans-nationalism and Telematics," sponsored by the United Nations University Institute for Advanced Studies, Vancouver, B.C., June.

Gereffi, Gary, and Miguel Korzeniewicz, eds. 1994. *Commodity Chains and Global Capitalism.* New York: Greenwood.

Graham, Steven. 1998. "Giant Invisible Cobwebs: On Global Cities, Optic Fibre Grids, and Planetary Urban Networks." Paper presented at a conference on "Global Cities and Their Cross Border Networks: The Impact of Transnationalism and Telematics," sponsored by the United Nations University Institute for Advanced Studies, Vancouver, B.C., June.

Gras, Norman S. B. 1922. *An Introduction to Economic History.* New York: AMS.

Gugler, Josef, ed. 1996. *The Urban Transformation of the Developing World.* New York: Oxford University Press.

———. 1997. *Cities in the Developing World: Issues, Theory, and Policy.* New York: Oxford University Press.

Gugler, Josef, and William Flanagan. 1977. "On the Political Economy of Urbanization in the Third World: The Case of Africa." *International Journal of Urban and Regional Research* 1:2:272–292.

Harvey, David. 1973. *Social Justice and the City.* Baltimore, Md.: Johns Hopkins University Press.

Hawley, Amos H., ed. 1968. *Roderick D. McKenzie on Human Ecology: Selected Writings.* Chicago: University of Chicago Press.

———. 1971. *Urban Society: An Ecological Approach.* New York: Ronald.

Hay, Richard. 1977. "Patterns of Urbanization and Socio-Economic Development in the Third World: An Overview." Pp. 71–101 in *Third World Urbanization,* edited by Janet Abu-Lughod and Richard Hay. Chicago: Maaroufa.

International Civil Aviation Organization. 1993. *On-Flight Origin and Destination: Year and Quarter Ending 31 December 1991.* Montreal: ICAO.

Kasarda, John, and Edward Crenshaw. 1991. "Third World Urbanization: Dimensions, Theories, and Determinants." *Annual Review of Sociology* 17:467–501.

King, Anthony. 1990. *Global Cities: Post-Imperialism and the Internationalization of London.* London: Routledge.

Knox, Paul, and Peter Taylor, eds. 1995. *World Cities in a World-System.* Cambridge: Cambridge University Press.

Lipietz, Alain. 1987. *Mirages and Miracles: The Crisis of Global Fordism.* London: Verso.

London, Bruce. 1980. *Metropolis and Nation in Thailand: The Political Economy of Uneven Development.* Boulder, Colo.: Westview.

———. 1985. "Thai City-Hinterland Relationships in an International Context: Development as Social Control in Northern Thailand." Pp. 207–230 in *Urbanization in the World-Economy,* edited by Michael Timberlake. New York: Academic Press.

London, Bruce, and David A. Smith. 1988. "Urban Bias, Dependence, and Economic Stagnation in Non-Core Nations." *American Sociological Review* 53:3(June):454–463.

Lozano, Wilfredo. 1997. "Dominican Republic: Informal Economy, the State, and the Urban Poor." Pp. 153–189 in *The Urban Caribbean: Transition to a New Global Economy,* edited by Alejandro Portes, Carlos Dore-Cabral, and Patricia Landolt. Baltimore, Md.: Johns Hopkins University Press.

McGee, Terence G. 1969. *The Southeast Asian City.* New York: Praeger.

———. 1976. "The Persistence of the Proto-Proletariat: Occupational Structures and Planning the Future of Third World Cities." Pp. 3–38 in *Progress in Geography,* vol. 9, edited by P. Haggart. London: Arnold.

McKenzie, Roderick D. 1968 (1926). "The Scope of Human Ecology." Pp. 19–32 in *Roderick McKenzie on Human Ecology: Selected Essays,* edited by Amos H. Hawley. Chicago: University of Chicago Press (reprint of 1922 original).

Meyer, David. 1986. "The World System of Cities: Relations Between International Financial Metropolises and South American Cities." *Social Forces* 64:3(March):553–581.

———. 1991. "Change in the World System of Metropolises: The Role of Business Intermediaries." *Urban Geography* 12:5(Sept.):393–416.

Mollenkopf, John, and Manuel Castells. 1991. *Dual City: Restructuring New York.* New York: Russell Sage.

Nemeth, Roger, and David A. Smith. 1985. "The Political Economy of Contrasting Urban Hierarchies in South Korea and the Philippines." Pp. 183–206 in *Urbanization in the World-Economy,* edited by Michael Timberlake. New York: Academic Press.

Oliveira, Orlandina de, and Bryan Roberts. 1996. "Urban Development and Social Inequality in Latin America." Pp. 253–314 in *The Urban Transformation of the Developing World,* edited by Josef Gugler. New York: Oxford University Press.

Palen, John. 1997. *The Urban World.* 5th ed. New York: McGraw-Hill.

Park, Robert, and Ernest Burgess. 1925. *The City.* Chicago: University of Chicago Press.

Perez-Sainz, Juan Pablo. 1997. "Guatemala: The Two Faces of the Metropolitan Area." Pp. 124–152 in *The Urban Caribbean: Transition to the New Global Economy,* edited by Alejandro Portes, Carlos Dore-Cabral, and Patricia Landolt. Baltimore, Md.: Johns Hopkins University Press.

Portes, Alejandro. 1985. "The Informal Sector and the World-Economy: Notes on the Structure of Subsidized Labor." Pp. 53–62 in *Urbanization in the World-Economy,* edited by Michael Timberlake. New York: Academic Press.

Portes, Alejandro, Carlos Dore-Cabral, and Patricia Landolt, eds. 1997. *The Urban Caribbean: Transition to the New Global Economy.* Baltimore, Md.: Johns Hopkins University Press.

Portes, Alejandro, Jose Itzigsohn, and Carlos Dore-Cabral. 1997. "Urbanization in the Caribbean Basin: Social Change during the Years of the Crisis." Pp. 16–54 *in The Urban Caribbean: Transition to the New Global Economy,* edited by Alejandro Portes, Carlos Dore-Cabral, and Patricia Landolt. Baltimore, Md.: Johns Hopkins University Press.

Portes, Alejandro, and Alex Stepick. 1993. *City on the Edge: The Transformation of Miami.* Berkeley: University of California Press.

Reich, Robert. 1991. *The Work of Nations: Preparing Ourselves for Twenty-first Century Capitalism.* New York: Knopf.

Roberts, Bryan. 1978. *Cities of Peasants: The Political Economy of Urbanization in the Third World.* Beverly Hills: Sage.

Ross, Robert, and Kent Trachte. 1990. *Global Capitalism: The New Leviathan.* Albany: State University of New York Press.

Salau, Ademola. 1978. "The Political Economy of Cities in Tropical Africa." *Civilizations* 28:281–290.

Sassen, Saskia. 1991. *The Global City: New York, London, Tokyo.* Princeton, N.J.: Princeton University Press.

———. 1994. *Cities in a World Economy.* Thousand Oaks, Calif.: Pine Forge.

———. 1996. *Losing Control? Sovereignty in an Age of Globalization.* New York: Columbia University Press.

Sassen-Koob, Saskia. 1985. "Capital Mobility and Labor Migration: Their Expression in Core Cities." Pp. 231–265 in *Urbanization in the World-Economy,* edited by Michael Timberlake. New York: Academic Press.

Scott, Allen. 1988. "Flexible Production Systems and Regional Development." *International Journal of Urban and Regional Research* 12:171–186.

Smith, Carol A. 1985. "Class Relations and Urbanization in Guatemala: Toward an Alternative Theory of Primacy." Pp. 121–167 in *Urbanization in the World-Economy,* edited by Michael Timberlake. New York: Academic Press.

Smith, David A. 1991. "Method and Theory in Comparative Urban Studies." *International Journal of Comparative Sociology* 32:1–2(Jan.):39–58.

———. 1995. "The New Urban Sociology Meets the Old: Rereading Some Classical Human Ecology." *Urban Affairs Review* 30:3(Jan.):432–456.

———. 1996a. *Third World Cities in Global Perspective: The Political Economy of Uneven Urbanization.* Boulder, Colo.: Westview.

———. 1996b. "Going South: Global Restructuring and Garment Production in Three East Asian Cases." *Asian Perspective* 20:2(Fall–Winter):211–241.

Smith, David A., and Bruce London. 1990. "Convergence in World Urbanization? A Quantitative Assessment." *Urban Affairs Quarterly* 25:4(June):574–590.

Smith, David A., and Michael Timberlake. 1995. "Conceptualising and Mapping and Structure of the World System's City System." *Urban Studies* 32:2(March):287–302.

———. 1997. "Urban Political Economy." Pp. 109–128 in *The Urban World*, edited by John Palen. New York: McGraw-Hill.

Smith, David A., and Douglas White. 1992. "Structure and Dynamics of the Global Economy: Network Analysis of International Trade 1965–1980." *Social Forces* 70:4(June):857–893.

Smith, Michael P. 1998. "The Global City—Whose Social Construct Is It Anyway? A Comment on White." *Urban Affairs Review* 33:4(March):482–488.

Snyder, David, and Edward Kick. 1979. "Structural Position in the World System and Economic Growth, 1955–1970: A Multiple Network Analysis of Transnational Interaction." *American Journal of Sociology* 84:5(March):1096–1126.

Soja, Edward. 1989. *Postmodern Geographies: The Reassertion of Space in Critical Social Theory.* London: Verso.

———. 1991. "Poles Apart: Urban Restructuring in New York and Los Angeles." Pp. 359–376 in *Dual City: Restructuring New York*, edited by John Mollenkopf and Manuel Castells. New York: Russell Sage.

Storper, Michael, and Richard Walker. 1989. *The Capitalist Imperative: Territory, Technology and Industrial Growth.* New York: Blackwell.

Tiano, Susan. 1994. *Patriarchy on the Line: Labor, Gender, and Ideology in the Mexican Maquila Industry.* Philadelphia: Temple University Press.

Timberlake, Michael, ed. 1985. *Urbanization in the World-Economy.* New York: Academic Press.

Timberlake, Michael, and Jeffrey Kentor. 1983. "Economic Dependence, Overurbanization, and Economic Growth: A Study of Less Developed Countries." *Sociological Quarterly* 24:489–507.

Timberlake, Michael, and James Lunday. 1985. "Labor Force Structure in the Zones of the World-Economy, 1950–1970." Pp. 325–349 in *Urbanization in the World-Economy*, edited by Michael Timberlake. New York: Academic Press.

Timberlake, Michael, and David A. Smith. 1998. "Global Cities, Transnational Networks: What We Know, What We Need To Know, and Why It Matters." Paper presented at a conference on "Global Cities and Their Cross Border Networks: The Impact of Transnationalism and Telematics" sponsored by the United Nations University Institute for Advanced Studies, Vancouver, B.C., June

United Nations. 1995. *1993 Demographic Yearbook.* New York: United Nations.

Wallerstein, Immanuel. 1974. *The Modern World-System: Capitalist Agriculture and the Origins of European World-Economy in the Sixteenth Century.* New York: Academic Press.

———. 1976. "Semi-peripheral Countries and the Contemporary World Crisis." *Theory and Society* 3:4:461–484.

———. 1979. *The Capitalist World-Economy.* New York: Cambridge University Press.

Walton, John. 1977. "Accumulation and Comparative Urban Systems: Theory and Some Tentative Contrasts of Latin America and Africa." *Comparative Urban Research* 5:1:5–18.

———. 1979. "Urban Political Economy: A New Paradigm." *Comparative Urban Research* 7:1:5–17.

————. 1981. "The New Urban Sociology." *International Social Science Journal* 33:2:374–390.

————. 1982. "The International Economy and Peripheral Urbanization." Pp. 119–135 in *Urban Policy under Capitalism,* edited by Norman and Susan Fainstein. Newbury Park, Calif.: Sage.

Walton, John, and David Seldon. 1994. *Free Markets and Food Riots.* Oxford: Blackwell.

Ward, Kathryn. 1985. "Women and Urbanization in the World-System." Pp. 305–323 in *Urbanization in the World-Economy,* edited by Michael Timberlake. New York: Academic Press.

Weber, Max. 1958. *The City.* New York: Free Press.

White, James. 1998. "Old Wine, Cracked Bottle? Tokyo, Paris, and the Global City Hypothesis." *Urban Affairs Review* 33:4(March):451–477.

Wolff, Goetz. 1992. "The Making of a Third World City? Latino Labor and the Restructuring of the Los Angeles Economy." Paper presented at the XVII Congress of Latin American Studies Association, Los Angeles.

9

World-Systems Theory in the Context of Systems Theory: An Overview

Debra Straussfogel

Over its more than twenty-year history, world-systems theory has been considered from many perspectives—the economic, political, cultural, and, more recently, the environmental. We have learned about and gained appreciation for a great many aspects of world-systems—core-periphery relations, commodity chains, hegemonic rises and falls, economic cycles and trends, and systemic logics, to name a few. Perhaps a little ironically, the one way world-systems have not been much considered is as a *system*. What does it mean to say we have a world-*system*? How can knowledge of other kinds of systems help increase understanding of the world-system today or of world-systems in the past? How can advances in the various areas of systems theory augment and improve *world*-systems theory? Some first steps toward answering these questions are the subject of this chapter.

SYSTEMS AND SYSTEMS THINKING: SOME BASIC CONCEPTS

No doubt, the reason a systems approach to world-systems study has been ignored is because systems-type analyses in the social sciences and humanities have historically had a (not entirely undeserved) bad reputation. The development and application of systems theories began in the "hard" sciences, in physics and in chemistry. The systems of study in these fields were traditionally simple, highly controlled, and, most significantly, closed systems. The quintessential example is Newtonian mechanics. In closed systems there are few types of elements, and these are structured by simple relations. An experimenter can easily intervene to change the character of these relations with predictable, replicable, and reversible results. Moreover, with closed systems, obtaining equilibrium means movement in the direction of maximum entropy, generally understood as a state of disorder, randomness, and "system death."

169

These qualities do not seem to apply to human and social systems. Human societies and their institutions are not simple, nor do they behave deterministically. While the truism exists that "those who do not learn from history are doomed to repeat it," the fact is that historical events do not reoccur. And human history most certainly cannot be reversed. Human systems persist, adapt, change, and grow. The thought of our sun burning out and the earth moving inexorably toward entropy, or complete system death, while conceivable and even predictable, remains mostly in the realm of science fiction novels and doomsday movies.

Then in 1968, Ludwig Von Bertalanffy published *General System Theory*. In this pivotal volume, Von Bertalanffy laid out the most basic and general principles regarding systems and proposed how some of these principles had analogs across a great many kinds of systems sciences, including social science. Of particular importance was Von Bertalanffy's distinction between closed systems and open systems. Unlike the closed systems described above, open systems do not tend toward maximum entropy. They exist in relationship to an environment and have dynamic and complex structures. Open systems seem to be more analogous to human and societal systems, and many social scientists adapted the tenets of general systems theory to build theories in their own disciplines. But this approach had a limited acceptance. And even among the adherents, the practicality of methodologies that used a systems approach was constrained by analytical tractability. Because of the complexities of open systems, nonlinear relations are the norm. But because nonlinear solutions are difficult to solve and produce multiple possible outcomes, systems analyses were, in earlier incarnations, linearized for computational ease. This made such systems models poor analogs for the more complex relations they were meant to estimate. Models that were not linear produced very complicated mathematical constructs that most social scientists and those in the humanities did not have the background or expertise to solve. Furthermore, the mathematization of human issues was believed by many across disciplines as inappropriate and dehumanizing. Despite Von Bertalanffy's exposition of the differences between open and closed systems, these computational issues reinforced the view among many that systems are constrained by their functionalism. This functionalism implies an inherent conservatism that in fact, in open systems, is not the case. Nonetheless early views of systems approaches saw them as prologue to social engineering where individual choice is abstracted and human beings are relegated to the role of molecules bouncing around in a social petri dish.

Over the last two decades, however, systems theories, analyses, and methodologies have both "broadened and deepened," to use terms common to students of world-systems. There is no one systems theory. The study of systems encompasses many subfields such as chaos theory, dissipative structures, soft systems science, organizational cybernetics, and others (see Suggested Readings). Systems methods have been facilitated by advances in nonlinear mathematics and especially by the development of computer technologies and computer software that allow those with limited mathematical training to produce systems models and run meaningful simulations. The kinds of systems studied are rarely assumed to be closed. In fact, even in the physical sciences there have been a revolution of thought and a realization that closed, linear, mechanistic systems represent only a small fraction of physical phenomena. The assumption of open systems has produced a wealth of new approaches dealing with their complexities. The "new science" is one of nonlinearity, complexity, incomplete knowledge, and limited predictability. These are the attributes of social systems with which the "soft sciences" have been dealing for some time and for which the "hard scientists" are gaining respect. Thus, the dialogue between the physical, biological, and human sciences has accelerated, especially within the last decade. And the new systems theoretic perspectives and methodologies emerging from the expanding scientific view are more easily crossing disciplinary boundaries.

Even among those less quantitatively oriented, the advancements in studies of systems have given rise to a perspective broadly known as *systems thinking*. This is an approach to problems and issues in the human and social sciences that is very holistic in nature. It takes a "big picture" approach, with the clear understanding that human issues are very complex, integrated, and unpredictable in the long term; that they contain both quantifiable and nonquantifiable elements; that interdependence is a key feature of the survival of human systems; and that to separate out subsystems to be studied in isolation changes the very nature of the system and the problem (Anderson and Johnson 1997). Systems approaches need not, therefore, involve quantitative analysis and modeling. A systems approach looks for the linkages and interactions between people, things, or events. It accepts the complexities inherent in studying the constructs (tangible and intangible) of human societies. It seeks to uncover the structure of these constructs and their functional dynamics in order to comprehend their genesis, current condition, and possibilities for the future.

DISSIPATIVE STRUCTURES

Among the many subareas within system studies, of particular relevance for the development of world-systems theory is the theory of dissipative structures. Ilya Prigogine received the 1977 Nobel Prize in physical chemistry for his theory of dissipative structures. The theory was revolutionary in the physical sciences and has slowly infiltrated the theoretical paradigms of other physical as well as social sciences since its inception. What was revolutionary was that it diminished the prevailing scientific view, that of Newtonian mechanics (with its extensions by Boltzman as statistical mechanics), as being applicable to only a very small subset of phenomena, those being closed systems.

As stated earlier, in a closed system, laws of structure and dynamics are observable. Relations are linear and stable, and therefore predictable. Processes are replicable and reversible. Prigogine's systems of study, instead, were open systems that exchanged matter and energy with their environments. His discovery was that this exchange produced structures that self-organized, utilizing the available matter and energy, and maintained themselves until the flow from the environment was somehow altered. In turn, this structural maintenance produced (dissipated) a degraded form of the matter or energy that was then returned to the environment.

Moreover, Prigogine observed that as the fluctuation in environmental inputs reached critical values, which he called bifurcation points, the dissipative structure would be unable to maintain itself and would spontaneously reorganize into a new and qualitatively different structure. Hence, a history is observable in these dissipative structures. Once formed, a dissipative structure remains in a stable state only as long as the environment in which it is embedded provides the necessary matter and energy for its maintenance. Should those values fluctuate outside critical limits, however, structural maintenance becomes impossible and structural change occurs. While elements of the prior structure may exist within the new, its form will be qualitatively different from the old. What that form will be is impossible to predict, as a range of possibilities is likely (hence, Prigogine's choice of the term "bifurcation point"). The possibilities are functions of the nonlinearities inherent in the existing system, the nature and degree of randomness of the fluctuations in the environment, and the interactions of environmental fluctuations with the internal structural nonlinearities (Allen 1982, 1983, 1992; Allen and Sanglier 1978, 1979, 1981; Allen, Sanglier, and Boon 1981; Nicolis and Prigogine 1977, 1989; Prigogine 1980, 1996; Prigogine and Stengers 1984; Prigogine, Allen, and Herman 1977).

Dissipative structures are far more complex than the closed systems describable by Newton's mechanical equations. The self-organizing behavior described produces a hierarchically organized dissipative structure that as a whole has identifiable and measurable characteristics (Auger 1986, 1990; Eldredge and Salthe 1984; Nicolis 1986; Pattee 1973; Salthe 1985, 1989). Each dissipative structure is a system unto itself, organized according to its own history of preceding dissipative structures and of the specific nature of the various fluctuations that have produced the particular structural sequence. This "holistic view" of the dissipative structure is its *macroscale*. The particular characteristics of the dissipative structure are known as its *macroscopic variables,* and the nature of change at this scale is known as the *macroscale dynamics*.

But because a dissipative structure is hierarchically organized, its structure is comprised of many subsystems. The internal subsystems of a dissipative structure may exist at several hierarchical levels themselves. That is, there may well be systems embedded in systems embedded in systems (such as cities embedded in states embedded in the world-system). Together all of the internal subsystems are known as the microscale, describable by *microscopic variables* and exhibiting *microscale dynamics*.

Dissipative structures also exist in nonequilibrium conditions. How far from equilibrium they are is indicated in the amount of energy they must consume from their environment in order to maintain themselves (Adams 1988). But also, "[t]he interaction of a system with the outside world, its embedding in nonequilibrium conditions, may become in this way the starting points for the formation of new dynamic states" (Prigogine and Stengers 1984, 143). That is, the organization of a dissipative structure includes the conditions for its own structural change.

While it is true that dissipative structures are "best described by macroscopic variables and by boundary conditions linking them to their environment . . . not by the properties of their individual components" (England 1994), this is not to say that the individual components should be discounted. In fact, the internal variability and the specific nature of the internal components play significant roles in the maintenance of any particular dissipative structure and in the particular succession of dissipative structures throughout the system's history (Allen 1988, 1990, 1994). Although the system as a whole may be described by macroscopic variables, microscopic processes are critical to defining what the values of these variables will be.

The interaction of macro- and microscales is fundamental to the dynamics of dissipative structures. The steady state exhibited by any dissipative structure is not an equilibrium state in that the system is at complete rest, as would be the case for closed systems with maximum entropy. Instead, in a steady state, the internal microscale dynamics of embedded systems continues to fluctuate. However, when the throughput of energy or material from the wider environment to the macrolevel system itself changes, the effects cascade through the nonlinear hierarchical structure causing the microscale fluctuations to amplify. If those amplifications cannot be damped within the relations defined by the existing structure, then the macroscale variables will be pushed to a bifurcation point, the internal relations will self-organize and the dissipative structure at the macroscale will have qualitatively changed to adapt to both the internal and external pressures, and a new dissipative structure will emerge (Straussfogel 1997a, 1997b, 1998).

The form of the new dissipative structure will therefore be a function of the microlevel structures as they recombine with both existing and novel structures introduced by the environmental shifts to become the source of new structural relations (Allen 1990, 1992). In this way, the microscale diversity of dissipative structures becomes a source of innovation in the future. "[A] system contains within itself the possibility of becoming something different, of 'adapting' of evolving" (Allen and McGlade 1987, 3). Hence, evolution in the theory of dissipative structures is a historical, creative process. The signposts along the course of this evolution are the bifurcation points in the macrolevel state variables indicating when internal instabilities and impinging environmental shifts effected evolutionary change. Because the evolutionary history of a dissipative structure is therefore continuous and unidirectional, each structure at any time it is observed is unique and represents a culmination of its past and a prologue to its future.

THE WORLD-SYSTEM AS A DISSIPATIVE STRUCTURE

Adams (1982a, 1982b, 1988) makes a strong case for the perspective that the theory of dissipative structures can offer to the study of large scale human systems, and the world-system in particular. He generalizes to make the case that "a dissipative structure is composed of a constantly changing flow that succeeds in maintaining a particular form by maintaining a particular input" (1988, 17). Adams underscores the point that it is the constant consumption of energetic inputs that comprise the structure, which is therefore potentially unstable, and will disintegrate if inputs cease to flow. Hence, a dissipative structure not only functions to increase the circulation of energy through itself but also will "use its own input to increase circulation through a larger structure that constitutes a part of its environment to assure itself of future inputs" (Adams 1988, 22). Here, then, is a theoretical basis for the broadening and deepening of the world-system.

Adams explicitly states that "societies operate as dissipative structures; there are continuities of form that are constituted by the very flow of energy that is expended (i.e., converted) in the process of acting out the behaviors and doing the work . . . that is carried on in the context of social relationships" (1988, 17). For societies, bifurcation occurs during the process of expansion when "the particular internal arrangements no longer suffice to handle the amount and kinds of energy flow that constitute them" (1988, 18). At a bifurcation point, the choices for a social system are for greater expansion, dissolution, decline, or maintenance of a steady state.

Adams makes the point that social dissipative structures differ from dissipative structures in other contexts on this issue of achieving a steady state. As he puts it, "there is nothing universally inherent in social organization that leads to a steady state" (Adams 1988, 18). A steady state for a social system exists when there is some restriction on inputs that results in the inhibition of continuing expansion. Furthermore, "whether a given dissipative structure is in a steady state depends on the actions of [internal] regulative mechanisms that may themselves be dissipative structures" (Adams 1988, 19). In the case of the modern capitalist world-system, those regulative mechanisms work to achieve continued expansion in order for the system to continue. Capitalism, in other words, "made possible by the conjunction of many economic, biomedical, technological, and social relational elements" is "[a] great, complex *catalytic mechanism* that releases flows of energy for expansion" (Adams 1988, 19). This means that a steady state is actually not a viable condition for a capitalist economic system. Linking the theoretical perspective with that of world-systems theory, Adams summarizes by saying:

> At every point in the history of capitalism thus far, its particular set of constituents has varied, always growing bigger. . . . The search for better investment opportunities, for ways to be more sure that energy will flow, always accompanies it. The shifting of people and goods from one continent to another is done in the same framework. (1988, 51)

Hence, as the theory of dissipative structures makes clear, if the flows necessary for capitalist expansion are restricted, the system's structure would by definition become unstable, potentially to the point of qualitative restructuring. According to Wallerstein, such a crisis constitutes "a structural strain so great that the only possible outcome is the disappearance of the system as such" (1991, 23). As to the nature of that outcome, Wallerstein later adds, "the indeterminacy of bifurcations is a central reality with which we must cope" (1996, 14). This latter point by Wallerstein reflects his own acceptance of at least the premise of Prigogine's theory, if not all the details (Wallerstein 1992).

What does it mean then to speak of a world-*system* in these terms? First, it means that the consideration of world-system structures must be enlarged to consider a global set of embedded systems that are functionally integrated and create the core-periphery structure we are used to considering. This does not preclude analysis at any of the many possible scales, from the household to the global, but neither does it violate the fundamental premise of world-systems theory—that a world-system exists whose structure and function are observable as a recognizable entity. In fact, to consider the world-system as a dissipative structure reinforces consideration of the world-system as the appropriate scale of analysis.

Moreover, the hierarchical nature of dissipative structures also allows the embedding of the world-system as a world-economy (if not also as a world-polity) into the global environment on which its structural and functional persistence depends (Straussfogel 1997a). This point is consistent with recent attention paid by some world-systems analysts to redress the lack of inclusion of the environment by world-systems theory up to the present (see chap. 16; Bergesen and Parisi 1997; Bergesen 1995; Chew 1995; Barbosa 1990; Straussfogel 1997a, 1997c; Straussfogel and Becker 1996). Seen as a multileveled complex system exhibiting the properties of a dissipative structure, the system-environment relationship looms as crucial. With increasing global environmental degradation, the available inputs to the world-system are decreased, creating a major source of fluctuation threatening the maintenance of the existing structure. In the past, the perpetrators of environmental degradation were able to relocate production within the existing world-system structure (i.e., the structure of capitalist relations) and shift the local impacts that capitalist production had on its natural resource base. However, as capitalism expanded and broadened to become global, so too were the environmental feedback effects exhibited at the corresponding global scale. So we have ozone layer depletion and global warming emerging from deforestation and desertification. We also have other feedback effects felt at a variety of scales, such as the impacts of environmental refugees.

Within the theory of dissipative structures one can validate the premise that the greatest threat to the persistence of the capitalist world-system is not from the traditional "antisystemic movements" discussed in world-systems literature. Instead, the accumulated dissipation of the degraded forms of energy inherent in capitalist production has finally impaired the environment's ability to provide the necessary inputs required for the maintenance of the capitalist world-system. As these environmentally induced fluctuations continue to amplify (as witnessed by environmental degradation at every geographical scale within the world-system's hierarchical and dissipative structure), maintenance, in fact, will become impossible and the system will reach a bifurcation point where it must contract or transform.

Second, the theory of dissipative structures provides world-system dynamics with a theoretical underpinning within which the "cyclical rhythms," such as hegemonic cycles and Kondratieff waves, and the "secular trends," those of broadening and deepening, may be understood and validated. Again, no prior assumptions are violated, only substantiated. The cycles and trends can now be understood as the regulatory mechanisms that have served to maintain and perpetuate the capitalist world-economy. They represent the adjustments made by and within the system in response to variations in the flow of goods, capital, people, and information through it.

Hence, even though since 1650 or so, the system (at the macroscale) has remained qualitatively unchanged, over time, the world-system has become more complex, with greater numbers and kinds of subsystems and many layers of relations and interdependencies. This makes implicit the assumption that with increasing complexity, the potential for a variety of longer- and shorter-term cycles consistent with the many scales of the system's hierarchy increases (Straussfogel 1997a). Each of these many cycles may itself become amplified to become a source of a destabilizing fluctuation pushing the system toward a future bifurcation point and qualitative system transformation. The same argument can be made looking toward the past. Considering earlier world-system types, the generalization would be that these particular systemic structures and relations, also being hierarchically organized dissipative structures, created, through their own self-organizing behavior, cyclical rhythms critical to maintaining these earlier structures (Chase-Dunn and Hall 1997).

This brings us to a third contribution of the dissipative structure approach to world-systems. In world-systems theory, the concept of "systemic logic" denotes the specific functional principles of a world-system. As new world-systems emerged in history, different systemic logics could be identified (Chase-Dunn 1988; Chase-Dunn and Hall 1992, 1993, 1994, 1997; Hall and Chase-Dunn 1994). According to dissipative structure theory, a new world-system would emerge when a prior world-system structure was unable to withstand amplifying fluctuations brought on by changes from its environment (an invasion or drought) or from within its subsystems (a revolution or famine) (Straussfogel 1997a). Resolution of these perturbations resulted in a reconfiguration of systemic components, the emergence of new subsystems, and the redefining of relations and interdependencies within the system and between the system and its environment. Hence, the evolution of world-systems from kin based to tributary to capitalist can be understood and analyzed according to this theory of system evolution. From this perspective, one can also consider the state of the current system and the many perturbations it is currently undergoing with the goal of creating scenarios for the structural form of the next world-system.

This underscores the fact that the emergence of new subsystems is a critical feature of the dissipative structure theory. The dynamics of transformation which is part of this theory is a creative dynamic. Criticisms based on functionalism can not stand up in the face of contemporary systems theory. The nonlinear nature of these systems allows for the possibility of novelty, variability, and therefore viability. As dissipative structures adjust to changes in environmental inputs at all scales, some internal subsystems may disappear (the former Soviet Union); others may recombine with novel inputs to become transformed (the United Nations from the League of Nations); or wholly new structures may emerge (the European Union, the Democratic Republic of the Congo, the International Labor Organization). The ability for dissipative structures to adapt in this way is the source of creativity that impacts the future of the system. At a bifurcation point, the future structure will (and has) emerged from the variability among the many microsystems combining with the particular nature of historical fluctuations. Successive world-systems have not sprung whole and all at once at some historical juncture. Elements of past structures existed in the new. Likewise, elements of the future exist in the present. World-system transformation (indeed *system* transformation) can occur as either a top-down or a bottom-up dynamic.

We can not know what the future will be. Dissipative structure dynamics illustrates that clearly. When a system is far from a bifurcation point, a certain degree of predictability is possible within a range of likely scenarios. But near to a bifurcation point, at a time of crisis, what will emerge on the other side is in large measure a function of chance. The possibilities are a combination of the resiliency of existing structures in the face of extreme perturbation, amplified through the complex web of nonlinear relations. This rather constrains the possibilities for planning for the future but does not make the ability to influence the future impossible (Straussfogel and Becker 1996). The source of the future structure will be an emergent property of the contemporary system. There is truth in the slogan "Think globally; act locally." The creative nature of system change embodied within the theory of dissipative structures implies that individuals are not necessarily powerless within the seemingly inexorable tide of global events. Grassroots efforts can effect large-scale change. Just as the flutter of a butterfly's wings on one part of the Earth can conceivably alter global climate patterns, so too can one person potentially change the world.

SUGGESTED READINGS

The following references are intended for general audiences interested in the systems sciences, especially in the subfields of self-organization, complexity, and chaos. Each has played a role in shaping and enlightening my own views.

Adams (1988), an anthropologist, applies the theory of dissipative structures to develop a theory of social evolution based on the capture and release of various forms of energy.

Briggs and Peat (1989) provide a very accessible and lavishly illustrated introduction to chaos theory and the holistic view provided by the systems approach.

Dawkins (1986) emphasizes the uncertainty inherent in the nonlinear relations of most kinds of systems and applies it to the interconnectivity of evolutionary processes.

Gleick (1987) offers a classic summary of the science of chaos.

Kauffman (1995) summarizes, in very readable fashion, the consistency of the self-organizing nature of complex systems across systems of all types and functions.

Prigogine and Stengers (1984) present Prigogine's exposition of the theory of dissipative structures and its applicability throughout the physical, natural, and social worlds.

Salthe (1985) provides an in-depth look at the structure of complex systems and how their hierarchical nature gives rise to emerging structures in the future.

Waldrop (1992) summarizes the work of the Santa Fe Institute, whose researchers, including Kauffman, have been at the forefront of discoveries in the sciences of complexity.

Additionally, the journals *Systems Research and Behavioral Science* (formerly *Systems Research*) and *Systems Practice* publish articles on the development and application of systems ideas and theoretical constructs.

REFERENCES

Adams, Richard N. 1982a. "The Emergence of Hierarchical Social Structure: The Case of Late Victorian England." Pp. 116–131 in *Self-Organization and Dissipative Structures,* edited by William C. Schieve and Peter M. Allen. Austin: University of Texas Press.

———. 1982b. *Paradoxical Harvest. Energy and Explanation in British History, 1870–1914.* Cambridge: Cambridge University Press.

———. 1988. *The Eighth Day.* Austin: University of Texas Press.

Allen, Peter M. 1982. "Evolution, Modelling, and Design in a Complex World." *Environment and Planning B* 9:1:95–111.

———. 1983. "Self-organization and Evolution in Urban Systems." Pp. 29–62 in *Cities and Regions as Nonlinear Decision Systems,* edited by Robert W. Crosby. Boulder, Colo.: Westview.

———. 1988. "Evolution: Why the Whole Is Greater than the Sum of the Parts." Pp. 2–30 in *Ecodynamics,* edited by Wilfred Wolff, Carl J. Soeder, and Friedhelm R. Drepper. Berlin: Springer.

———. 1990. "Why the Future Is Not What It Was: New Models of Evolution." *Futures* 22:6(July–Aug.):555–570.

———. 1992. "Modelling Evolution and Creativity in Complex Systems." *World Futures* 34:1–2:105–123.

———. 1994. "Coherence, Chaos and Evolution in the Social Context." *Futures* 26:6(July):583–597.

Allen, Peter M., and Jacqueline M. McGlade. 1987. "Evolutionary Drive: The Effect of Microscopic Diversity, Error Making, and Noise." *Foundations of Physics* 17:7(July):723–738.

Allen, Peter M., and M. Sanglier. 1978. "Dynamic Models of Urban Growth." *Journal of Social and Biological Structures* 1:1(July):265–280.

———. 1979. "A Dynamic Model of Growth in a Central Place System." *Geographical Analysis* 11:3:256–272.

———. 1981. "Urban Evolution, Self-organization, and Decision-making." *Environment and Planning A* 13:167–183.

Allen, Peter M., M. Sanglier, and F. Boon. 1981. *Modeling Adaptive Behavior.* U.S. Department of Transportation, Final Report Part IV, Contract TSC-1640. Washington, D.C.: U.S. Department of Transportation.

Anderson, B., and L. Johnson. 1997. *Systems Thinking Basics: From Concepts to Causal Loops.* Cambridge, Mass.: Pegasus Communications.

Auger, P. 1986. "Dynamics in Hierarchically Organized Systems: A General Model Applied to Ecology, Biology and Economics." *Systems Research* 3:1:41–50.

———. 1990. "Self-organization in Hierarchically Organized Systems." *Systems Research* 7:4:221–36.

Barbosa, Luiz C. 1990. "Dependencia, Environmental Imperialism and Human Survival: A Critical Essay on the Global Environmental Crisis." *Humanity and Society* 14:4(Nov.):329–344.

Bergesen, Albert. 1995. "Eco-Alienation." *Humboldt Journal of Social Relations* 21:1:1–8.

Bergesen, Albert, and Laura Parisi. 1997. "Editors' Introduction: Discovering the Environment." *Journal of World-Systems Research* 3:3(Fall):364–368 (electronic journal: http://csf.colorado.edu/wsystems/jwsr.html; ftp and gopher: csf.colorado.edu/wsystems/journals/).

Briggs, John, and F. David Peat. 1989. *Turbulent Mirror.* New York: Harper & Row.

Chase-Dunn, Christopher. 1988. "Comparing World-Systems: Toward a Theory of Semiperipheral Development." *Comparative Civilizations Review* 19(Fall):39–66.

Chase-Dunn, Christopher, and Thomas D. Hall. 1992. "World-Systems and Modes of Production: Toward the Comparative Study of Transformations." *Humboldt Journal of Social Relations* 18:1:81–117.

———. 1993. "Comparing World-Systems: Concepts and Working Hypotheses." *Social Forces* 71:4(June):851–886.

———. 1994. "The Historical Evolution of World-Systems." *Sociological Inquiry* 64:3(Summer):257–280.

———. 1997. *Rise and Demise: Comparing World Systems.* Boulder, Colo.: Westview.

Chew, Sing. 1995. "On Environmental Degradation: Let the Earth Live. An Introductory Essay." *Humboldt Journal of Social Relations* 21:1:9–13.

Dawkins, Richard. 1986. *The Blind Watchmaker: Why the Evidence of Evolution Reveals a Universe without Design.* New York: Norton.

Eldredge, N., and Stanley Salthe. 1984. "Hierarchy and Evolution." *Oxford Survey in Evolutionary Biology* 1:184–208.

England, Richard W. 1994. "On Economic Growth and Resource Scarcity: Lessons from Nonequilibrium Thermodynamics." Pp. 193–211 in *Evolutionary Concepts in Contemporary Economics,* edited by Richard W. England. Ann Arbor: University of Michigan Press.

Gleick, James. 1987. *Chaos: Making a New Science.* New York: Viking.

Hall, Thomas D., and Christopher Chase-Dunn. 1994. "Forward Into the Past: World-Systems before 1500." *Sociological Forum* 9:2:295–306.

Kauffman, Stuart. 1995. *At Home in the Universe: The Search for the Laws of Self-Organization and Complexity.* New York: Oxford University Press.

Nicolis, John S. 1986. *Dynamics of Hierarchical Systems: An Evolutionary Approach.* Berlin: Springer.

Nicolis, Gregoire, and Ilya Prigogine. 1977. *Self-Organization in Non-Equilibrium Systems.* New York: Wiley.

———. 1989. *Exploring Complexity.* New York: Freeman.

Pattee, Howard H. 1973. *Hierarchy Theory: The Challenge of Complex Systems.* New York: Braziller.

Prigogine, Ilya. 1980. *From Being to Becoming: Time and Complexity in the Physical Sciences.* San Francisco: Freeman.

———. 1996. "The Laws of Chaos." *Review* 14:1(Winter):1–9.

Prigogine, Ilya, Peter M. Allen, and R. Herman. 1977. "Long Term Trends and the Evolution of Complexity." Pp. 1–63 in *Goals in a Global Community: the Original Background Papers for Goals for Mankind, a Report to the Club of Rome,* edited by Ervin Laszlo and Judah Bierman. London: Pergamon.

Prigogine, Ilya, and Isabelle Stengers. 1984. *Order Out of Chaos.* New York: Bantam.

Salthe, Stanley N. 1985. *Evolving Hierarchical Systems: Their Structure and Representation.* New York: Columbia University Press.

———. 1989. "Self-Organization of/in Hierarchically Structured Systems." *Systems Research* 6:3:199–208.

Straussfogel, Debra. 1991. "Modelling Suburbanization as an Evolutionary System Dynamic." *Geographical Analysis* 23:1(Jan.):1–24.

———. 1997a. "World-System Theory: Toward a Heuristic and Conceptual Tool." *Economic Geography* 73:1(Jan.):118–130.

———. 1997b. "A Systems Perspective on World-Systems Theory." *Journal of Geography* 96:2(March/April):119–126.

———. 1997c. "Redefining Development as Humane and Sustainable." *Annals of the Association of American Geographers* 87:2(June):280–305.

———. 1998. "How Many World-Systems? A Contribution to the Continuationist/Transformationist Debate." *Review* 21:1(Winter):1–28

Straussfogel, Debra, and Mimi L. Becker. 1996. "An Evolutionary Systems Approach to Policy Intervention for Achieving Ecologically Sustainable Societies." *Systems Practice* 9:5(Oct.):441–468.

Von Bertalanffy, Ludwig. 1968. *General System Theory.* Rev. ed. New York: Braziller.

Waldrop, M. Mitchell. 1992. *Complexity: The Emerging Science at the Edge of Order and Chaos.* New York: Simon & Schuster.

Wallerstein, Immanuel. 1991. "Crises: The World-Economy, the Movements, and the Ideologies." Pp. 23–37 in *Unthinking Social Science: The Limits of Nineteenth Century Paradigms,* edited by Immanuel Wallerstein. Cambridge: Polity.

———. 1992. "The New Science." *Review* 15:1(Winter):entire issue.

———. 1996. "History in Search of Science." *Review* 19:1(Winter):11–22.

10

Postmodernism Explained

Albert J. Bergesen

I propose a world-system explanation for the cultural movement known as post-modernism (Jencks 1992; Rosenau 1992; Lasch and Friedman 1992; Connor 1989; Harvey 1990). This current in contemporary art and thought is international in scope and, I will argue, produced by the capitalist world-economy. The key sociological observation is that the emergence of a belief that we are in a postmodernist phase in culture does not occur within a social vacuum. It appears precisely at the time when the larger world economy is shifting from the dominance of a single hegemon (the United States) to a world characterized by a growing plurality of power and heightened economic competition in what could be called a balkanizing capitalist world economy. It is not an accident, then, that postmodernism appeared in the 1970s, precisely when the United States begins to decline, for postmodernism is the international culture of hegemonic rivalry seen in the growing plurality of national power centers within the world-system.

From the perspective of the history of the modern world system since the sixteenth century, there have been alternating periods of hegemonic domination by a single state (the Spanish Hapsburgs, Britain, the United States) and periods of a more plural competitive world composed of conflict and war (from the wars of the absolutist states to World Wars I and II). What is important to realize is that each of these phases produces the other in what has been an unending global history of political conflict/economic crisis and cyclically followed by periods of economic expansion/ imposed peace by a dominate state (Bergesen 1992; Bergesen and Schoenberg 1980; Bergesen, Fernandez, and Sahoo 1987; Bergesen and Fernandez 1998). Each phase of these cycles has been accompanied by a different framework in world culture that promises liberation and superiority over the cultural framework of the previous cyclic phase. During each historical moment each frame is believed, as the world now endorses contingent postmodernism and condemns absolutist modernism. What must be remembered, though, is that these cultural stances are not permanent but swing back and forth in sync with the hegemony/rivalry cycle of the world political economy.

There have been earlier cultural frameworks similar to postmodernism that appeared during earlier periods of hegemonic decline. The cultural expression called "mannerism" in the later sixteenth century accompanying the decline of Spanish Hapsburg power and the mannered academic art of the salons accompanying British decline were earlier versions of postmodernism following earlier periods of generalized/universal discourse—the sixteenth-century High Renaissance and the late eighteenth-/early nineteenth-century Enlightenment/neoclassicism/romanticism (Bergesen 1996). Postmodernism, then, is not a new cultural frame but a recurring set of assumptions that always accompanies hegemonic decline. The world-system has not only political/economic but also cultural cycles, and the swings in world culture between absolute universalism and contingent particularism are one such cycle.

Thus, the political economy of posthegemony is not a void, and neither, therefore, is the cultural condition known as postmodernism. This is why the critique of postmodernity as an expression of cultural nihilism is wrong. Following hegemony is the world-system's cyclical return to accelerated interstate rivalry and a growing struggle for hegemonic succession, and the appearance of postmodernism represents an early manifestation of heightened intercapitalist rivalry. This growing competition between countries heightens nationalism within countries. The result is a turn in culture away from abstract and general art and social theory. The post-1970s assertion of a multicultural flowering of many national, ethnic, racial, and gendered voices—that is, of difference—reflects the decline of the material base of the hegemon's once totalizing abstract and universal cultural categories that constituted the framework within which concern with the human condition was expressed during the cold war years.

The transsocietal scope of American economic and political domination took the cultural form of transsocietal, transgroup, nongrounded, abstract painting in art, the glass box formalism of the international style in architecture and abstract theory in social science. In this cultural milieu in which only the general or universal exists, particular social, class, gender, and national identities are masked, suppressed, or simply disappear from national consciousness. Then, with hegemonic decline, various nationalities, groups, genders, and ethnicities begin to find cultural space as they can once again assert their own identities in the cultural practices of art, architecture and cultural/social theory. With American decline and a growing plural competitive world, we see a growing praise of the appearance in world culture of multiple competing voices.

In short: the absence of a dominant hegemon in political economy; the absence of a dominant master narrative in ideological and cultural discourse; no hegemon, no theoretical perspective that takes priority over any other framework; no hegemon, no privileging of any one discourse or national expression above any other. These are the cultural assumptions of postmodernism. The world-systems point is that this change in world cultural consciousness reflects the plurality of economic power that is the condition of posthegemony in the capitalist world-economy. There is, then, a clear map between the new multicentric world political economy and the new multicultural postmodernist philosophy.

Postmodernity is the culture of a world-system entering a phase of heightened national competition and rivalry. The material base of the 1990s cultural world is a global political economy without an American hegemon ordering international politics and economic order. Such political multicentricism creates space for alternative positions and results in a new world ideology that emphasizes the equality of all positions, all voices, all groups, and all modes of expression. This is a philosophical stance in the human sciences that claims social life has no deep structural determination or ordering principle. Traditional twentieth-century modernism in art and generality in social theory are indicted as a mask covering and repressing identities of nation, race, gender, ethnicity, and sexual preference. Postmodern expression may be an advance over the universal grid of cultural space produced under the condition of the single American hegemony, but it is also the early phase of what will become a vicious 1930s-like nationalistic love of race and country when the world-system passes to the later stages of the rivalry cycle characterized culturally by national hatreds.

The process of global reorganization beginning at the end of the twentieth century is seen in economic restructuring forced by the rise of new Asian centers of economic strength accompanied by autonomous American industrial decline. In past world-system cycles this phase has produced a particularism and realism in culture and an intense nationalism as nation-states use love of country to mobilize their populations in the frenzied competition and conflict of economic downturn and crisis. While the late twentieth-century multicultural emphasis on expressing separate group identities is refreshing, if it turns to race and national hatred and is pressed into service for interstate rivalry, then it is something quite different. This point is not to demean today's multicultural surge but to try to understand our own cultural discourse by identifying its place within the long-term trends of world-system dynamics.

GLOBAL FRAME ANALYSIS

One way to understand this cyclic repeat of cultural assumptions is to view it from a frame analysis point of view, which arose in Erving Goffman's social psychology to account for the fact that people use the framing of problems, issues, and social reality itself to further their interests. More recently it has been used to account for how social movement organizations mobilize new members, identify goals, and advance their political agenda (Goffman 1974; Snow and Benford 1992; Hunt, Benford, and Snow 1994). While the idea of cultural frames has been used at the microlevel of social analysis, it also has a counterpart at the most macro- (i.e., world-systemic) level of analysis, where the cultural outlooks such as postmodernism are also frameworks used by different contending groups, classes, and nation-states to advance their interests.

Postmodernism, as an identifiable set of assumptions about social life, can be investigated like any other cultural frame. Where does it originate? Whose interests does it serve? Whose power base does it legitimate or mask, and how is such knowledge also power? Such questions invert the usual way the postmodern sensibility is approached, as it is traditionally used to analyze the world of power relations. But here we are going to take the world-system power relations and use them to analyze postmodernism. In the language of social science, postmodernism has usually been an independent variable—a framework used to analyze the power and privilege of social groups. Here we will study it as a dependent variable—an ideological frame that is used by social powers to advance their interests.

Three assumptions and a conclusion constitute the postmodern frame:

1. There is no valid general theory in the human sciences; and when such generality exists, it is but a mask for oppression.
2. In social, linguistic, and literary realities, only parts exist, not wholes, whether this be signifiers in narration or actors/identities in social structures.
3. In this universe of parts, there is an equality among all voices, interests, and identities. There is no hegemony, or if there is, it is to be opposed.
4. Hence, all is in flux; whether the infinite regression of signifiers or the unsutured formation of society, there is no essentially fixed order within human affairs.

All is contingent and open, and what order exists is the product of struggle among constituent identities and actors. The human world, then, is composed of a constant combination/recombination of signs, identities, and social relations. Nothing is permanent or fixed, all is open to contestation.

Now the question: Whose material interest would such a framework benefit? There are two answers, and we will focus on the second, although it has implications for the first. The first answer is groups within social formations: genders, races, classes, sexual preferences, whose interests are increasingly spoken for by intellectuals using a postmodern framework. A second perspective is from the world-system as a whole, examining relations between, rather than within, national formations. Here nation-states compete and jockey for position within the larger world-economy. The world shifts back and forth between the dominance of a single state and a more plural distribution of power amongst states in what appear to be cycles of hegemony and rivalry.

From this point of view, what is the condition of the world when the postmodern way of framing reality arose in the 1970s? American economic hegemony was beginning to decline, and with the overarching power of the United States contracting, other states had more room to assert their interests and demand recognition for their point of view. This means when the material world shifts from hegemony to rivalry there is a cultural shift in how the world is viewed, and the tenets of postmodernism capture the very changes that are occurring in the world political economy at the end of the twentieth century.

It's not an accident, then, that postmodern ideas arise and have their most forceful champions among nonhegemonic states like France. Paris seems to recurringly ascend in world culture between Anglo-American hegemonies. There is French dominance in art throughout the pluralist period from the decline of Britain in the mid–nineteenth century until the ascendancy of New York in the mid–twentieth century. The process seems to repeat today: American decline and French initiated structural, poststructural, semiotic/discursive postmodern, theories of human culture (Althusser, Foucault, Derrida, Lyotard, Kristeva) capture the imagination of the world. General theory's dissolving universalism is now reified as the new cultural orthodoxy. Cultural collapse by the hegemon championing universalist theory now becomes cultural victory for lesser core states, as the French now praise nonsutured social formations, infinitely regressing signifiers, irony, play, decentered meanings, and all the other tools of postmodernism as the very heart of a posthegemonic cultural victory. French thought presents pieces and snippets of past general theory as the new theory, as the way the world is—no order, no theory, no logocentrism, no privileged voice, no master theory. In a world where all nation-states are contending, we see a corresponding cultural program claiming that all intellectual positions are as equal as all others. Unified theory is no longer seen as an advance but as a tyranny of the past; and now, nontheory is proposed as the theory of the day.

Such postmodernism was the ideology of the emerging rivalrous international system of the last quarter of the twentieth century. It is taken by adherents as a new form of truth, an ascent over general theory which prevailed during the period of American hegemony. But it is nothing but the frame used by rivalrous states to depict the world in terms favorable to their growing power and ascension vis-à-vis American hegemonic decline. In a world with a center in decline, what cultural frame is better suited than one that claims there is no center, no transcendent truth, no absolutes, and only the contingent, agreed-on truths resulting from the struggles of the contending parties?

From this perspective, we can now better understand the ideological role of the universality in theory and pictorial expression that was the "modernism" of the twentieth century and the "high modernism" of the pinnacle of American hegemony between 1945 and 1975. Again, the match between the material condition and culture. America exercises power worldwide, and historical events, crises, and stalemates are viewed as the product of that power, by itself or in gridlock engagement with the Soviet Union. In such a world, theoretical frames in the human sciences are abstract and universal, parts are derived from the whole, and the intellectual world is dominated by general theory. In sociology it is the general theory of Talcott Parsons; in linguistics it is the idea of a universal grammar of Noam Chomsky; and in economics the generalizing, standardizing overview of neoclassical theory of Paul Samuelson.

The great works of social science in the era of American hegemony are all concerned with general theory, and all arise, by and large, in the United States. Whether theories of society, economy, or language, such generality reflects the generality of American power. Then, with the start of American decline, comes a shift in cultural frames from seeing everything as an instance of a general principle, to seeing things as separate, unique, partial, without general ordering principles—that is, seeing things through the postmodern frame.

Culture and material condition co-vary. A general exercise of power by one nation across all others in material life produces a particular global life-world, that when reflected upon in the act of creating explanatory theory produces a theoretic imagination that deeply believes in the power of general theory from which the behavior all aspects of society can be derived. The decline in that power creates a new world situation and accompanying habitus and a new consciousness that now believes all is relative and in flux—that is, the postmodern mind. It is simply not an accident that changes in the material world are mirrored by changes in the ideational world of theoretical outlooks where it is now believed that no general ordering principle— be it Karl Marx or Talcott Parsons—can possibly explain all behavior. Parts are no longer derivations of wholes but are now seen as having a life of their own, and what social wholes that do exist are theorized to be momentary contingencies produced by the day to day struggles between actors and identities.

Again, it is not an accident that with American decline comes the decline of generalizing culture and universal standards of which New York was the center. It is not an accident that, with the rise of other centers of material power, there is a rise of other centers of cultural analysis, and therefore that such an other center as France produces the intellectual leaders of postmodern thought. Americans may think in a postmodern way and may apply the postmodern frame, but its most primal assumptions are generated abroad, principally in France.

PICTURES AND POWER

This shift in cultural frames between periods of rivalry and hegemony is general and applies to art as well as social science theory. Hegemonies have risen and fallen before the United States and postmodern-like frames have in turn succeeded universal/general outlooks. One way to see this is to briefly examine the synchronization of hegemony/rivalry cycles in political economy and abstraction/realism cycle in art history. In general, periods of a single hegemony in the world-system produce a cultural frame of ideal, universal generalities, whether this be Renaissance classicism under the Spanish Hapsburg hegemony of the early sixteenth century, neoclassicism/ romantic frames under the British hegemony of the early nineteenth century, or the abstraction in art and architecture under American hegemony in the middle of the twentieth century. Conversely, periods of a more multicentric distribution of national power produce world art emphasizing the specific, concrete, particular, and realistic frames, whether this be the baroque of the seventeenth century, the realism/Impressionism of the post-British hegemony of the nineteenth century, the cubism/social realism of the first half of the twentieth, or the growing multicultural postmodernism of the late twentieth century.

A world under the domination of a single hegemonic state produces an artistic style that is general and universal. This serves the hegemon's power by suppressing the identities of nations, women, races, ethnicities, and genders. Realism and multicultural identity art, on the other hand, are used to mobilize group against group and nation against nation. Neither type of artistic frame, then, is devoid of a social base in real-world power politics. They are just the appropriate frame for the next phase of the larger world political economy's cyclical swings. We shift back and forth between periods of expanded economic growth under the domination of a single hegemon and periods of rivalry, conflict, and economic downturn, with no hegemonic leadership. During each phase we endorse the cultural logic of our time as superior to that of the preceding phase. Economic expansion and political hegemony are associated with universalism/idealism in culture; economic contraction and hegemonic decline are associated with particularism, social realism, multiculturalism, and postmodernism. In the 1950s the universalism of abstract art was not seen as a means of American hegemonic domination but of personal freedom and a liberation from the stifling national conformity and political agenda that was the art of social realism. But now a similar social realism—not of the working class, or of Nazi and Stalinist national heroics—but a postmodern multicultural realism, heroically portraying gender, race, ethnicity, is seen as liberation from the purported tyranny of generalization, universalism, and totalizing abstraction in pictorial representation.

To advance beyond this dualism of abstract high modernism versus postmodern realism, we must see both as different master frames utilized by different states at different times to advance different configurations of power. Abstract and classical art serves the purposes of a dominant hegemonic state; realism serves the purposes of states in rivalry. Abstraction/classicism is an image anticoagulant, preventing the emergence of coherent national identities, for by definition abstract art doesn't signify national origin, whereas realism allows people, places, landscapes, and other national identifications to be realized.

In periods of early hegemonic decline the emerging realism, like multicultural art, pictorially represents more intrasocietal groupings—race, gender, ethnicity, age, class—as seen in the 1980s–1990s rise of feminist, African-American, Chicano/a, Asian-American, and environmental art, with the social background of the artist also being defined as part of the overall interpretative process. The visual articulation of the social starts locally, but as hegemonic decline continues and rivalry between competing states heightens, the portion of representation dealing with national themes grows. Nationalism grows in art as national populations are mobilized to support their state in struggle with other states. This is 1930s Nazi, fascist, and Soviet socialist realist art and American social realists and regionalists, who played to love of country and other patriotic themes.

At the most general theoretical level, then, both abstract modernism and realistic postmodernism partake in framing enterprises; they just represent different sources of power. Realism is national art, mobilizing domestic populations for service to the nation, symbolically wedding citizenry to state in the struggle between states that is a multicentric world-system. Abstraction suppresses these very national identities in a frame of totalizing abstract art initiated by the hegemonic state. For hegemonic domination, the social agenda of nations is reduced and the focus is placed on the individual artist and her or his genius, with universal standards being held high. Here art is seen as more personal and private expression, devoid of the politics of Stalinist realism, American regionalism, Nazi or fascist art. But it is a mistake to only associate the political in art with this thirties type social realism, for there is also an ideological purpose to the universalism of abstraction/classicism. As a frame it suppresses national identities that might challenge the hegemony of the single dominant state, which is why hegemonic states so enthusiastically champion abstract art (Guibaut 1983; Frascina 1985).

The ideological, therefore, exists in the framing styles that are social realism and abstraction. They are simply styles for different levels of political power within the world system. Social realism allows the freedom for states to express their national identity vis-à-vis each other, but such states tend to control the expression of individual artists during these periods. Ironically, then, social realism is freedom of expression for the corporate state, while simultaneously control over the individual artist. Abstraction is just the opposite. It appears when there is less state control over individual expression and so there is freedom for the artist, but such nonnational looking art exercises tacit control over the state's freedom to express its national identity. Ironically, then, abstraction is control over the corporate state's ability to express its collective identity, but freedom for the individual artist to express her or his personal identity.

Abstraction and realism, then, produce unity and antiunity at different levels of the world-system. Realism with a social bent erodes the power of identity-dissolving abstraction and as such advances national solidarity at the expense of the hegemon's imposed global order. Emphasizing national, local, regional, and social themes in art raises national consciousness and helps build patriotism, mobilize the populace, and bind the nation together. In this way 1980–1990s multicultural realism erodes the sense of global unity and international conformity that existed under American hegemony since the 1950s. Realism produces solidarity within states, but that weakens the solidarity of the larger international system that has been enforced by the power of the single hegemonic state. Conversely, the general themes of more classic, universal, and abstract art erase national identities and, as such, work against a sense of national solidarity. The splashes and dabs of abstract expressionism under American hegemony say little about the national conditions under which they were produced. Such abstraction helps produce the idea that the whole world is working as a singular artistic project solving the problems of modern art, which, by definition, has nothing to do with national pictorial space. On the other hand, postmodernism's call for reintroducing the social context of artistic production is the first stages of the cyclic renationalization of art.

ART SINCE BRITISH HEGEMONY

Since the mid-1970s, the art world has begun to move away from the universalism of American led abstraction and assert the particularism of genders, races, and ethnicities, the multicultural art of "other voices" (Wallis 1984; Lippard 1990; Robins 1984; Gablik 1991). What such postmodern multicultural art evolves into remains to be seen, but the stylistic evolution of world art after the decline of British hegemony in the mid–nineteenth century suggests a series of stages that art styles pass through.

The ascending hegemony of Britain starts in 1763 with the Treaty of Paris and the defeat of the remaining serious challenger, France. At about the same time there is the rise of a new era of generalized universal frames for discourse, the Enlightenment, and in art there is an emerging neoclassicism seen in the idealized modeling of figures that had not been seen since the last surge of universalism, the High Renaissance under the Spanish Hapsburg hegemony. Also under British hegemony in the early nineteenth century, there is the appearance of a romantic idealization of nature with the transsocial ideal of "sublime" experience that was thought to cross class, region, ethnicity, and nation, making romanticism, like neoclassicism, a form of transsocial generalized styling. Romanticism is another manifestation of a realism-denying visual framing that accompanies political/economic hegemony.

Around the middle of the nineteenth century Britain starts to decline, and with this we see a lessening of the grip of romanticism and a decaying of neoclassicism into the formal academic art of the Salons. At the same time we see the rise of realism in art, of Millet and Courbet painting stone breakers and agricultural workers in fields with their lumpy nonclassical bodies bathed in real light. This early realism represents a *neutralist nonideological realism* that will eventually become a fully social realism in the full service of the state in the 1930s. Such postneoclassicism/romanticism, an earlier version of postmodernism, continues with the rise of Impressionism, a visual frame that adds more institutional, class, and gender images that reflect national social infrastructure by painting urban scenes, cafe society, boulevards, landscape, people in classes, and women in gendered positions (Broude and Garrard 1982; Pollock 1988; Raven, Langer, and Frueh 1988). The breakup of British economic hegemony and the growing competitiveness of the world economy finds cultural expression in the growing portrayal of national difference in art. This constitutes a more *institutional realism*—not of classically eternal figures under neoclassicism, or quivering states of sublime ecstasy in the face of the universal power and force of nature in Romanticism, but socially situated people doing very ordinary things, as there is nothing classical or universal in Impressionism or post-Impressionism. There is only the day to day present as seen by ordinary imperfect, nonclassical eyes. From realism to Impressionism, localism returns to art. The post-Impressionism of Gauguin, Matisse, and Van Gogh continues the trend of posthegemony realism, and even cubism, while seen as abstract, is but another way of getting at the underlying structures of reality.

Entering the twentieth century, realism begins to turn more collective and nationalistic, taking on distinctive social programs, and having more multiple national origins. Futurism, dadaism, surrealism all have social/radical programs and a separate national origin: futurism from Italy; expressionism in Germany; dadaism/surrealism in New York, Paris, Zurich; constructivism in the Soviet Union. The balkanization of the world economy continues during the interwar years and is culturally expressed in the balkanization of modern art with these new, and nationally specific, art movements.

By the 1930s the realism that has been evolving since the mid–nineteenth century has now turned to the complete service of the nation. Realism is now fully social, having gone from neutral in mid–nineteenth century, through institutional in the late nineteenth, to being accompanied by social programs in the early twentieth, to fully praising country in the 1930s. This is the period of *social realism*. From the Mexican muralist Rivera, Orosco, Siquerios, to American regionalist and urban social realists Ben Shawn or Jack Levine, to Nazi, Stalinist, and fascist art, painting turns toward love of country, in representing people at home and work, rural and urban.

This social realism, though, abruptly ends with the appearance of the next hegemony, the United States after 1945. Art is once again dominated by generalized universal themes, now as avant garde abstract art. Again hegemony. Again the resurrection of the isolated individual artist and the denial of realism, social context, folk art, crafts, regionalism, and the social in art in general. From abstract expressionism through minimalism to conceptualism, the abstract and the modern hold sway in world consciousness as long as there is a clear hegemon. But when hegemony falters, when the United States starts to decline, so does the modernism of abstract art. The mood since the 1970s is again toward realism and the return of social subjects. Instead of Millet's sowers and Courbet's stone breakers it is now women, ethnicities, and races as the social face of voices left out again takes center stage and abstract modernism recedes under attack from the new realistic particularists. First the neutral realism of the photorealism of the 1970s, then more institutional realism of race, gender, ethnicity, in the 1980s and 1990s. This is where we are now. The future, of course, remains open. But if the cycles of world-systems political economy continue, cycles of global culture should follow suit, and the twenty-first century should see the rise of art equivalent to manifest realism—neosurrealism/dadaism. Then, as American decline continues, balkanization increases, and trade war turns to the possibility of World War III over who will succeed the

United States as the next hegemon, art should turn to 1930s-like social realism in praise of country and against the other.

So, in sum, changes in forms of art and forms of social theory vary at the same time. Abstraction in art is, then, similar to generality in cultural and social theory. Both appear when a single state dominates the world economy. Realism in painting and focus on specific facts, historical episodes, and the interests of specific social groups predominate in social science inquiry during periods when no single state dominates the world-system.

REFERENCES

Bergesen, Albert. 1992. "Communism's Collapse: A World-System Explanation." *Journal of Political and Military Sociology* 20:1(Summer):133–155.

———. 1996. "The Art of Hegemony." Pp. 259–278 in *The Development of Underdevelopment: Essays in Honor of Andre Gunder Frank,* edited by Sing Chew and Robert Denemark. Thousand Oaks, Calif.: Sage.

Bergesen, Albert, and Ronald Schoenberg. 1980. "Long Waves of Colonial Expansion and Contraction, 1415–1969." Pp. 231–277 in *Studies of the Modern World-System,* edited by Albert Bergesen. New York: Academic.

Bergesen, Albert, Roberto Fernandez, and Chintamani Sahoo. 1987. "America and the Changing Structure of Hegemonic Production." Pp. 157–175 in *America's Changing Role in the World-System,* edited by Terry Boswell and Albert Bergesen. New York: Praeger.

Bergesen, Albert, and Roberto Fernandez. 1999. "Hegemonic Rivalry between Multinational Corporations, 1956–1989." Pp. 151–173 in *Future of Global Conflict,* edited by Volker Bornschier and Christopher Chase-Dunn. London: Sage.

Broude, Norma, and Mary D. Garrard, eds. 1982. *Feminism and Art History: Questioning the Litany.* New York: Harper & Row.

Connor, Steven. 1989. *Post-Modernist Culture: Introduction to Theories of the Contemporary.* Oxford: Blackwell.

Frascina, Francis, ed. 1985. *Pollock and After: The Critical Debate.* New York: Harper & Row.

Gablik, Susi. 1991. *The Reenchantment of Art.* New York: Thames & Hudson.

Goffman, Erving. 1974. *Frame Analysis.* Cambridge, Mass.: Harvard University Press.

Guibaut, Serge. 1983. *How New York Stole the Idea of Modern Art.* Chicago: University of Chicago Press.

Harvey, David. 1990. *The Condition of Postmodernity.* Oxford: Blackwell.

Hunt, Scott A., Robert D. Benford, and David A. Snow. 1994. "Identity Fields: Framing Processes and the Social Construction of Movement Identities." Pp. 185–208 in *New Social Movements: From Ideology to Identity,* edited by Enrique Larana, Hank Johnston, and Joseph R. Gusfield. Philadelphia: Temple University Press.

Jencks, Charles, ed. 1992. *The Post Modern Reader.* London: Academy Edition.

Lasch, Scott, and Jonathan Friedman, eds. 1992. *Modernity and Identity.* Oxford: Basil Blackwell.

Lippard, Lucy R. 1990. *Mixed Blessings: New Art in a Multicultural America.* New York: Pantheon.

Pollock, Griselda. 1988. *Vision and Difference: Femininity, Feminism, and the Histories of Art.* London: Routledge.

Raven, Arlene, Cassandra Langer, and Joanna Frueh, eds. 1988. *Feminist Art Criticism: An Anthology.* Ann Arbor, Mich.: UMI Research Press.

Robins, Corrine. 1984. *The Pluralist Era: American Art 1968–1981.* New York: Harper & Row.

Rosenau, Pauline Marie. 1992. *Post-Modernism and the Social Sciences.* Princeton, N.J.: Princeton University Press.

Snow, David, and Robert D. Benford. 1992. "Master Frames and Cycles of Protest." Pp. 133–155 in *Frontiers in Social Movement Theory,* edited by Aldon D. Morris and Carol McClurg Mueller. New Haven, Conn.: Yale University Press.

Wallis, Brian, ed. 1984. *Art after Modernism: Rethinking Representation.* New York: New Museum of Contemporary Art.

Part IV

Gender, Urbanism, Cultures,
Indigenous Peoples, and Ecology

11

Women at Risk: Capitalist Incorporation and Community Transformation on the Cherokee Frontier

Wilma A. Dunaway

THE HISTORICAL SETTING

Between 1600 and 1750, England, France and Spain competed for the position of hegemonic world power. As part of that international rivalry, a major segment of the New World was absorbed into the capitalist world-economy. This new periphery was the *extended Caribbean,* stretching from northeast Brazil to Maryland. The colonizations of Virginia, Carolina, Florida, and Georgia and the subsequent incorporation of their mountainous hinterlands ensued as part of the creation of this large new peripheral region. By 1700, the southern Appalachians and their eight indigenous populations formed a buffer zone between British, French and Spanish settlements in the North American Southeast (see fig. 11.1). Each of the colonizers sought to take hold of the western frontier, out of fear that one of the other powers would capture those crucial mountains; for this vast region formed a geographical barrier between the Atlantic coast and the rich inland valleys of the Ohio and Mississippi Rivers. All three colonizing powers knew that whoever was master of the Appalachians might hold the key to further advancement into the continent, for the Cherokees occupied sixty towns of twenty-two thousand people and claimed the land area from northern Alabama to east Kentucky and the Shenandoah Valley. Indeed, Cherokee country was "the key" to the continued existence of four of the British colonies.

Incorporation into the capitalist world-economy bore a high cost for indigenous Appalachians. The Cherokee economy was transformed into a putting-out system that generated dependency on European trade goods and stimulated debt peonage. Within a few decades, Cherokee village activities were restructured from subsistence production into an export economy in which hunting for slaves and deerskins and gathering marketable herbs assumed primacy. Beginning in the late seventeenth century, slave raiders, fur traders, and land speculators carried capitalism into the in-

195

Figure 11.1   Southern Appalachia in the Early 1700s

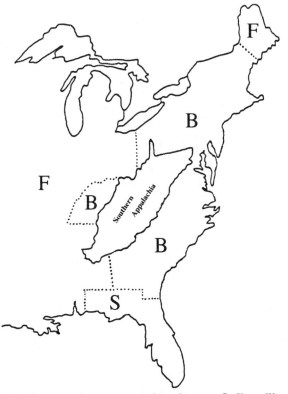

Nation holding colonies, pre-empted territory, or Indian alliances:
B = British   F = French   S = Spanish

land southeastern mountains. In the early 1700s, Charleston merchants exported as many as 121,355 skins annually; and that number rose steadily to 255,000 skins by 1730. Within less than fifty years, the Cherokees lost economic and political autonomy and became dependent on the commodities they obtained through trade with the European nations. The cultural collision between the Cherokees and the Europeans stimulated serious repercussions for the communal-subsistence economy. Looking from the vantage point of indigenous women, this chapter will explore the dramatic political and cultural transformations of Cherokee society that were caused by the absorption of this indigenous people into the capitalist world-system (Wallerstein 1989; Dunaway 1994).

CAPITALIST BIFURCATION OF ECONOMIC PRODUCTION

Prior to European trade, women's economic activities formed the essential basis for community survival. Women produced, preserved, and prepared the majority of the village food supply and controlled the households that supplied every other aspect of livelihood. Because they contributed so much labor to the survival of their villages, women held high status within traditional customs. Children were recognized through matrilineal lineage, so women were given control over the agricultural fields and family homes. During spring planting, all residents of Cherokee villages worked together in the fields. Moreover, Cherokee laws ensured women control over their food production; and women's labor in subsistence agriculture was celebrated in annual ceremonies (Perdue 1985; Hudson 1976; Fogelson 1990).

During the eighteenth century, Cherokee villages paid for imported European commodities by expanding their deerskin output. Consequently, the Cherokee economy was transformed into a *putting-out system* that required the redistribution of half the village's collective labor to the fur trade. As Cherokee households were reshaped around export activities, communal labor arrangements were replaced by a new gender bifurcation of tasks. Cherokee males were exclusively engaged in export and diplomacy activities, leaving women responsible for all the subsistence production. As villages became dependent on European commodities purchased with male-controlled skins, the status of women diminished—at the same time that their work load intensified.

Increasingly, the deerskin trade and the European impetus to war became the central economic and political foci. The resultant new division of labor disrupted traditional production of survival essentials and intensified Cherokee dependency on expensive core-manufactured commodities. Because hunting, diplomacy, and warfare consumed the entire year, a new gender bifurcation of tasks emerged. Emphasis on hunting and the trading-diplomacy process siphoned Cherokee males away from the seasonal rhythm of economic production. In addition to commercial hunting, the male labor force was proletarianized as burdeners, canoers, or pack horsemen for traders, to help build forts, to engage in slave raids, and to fight wars. The new emphasis on hunting and warfare also necessitated greater male labor time for the production of weapons and canoes. No longer were men periodically summoned to assist with spring planting. By the mid-1700s, British observers reported that the Cherokee "women alone do all the laborious tasks of agriculture" (Williams 1927, 68; Dunaway 1994; Fogelson and Kutsche 1961).

Export production also drained away six months or more of female labor time. Between November and March, women accompanied men on the annual hunts that might range as far as three hundred miles from their home villages. During the early years of the fur trade, Cherokee males had not yet adapted European horses. Thus, women walked fifteen or twenty miles a day, carrying sixty to eighty pounds of skins. After men killed the animals and dressed the hides in a preliminary fashion, women completed two tedious tasks. They dried, smoked, and packed the meat; and they cured the skins. At the peak of the Charleston fur trade, Cherokee males were marketing 255,000 skins per year. Consequently, village women must have invested more than one million labor hours to subsidize male commodity production (Hudson 1976; Catesby 1683–1749:1, ix; Dunaway 1994).

GENDER BIFURCATION OF TRADE

Despite the heightened demands on their labor, women did not participate equitably with men in the fur trade. Prior to capitalist incorporation, women engaged freely in long-distance exchange between the Cherokees and other southeastern groups; and their exchanges of food, clothing and decorative items were just as valued as male items. After European linkages, *formal trading* changed from the communal activity it had once been into a male-dominated enterprise. Because European trade was linked to political alliances, Cherokee hunters were also the warriors among the villages; and women were almost totally excluded from the male-dominated fur-trading process (Reid 1976; Gearing 1956; Dunaway 1994).

Capitalist trading evolved into a bifurcated system in which males dominated trade and diplomacy. Because of their own sexism, Europeans restricted Cherokee males to transact business within the structured networks of the trading companies. European traders advanced commodities to villages for which males paid with their annual deerskin production; thus, the Cherokees became perennial sharecroppers indebted to the traders who staked them in their winter hunts. Pushed out of the capitalist market, Cherokee women operated *outside the reach of formalized debt* by marketing surpluses they produced as an extension of household subsistence. Women lacked control over the fur trade, but they maintained dominance over their own commodity production. They hawked their crafts and foodstuffs through the informal economy, typically to British forts or to white urban households. At British forts, Cherokee women peddled agricultural commodities in such numbers that the facilities periodically had "the Appearance of a Market." In addition to their exchanges with nearby forts, Cherokee women traveled to the coastal settlements to peddle their mountain herbs, carpets, turkey feather blankets, colorful pottery, and intricately designed baskets (Sellers 1934; Corkran 1962; McDowell 1958–72:2, 218; Carroll 1836:2, 482; Bartram 1791, 53; Adair 1775, 388, 424–425).

Because Europeans accorded them lower status than male deerskins, women's commodities were exchanged outside the networks of formalized trade. In an attempt to bring Cherokee women under the influence of the capitalist market and make them conform to its rules, Europeans began to ban their commodities from white consumers. The British Trade Commissioners prohibited traders from accepting women's baskets, pottery, and mats because these items were cheaper than imported manufactured household goods. Moreover, the British put in place a "pass" system to regulate the peddling of women in the streets and neighborhoods of the coastal cities. Fort garrisons were instructed not to substitute indigenous crafts for manufactured household goods available through European trading companies. By 1730, male deerskins were still in demand on the world-market because they were essential to the emergent leather industry of western Europe. In fact, England prohibited the flow of deerskins from British colonies to any locations except London and Bristol. However, the trading companies identified many female commodities as economic threats because they lowered demand for European manufactured goods. European traders even began raising their own hogs and chickens, so Cherokee women no longer could dispose of those commodities in the informal sector. Consequently, British competition with women's informal trade deepened village dependence on the male-dominated fur trade (McDowell 1955, 126–128, 131–132, 201; McDowell 1958–72:2, 137–139, 150, and vol. 3, 346; Adair 1775, 240; Dunaway 1994).

THE MALE ROLE IN ECONOMIC DEPENDENCY

As the fur trade expanded, Cherokee villages were "deindustrialized" of traditional male crafts that did not contribute directly to fighting wars or killing deer. Even specialized copper work and body paint preparation were displaced by European commodities. Once the males stopped their indigenous salt manufacturing, Cherokee women were compelled to exchange their precious corn for European salt. By the mid-1700s, the British could report, "The Indians, by reason of our supplying them so cheap with every sort of goods, have forgotten the chief part of their ancient mechanical skill, so as not to be well able now, at least for some years, to live independent of us" (Williams 1928, 112; Williams 1927, 77–78; McDowell 1958–72:3, 344; Adair 1775, 456).

Only twenty-five years after organized trading had begun, a new generation of young Cherokees had "been brought up after another Manner than their forefathers"; and their head warriors taught them that "they could not live without the English." Commercial hunting, population declines and frequent warfare resulted in lowered production in those agricultural and craft functions that were essential to the survival of the villages. At the same time, Cherokee males expanded their consumption of European luxury goods. The native clothing and decorative items produced by Cherokee women were replaced by European beads, metal necklaces, and British shirts and match coats. Indigenous tobacco was displaced by West Indian imports. To demonstrate their new economic status, many Cherokee males even began to have European goods buried with them (Williams 1928, 112; McDowell 1958–72:l, 255; Williams 1927, 76–77; Sellers 1934; Corkran 1969, 26).

Because demand for manufactured goods was relatively inelastic, British traders identified a commodity that would be in constant demand. Introduced to Cherokee men by 1700, rum became the trade good that spurred abandonment of subsistence production and intensified dependency on deerskin exporting. The British Trade Commissioner reported in 1730 that white traders

> made a constant Practice of carrying very little Goods, but . . . Rum. . . . [S]ome of those Rum Traders place themselves near the Towns, in the way of the Hunters returning home with their deer skins. The poor Indians in a manner fascinated, are unable to resist the Bait; and when Drunk are easily cheated. After parting with the fruit of 3 or 4 Months Toil, they find themselves at home, without the means of buying the necessary Clothing for themselves or their Families. (Jacobs 1954, 35)

While women were more reticent about altering traditional customs, they could not escape the trade dependency that male economic decisions entailed (Williams 1927, 77).

DEBT PEONAGE AND DISEMPOWERMENT OF WOMEN

The average European trading company received a 500 to 600 percent profit on deerskins, yet the Cherokees "roamed the forests almost as employees of a trading system built around the faraway demands of European society." Male rum addiction, horse stealing, and luxury-good consumption exacerbated Cherokee debt peonage; for the Europeans made the unpaid debts of any single member of the town the obligation of the entire village. In addition, traders seized assets from the clan who were kin to a deceased debtor. Consequently, women were responsible for the indebtedness of male members of their clans. To prevent transfer of matrilineal clan lands to white traders, Cherokee women were forced to provide more labor to male-dominated tasks and to sacrifice accumulated household assets (Corkran 1962, 6, 26; Jacobs 1954, 35; Williams 1927, 76–77; Salley 1928–47:1, 188; Hudson 1976; Reid 1976).

To meet village debt obligations, women increased their allocation of labor to deerskin processing. As a result, their subsistence agricultural cultivation and their craft production became more erratic. Villages that once marketed surpluses now purchased British foodstuffs at exorbitant prices. Women had avoided indebtedness for nonfood household essentials through their craft production and through exchanges in the informal sector. However, that form of household subsidy disappeared, as their commodity production declined. Women had traditionally produced pottery, baskets, carpets, mats, and blankets during the winter months. By 1715, women spent November through April hunting and skin processing. In short, subsidization of the male-dominated fur trade locked women into an inescapable circle of deepening dependency on imported commodities (Adair 1775, 414, 422–425; Hudson 1976).

EXPANSION AND DEVALUATION OF WOMEN'S WORK

While their traditional work load may have been inequitable, Cherokee women occupied a pivotal position within their communities. In precapitalist villages, women produced more of the subsistence requirements than males, and their work was as respected and as valued as that of males. What was new under historical capitalism was the correlation of a gendered division of labor and valuation of work. Under historical capitalism, there has been a steady devaluation of the work of women and a corresponding emphasis on the value of the adult male's work (Perdue 1985; Mooney 1900; Wallerstein 1983, 25).

As the male-dominated trade with Europeans assumed primacy, Cherokee men gradually reflected European sexism in their devaluation of women's contributions. Before capitalist incorporation, women's farming and gathering were subsistence functions equal in status to male meat production. However, trade relations with the Europeans restructured hunting from a part-time subsistence function into the central economic focus of most villages. Because Cherokee household production was not part of the cash export economy, traditional respect for women's contributions declined. The economic status of men was linked to hunting, trade, and warfare, and those activities provided the direct connection to the capitalist world-economy. In traditional villages, control over resources and distribution of subsistence goods supported women's high social prestige and authority. As trade dependency deepened, men acquired and redistributed the imports on which the villages depended. Moreover, males disdained women's household crafts for imported commodities. The male devaluation of women's subsistence production is clear in these remarks of Cherokee Chief Skiagonota: "My people cannot live independent of the English. The clothes we wear we cannot make ourselves. They are made for us. . . . Every necessity of life we have from the white people." Despite the crucial role that agriculture and exchanges in the informal sector played in village survival, males almost never engaged in those forms of economic production. Cherokee women did the farming and produced all the household items (McDowell 1958–72:1, 45, and vol. 3, 321–322; Hudson 1976; Dunaway 1994; McDowell 1955, 127; Blumberg 1978; Adair 1775, 422–425).

On the one hand, women's household subsistence and production for the informal economy were devalued. On the other hand, women's labor contribution to skin exports remained hidden behind that of their husbands. Still the Cherokee woman was intricately controlled by the village's contract to pay its indebtedness for trade goods. To produce the deerskins needed, she was expected to become an "unpaid employee" of the men of her clan. However, those labor outputs for deerskin production remained socially invisible. Women carried loads, assisted with burnings, prepared meals, and cured skins during annual hunts—thereby contributing more labor-hours than men. However, it was the male act of deer killing that received acknowledgment from the Europeans, and women's inputs were seen as secondary

subsidization of male production. Because this part of her work was now an extension of male work, the Cherokee woman lost the separate labor identity she had traditionally held. While women maintained control over their agricultural crops and
their crafts, they had no similar jurisdiction over the deerskins they helped to produce (Corkran 1969; Hudson 1976; Reid 1976; French and Hornbuckle 1981).

WOMEN'S WORK AND ECOLOGICAL DEGRADATION

Economic *disarticulation* between women's subsistence activities and the male-dominated export sector brought serious repercussions for Cherokee villages. The fur trade
induced ecological change in the southern mountains, and that environmental degradation deepened Cherokee dependence on European trade commodities and wors-
ened the lot of village women. Greater risk of external attack triggered the movement of Cherokee towns to marginal sites that provided easier defense. Some villages
were moved to more mountainous sites where crops were limited to narrow river
bottoms. Thus, the Cherokees abandoned the fertile outfields they had once kept
at a distance from their towns and shifted to smaller garden plots inside the villages.
Thus, women no longer had enough space to produce sufficient corn, and repeated
warfare destroyed women's crops (Amin 1974, 390–394; French 1977, 297–298;
Bartram 1791, 284; Adair 1775, 94; Grant 1933).

As women used more of their time and energy to subsidize the labor of males in
commercial hunting, the Cherokees encountered more frequent famines. By 1756,
deer had grown so endangered by the fur trade that women's hogs were being sub-
stituted for venison in the Cherokee diet. To complicate matters, male leaders neglected the public granaries that were once filled with women's grain surpluses because the villages could now depend on British corn, pork, and beef to overcome
shortages. By the mid-1700s, some Cherokee towns broke up and combined with
other villages because they could no longer survive the food scarcities (Thornton
1990; McDowell 1958–72:2, 264, 118–119, 151; Williams 1927, 179–181, 67;
French 1977).

To facilitate deer hunting, men engaged in annual burning of meadows and
wooded areas, and that deforestation complicated female work. Traditionally, women
had fuelled the village fires with underbrush and dead trees. By 1740, those sources
had been destroyed, so women were walking many more miles each day to gather
firewood. Deforestation also eliminated several nuts, herbs, roots, fruits, and berries that women had gathered to supplement the household diet. As the number of
trees declined near the villages, women produced less natural sweetener—a household commodity they had once collected from native maples (Hudson 1976, 308;
Adair 1775, 407–410, 416; Williams 1928, 477–478, 490–491).

As villages imported new species of European livestock, women's subsistence production came into direct conflict with the male-dominated trade with Europeans.
Women readily adapted swine and chickens because they could be raised in pens.

However, women resisted male production of large numbers of export cattle because they were afraid the animals would damage their open corn fields. By the mid-1700s, Cherokee males were importing large numbers of horses to facilitate warfare, annual hunts, and the transport of deerskins to Charleston. In an attempt to protect their crops, the women tethered the horses and surrounded their garden plots with stakes. The men were less concerned about restraining their animals, so the horses crept through the garden fencing, destroying family gardens (Williams 1928, 257; Adair 1775, 138–139, 240, 263, 406; Silver 1990).

In addition to risks to crops, hooved livestock endangered the wild flora on which the women depended for craft production and supplementary foods. Women lost the mulberry trees they once cultivated at the peripheries of their towns. Mulberry bark once supplied a natural resource used to produce blankets, carpets, aprons, and mats—items that were replaced by European manufactured items by 1730. Traditionally, women had encouraged and protected river cane and several berries, barks, and roots because they supplied natural materials for craft production. Within twenty years after their introduction to the southern Appalachians, hooved livestock devastated those dense canebrakes on which women depended for reeds and dyes to produce baskets, carpets, and mats (Silver 1990; Bartram 1791, 185–187; Jacobs 1954, 49; Williams 1928, 478).

WOMEN AND CHILDREN PLACED AT RISK

In addition to the stresses associated with famines and ecological change, Cherokee women were impacted by several other economic and physical risks that accompanied the fur trade. Because of the demands for labor in the West Indies and in the emergent North American colonies, Indian slaves were the first profitable commodity to attract the interest of the world-economy in southern Appalachia. By 1681, the capture and selling of Cherokee slaves had begun. Indeed, the Cherokees' initial diplomatic mission to Charleston in 1683 was aimed at seeking relief from slavery raids. In the early 1700s, 70 percent of the Indian slaves in South Carolina were women and children. Because of increased slave raiding and intensified warfare, the daily lives of Cherokee women were filled with new physical dangers. Women engaged in work that made them easier targets than men for intruders. As the woodlands around the villages were deforested, women walked longer distances to retrieve daily firewood. Women did most of their agricultural work in outfields distant from the villages, and enemy raiders sometimes killed them before they could run to safety (Crane 1929; Snell 1972, 96; Salley 1928–47:5, 197; Dunaway 1994; Hudson 1976; Adair 1775, 407–408).

In addition, Cherokee women were required by purity taboos to isolate themselves in separate huts during and after menstruation or childbirth. Violation of these purity taboos was equal to murder in the severity of sanction. Such isolation made women particularly vulnerable to sudden attacks by intruders. Cherokee villages insisted that their untouchable women segregate themselves from public interaction in small huts at a considerable distance from their households. Cherokee males believed that anyone who touched or walked near these women would be polluted. Because these women were not situated within defensible range, they stayed in their purity huts at the risk of their lives (Hudson 1976; Adair 1775, 125–126).

Capitalist incorporation of Cherokee villages brought additional health dangers to women. During the 1700s, a devastating disease was transmitted to the Indian towns about every four years, spreading ninety-three epidemics from the coastal European settlements. In addition to the greater risk of death, women were often blamed when European diseases struck Cherokee villages. When the 1738 smallpox epidemic was transmitted from Charleston in trade goods, Cherokee medicine men contended that the sickness had been sent among them as punishment for the adulterous behavior of the village's married women. When the male doctors could not cure the villagers, they destroyed their "physic pots" believing that they had lost their divine power by being polluted by women who had broken marriage laws and purity taboos. The punishment for such violations could include stoning, mutilation, whipping, or death of the accused women (Thornton 1990; Mooney 1900; Adair 1775, 232; Reid 1976).

As direct outcomes of capitalist incorporation, women and their children were placed at risk in three other ways. Male rum drinking not only expanded the work load of women, but alcohol addiction also created an upsurge in domestic violence. Under Cherokee law, women could kill or abort infants; but men who engaged in these practices were charged with murder and could expect the death penalty. Even in the face of declining populations, women engaged more often in abortion and infanticide, in order to accompany their husbands on hunting expeditions. According to Cherokee custom, women kept their infants inside a cradleboard while they were breast-feeding the first year. As women spent more time on hunts and in processing deer skins, they left their babies inside the cradleboards for longer periods. As a result, the women unintentionally deformed the skulls of their children (Adair 1775, 5, 198; Jacobs 1954, 26; Williams 1927, 60, 89–91; Corkran 1969; Owsley and Guevin 1982; Reid 1976).

POLITICAL DISEMPOWERMENT OF WOMEN

Before capitalist incorporation, the Cherokees were an agglomeration of independent villages, without any unifying structure to facilitate coordination of all the dispersed settlements. Each separate town was populated by members of seven matrilineal clans, and every individual Cherokee derived his or her political alignment from membership in one of the clans. To be without a clan was to be without rights. Although women could not hold office, they exercised authority over the clans. Married women of child-bearing age held a council that nominated candidates for chief and subchief of each clan. These males, then, comprised two hierarchical councils of red and white clans who alternately directed village affairs. Headed by the most prominent warrior, the red organization functioned during three significant dilemmas faced by the town: external warfare, external diplomacy, and trade alliances with outsiders. In cases of unresolved disagreements, the white council made the final decision (Corkran 1962; Perdue 1980, 20–21; Foreman 1954, 7; Gearing 1956; Reid 1976).

Articulation with the world-system necessitated a political structure that permitted the Europeans to negotiate with the Cherokees as a single corporate entity. It was more *rational* and more *efficient* to collect trade debts, make treaties, engineer war alliances and seek reparations from one leader who was the "Mouth of the Nation" and who could enter into agreements that "should be binding upon him and all the Nation." Thus, the British pressured the loosely-knit Cherokee towns toward *secularization* and *centralization* of their nonstate political mechanisms, culminating in a regional "priest-state" comprised of representatives from all the towns. Now the Cherokees abandoned their ultrademocratic methods to support, instead, a single tribewide sentiment and to legitimate that decision with universally applied sanctions against violators. Because of the distance between settlements, selected spokesmen acted at the regional capitol on behalf of their villages, establishing a system through which the Europeans could easily manipulate a small number of elite males (McDowell 1955, 188; Gearing 1956; Corkran 1962).

As the Europeans expanded their diplomatic and trade manipulation of Cherokee males, the political participation of women eroded. Traditionally, matrilineal clans played a key role in village politics, and a seven-member advisory panel represented all the divergent interests. With the advent of regional council meetings, however, clans were no longer equitably represented, and women had fewer avenues through which they could bring pressure to bear on the males. After the emergence of the "tribal half-government," the white clans no longer shared village leadership—a change in traditional governance that silenced the political voices of half the female population. Cherokee settlements diminished in number from sixty towns in 1715 to thirty-nine towns in 1755. When towns died or relocated, women's positions within their clans shifted. While males retained their war and trade reputations, women lost any status or political clout they may have held in the old town and had to assimilate themselves into the new community (Gearing 1956; Williams 1927; Thornton 1990; Adair 1775; Hudson 1976).

As political deliberations were centralized, women's sphere of influence contracted. As the role of clans and towns narrowed, so did the political participation of women. Traditionally, each matrilineal clan had chosen a "Beloved Woman" as its leader; and these seven women formed the Women's Council, headed by the "War Woman." These women leaders were selected because of their long service or sacrifice for the village. The War Woman was the most highly respected because she had either lost many male family members to war or had shown great courage on the battlefield. Nancy Ward was the most famous eighteenth-century Cherokee War Woman. The Women's Council selected the candidate for town chief, and their nominations were usually approved by the male-dominated town council. Women lost this power when the British began to select Cherokee leaders through war commissions or through trade co-optation of elites (Foreman 1954; Gearing 1956; Corkran 1962; Mooney 1900).

As instances of warfare multiplied, so did the British demand for Cherokee warriors, and, in the minds of the Europeans, women should have no place in decision making about war. Even though a few male Cherokees defended the right of women to participate in such councils, the regional government eventually relented to European custom. Ultimately, the British imposed their sexist prejudices by excluding women from their deliberations with Cherokee males. Women even lost the right to determine the fate of war captives, once males began to sell such unfortunates to the British for slave exports to the Caribbean (Hamer and Rogers 1972:3, 279–280; Reid 1976, 69; Adair 1755, 152–153; Snell 1972).

In the traditional Cherokee way of life, she who controlled essential resources garnered power. Because farming and child rearing were primarily their responsibility, Cherokee women controlled the means of production: village lands. By the mid-1700s, however, Cherokee men were ceding lands to Europeans to settle trading debts, without the input of women. Within little more than fifty years, the British extinguished Cherokee claims to nearly forty-four million acres—more than half their ancestral territory. Moreover, the Cherokees wasted more than half of those natural resources to pay male debts to European traders. In treaties that parallel contemporary third world "debt for nature swaps," Cherokee lands were transferred from local villages to distant capitalists, forever altering the central roles of women in this indigenous society (Palmer 1875–93:1, 291; DeVorsey 1961, 162–163; Royce 1884; Mies and Shiva 1993).

THEORETICAL REPRISE

When the capitalist world-system incorporates a new frontier, dramatic social changes are set into motion. If we explore this process from the perspective of the rich core nations, we see that incorporation is the long-range civilizational project of capitalist colonizers. Driven by the economic and cultural logic of historical capitalism and by their own ethnocentric sense of superiority, the intruders mythologize their domination as a lofty mission to implant "civilization" and "progress" on backward barbarians. Frontier histories have typically been constructed from the viewpoint of the colonizer, the indigenous people being transformed into a "ghost" from the past of the land area captured by the invader. If we view the process from the bottom up, as this research does, we ask questions from the perspective of the indigenous people; and we attempt to "represent that ghost" and to give voice to a people whom colonizers have obliterated from written history (Dunaway 1996; Trouillot 1995, 147; Wolf 1982).

Like indigenous peoples, women's voices have also been silenced from official world history. This investigation of the Cherokee frontier calls attention to our need to examine more closely the unusual impacts of capitalist incorporation on women. Historically and in the contemporary period, the intrusion of capitalism has generated many more negative side effects for women than for males (Mies and Shiva 1993). Even when Cherokee males benefited economically and politically, women lost their pivotal roles in the village economy, polity, and culture. As Cherokee households were reshaped around export activities, communal labor arrangements were replaced by a new gender disarticulation between women's subsistence activities and the male-dominated export sector. Under historical capitalism, there was a steady devaluation of the work of Cherokee women and a social inflation of the value of the men's labor. While females had done an inequitable share of precapitalist labor, women's work held high status within traditional customs. As villages became dependent on European commodities purchased with male-controlled skins, the work load shifted even more inequitably on women. However, traditional respect for women's subsistence production declined, and women's contributions to the export economy were disguised as minor subsidies to male labor.

Women's work load may have been inequitable in primitive horticultural societies, but that work was respected and publicly acknowledged. Moreover, women owned the means of production and controlled the products of their labor. In capitalist societies, women no longer own production resources, their work is devalued, and they do not control what they produce. As it incorporates new zones of the globe, capitalism embraces two antithetical labor recruitment mechanisms: (1) a historical proletarianizing of males into laborers who produce surplus commodities for the market and (2) a simultaneous historical concentration of women's labor into arenas that are never fully or fairly remunerated. Thus, a society undergoing the transition to capitalism experiences a realignment of labor so that the economic contributions of women are devalued. Women are transformed into "the last link in a chain of exploitation," permitting by their unpaid labor the reproduction of the work force and the unrewarded subsidization of male-dominated labor (Blumberg 1978; Hopkins and Wallerstein 1987; Mies, Bennholdt-Thomsen, and Werhof 1988).

In contemporary peripheries of the capitalist world-system, poor women must move back and forth continually between household subsistence, assistance in export production, and activities to generate a few commodities that will be exchanged in the informal sector. Despite their complex interlinkages across several economic arenas, however, such women are less likely than males to receive remuneration or acknowledgment for their labor. Essentially, "women's work" is socially invisible because much of it is unpaid or rewarded only indirectly through male intermediaries, and there is a tendency to view it as a form of domestic labor without economic value. However, females subsidize the production of capitalist commodities, like the Cherokee deerskins, through devalued household labor and through their unpaid assistance to spouses.

This research also calls our attention to another aspect of the incorporation process that is usually ignored. Significantly, there could be no expansion of the capitalist world-system without rivalry over environment. Land and territory are the basic impetus for exploitation of the peripheral zones, for it is the capture of their raw materials that propels the growth of core manufacturing. Thus, economic restructuring around commercial hunting endangered or wasted the pool of natural resources on which the Cherokees relied. However, capitalism externalizes to women several ecological costs that males do not experience. After the fur trade ecologically degraded the southern mountains, women's natural resources diminished—thereby deepening Cherokee dependence on European trade commodities. To meet village debt obligations, women increased their allocation of labor to deerskin processing. As a result, their subsistence agricultural cultivation and their craft production fell sharply, weakening their control over village production even further. Women engaged in work and purity taboos that made them easier targets than men for warring intruders or slave raiders, and women were disproportionately affected by the frequent epidemics that struck Cherokee villages. Articulation with the capitalist world-system also necessitated a political structure that permitted the Europeans to negotiate with the Cherokees as a single corporate entity. As governance was centralized, males assumed primacy in decision making about the natural environment.

For women, the most dooming articulation between Cherokee environment and European world-system was the commodification of land that accompanied capitalist expansion. Ultimately, women even lost their traditional control over the means of household production, as males settled their trading debts by ceding more than half of Cherokee ancestral lands to Europeans. We can see direct parallels between eighteenth-century Cherokee women and today's peripheral nations. To pay international debts, third world countries initiate capitalist development projects that cause ecological devastation and waste the natural resources from which women have drawn subsistence resources. Moreover, recent "debt-for-nature swaps" shift ecological control from indigenous women to core corporations and to foreign nongovernmental organizations (Smith 1984; Mies and Shiva 1993).

REFERENCES

Adair, James. 1775. *The History of the American Indians.* Reprint, Johnson City, Tenn.: Watauga, 1930.

Amin, Samir. 1974. *Accumulation on a World Scale: A Critique of the Theory of Underdevelopment.* New York: Monthly Review Press.

Bartram, William. 1791. *Travels through North and South Carolina, Georgia, East and West Florida, the Cherokee Country.* Philadelphia: James & Johnson.

Blumberg, Rae Lesser. 1978. *Stratification: Socioeconomic and Sexual Inequality.* Dubuque, Iowa: Brown, Boserup, & Ester.

Carroll, Bartholomew R., ed. 1836. *Historical Collections of South Carolina.* New York: Harper.

Catesby, Mark. 1683–1749. *The Natural History of Carolina, Florida, and Bahama Islands, 1683–1749.* Reprint, Spartanburg, S.C.: Beehive, 1974.

Corkran, David H. 1962. *The Cherokee Frontier: Conflict and Survival, 1740–1762.* Norman: University of Oklahoma Press.

———, ed. 1969. "Alexander Longe's 'A Small Postscript of the Ways and Manners of the Indians Called Charikees, 1725.'" *Southern Indian Studies* 21:1:3–49.

Crane, Verner. 1929. *The Southern Frontier, 1670–1732.* Ann Arbor: University of Michigan Press.

DeVorsey, Louis. 1961. *The Indian Boundary in the Southern Colonies, 1763–1775.* Chapel Hill: University of North Carolina Press.

Dunaway, Wilma A. 1994. "The Southern Fur Trade and the Incorporation of Southern Appalachia into the World-Economy, 1690–1763." *Review* 17:2(Spring):215–241.

———. 1996. "Incorporation as an Interactive Process: Cherokee Resistance to Expansion of the Capitalist World-System, 1560–1763." *Sociological Inquiry* 66:4(Fall):455–470.

Fogelson, Raymond D. 1990. "On the Petticoat Government of the Eighteenth-Century Cherokee." Pp. 161–181 in *Personality and the Cultural Construction of Society: Papers in Honor of Melford E. Spiro,* edited by David K. Jordan and Marc J. Swartz. Tuscaloosa: University of Alabama Press.

Fogelson, Raymond D., and Paul Kutsche. 1961. "Cherokee Economic Cooperatives: The Gadugi." *Bureau of American Ethnology Bulletin* 180:87–123.

Foreman, Carolyn T. 1954. *Indian Women Chiefs.* Norman: University of Oklahoma Press.

French, Christopher. 1977. "An Account of Towns in the Cherokee Country with Their Strengths and Distance." *Journal of Cherokee Studies* 2:3:34–46.

French, Laurence, and Jim Hornbuckle, eds. 1981. *The Cherokee Perspective: Written by Eastern Cherokees.* Boone, N.C.: Appalachian Consortium.

Gearing, Frederick O. 1956. "Cherokee Political Organizations: 1730–1775." Unpublished Ph.D. dissertation, University of Chicago, Chicago.

Grant, James. 1933. "Journal of Lieutenant-Colonel James Grant, Commanding an Expedition against the Cherokee Indians, June–July, 1761." *Florida Historical Quarterly* 12:1(July):22–51.

Hamer, Philip M., and George C. Rogers, eds. 1972. *The Papers of Henry Laurens.* Columbia: University of South Carolina Press.

Hopkins, Terence K., and Immanuel Wallerstein. 1987. "Capitalism and the Incorporation of New Zones into the World-Economy." *Review* 10:5/6(Summer/Fall):763–780.

Hudson, Charles. 1976. *The Southeastern Indians.* Knoxville: University of Tennessee Press.

Jacobs, Wilbur R., ed. 1954. *Indians of the Southern Colonial Frontier: The Edmund Atkin Report and Plan of 1755.* Columbia: University of South Carolina Press.

McDowell, William L., ed. 1955. *Colonial Records of South Carolina: Journals of the Commissioners of the Indian Trade, September 20, 1710–August 29, 1718.* Columbia: South Carolina Archives Dept.

———. 1958–72. *Colonial Records of South Carolina: Documents Relating to Indian Affairs, 1750–1765.* Columbia: South Carolina Archives Department.

Mies, Maria, Veronika Bennholdt-Thomsen, and Claudia von Werholf. 1988. *Women: The Last Colony.* London: Zed.

Mies, Maria, and Vandana Shiva. 1993. *Ecofeminism.* Melbourne: Spiniflex.

Mooney, James. 1900. "Myths of the Cherokee." *Bureau of American Ethnology Annual Report* 19.

Owsley, Douglas W., and Bryan L. Guevin. 1982. "Cranial Deformation: A Cultural Practice of the Eighteenth-Century Overhill Cherokee." *Journal of Cherokee Studies* 7:2:79–91.

Palmer, William L., ed. 1875-93. *Calendar of Virginia State Papers and Other Manuscripts.* Richmond: State of Virginia.

Perdue, Theda. 1980. "The Traditional Status of Cherokee Women." *Furman Studies* 26:(Dec.):19–35.

———. 1985. "Southern Indians and the Cult of True Womanhood." Pp. 35–57 in *The Web of Southern Social Relations,* edited by Walter J. Fraser. Athens: University of Georgia Press.

Reid, John P. 1976. *A Better Kind of Hatchet: Law, Trade, and Diplomacy in the Cherokee Nation during the Early Years of European Contact.* University Park: Pennsylvania State University Press.

Royce, Charles C. 1884. *Cherokee Nation of Indians.* Washington, D.C.: Bureau of American Ethnology.

Salley, Alexander. S., ed. 1928–47. *Records in the British Public Record Office Relating to South Carolina, 1663–1684.* Columbia: Historical Commission of South Carolina.

Sellers, Leila. 1934. *Charleston Business on the Eve of the American Revolution.* Chapel Hill: University of North Carolina Press.

Silver, Timothy. 1990. *A New Face on the Countryside: Indians, Colonists and Slaves in South Atlantic Forests, 1500–1800.* New York: Cambridge University Press.

Smith, Joan. 1984. "Nonwage Labor and Subsistence." Pp. 23–36 in *Households and the World-Economy,* edited by J. Smith, I. Wallerstein, and Hans-Dieter Evers. Beverly Hills: Sage.

Snell, William R. 1972. "Indian Slavery in Colonial South Carolina, 1671–1795." Unpublished Ph.D. dissertation, University of Alabama.

Thornton, Russell. 1990. *The Cherokees: A Population History.* Lincoln: University of Nebraska Press.

Trouillot, Michel-Rolph. 1995. *Silencing the Past: Power and the Production of History.* Boston: Beacon.

Wallerstein, Immanuel. 1983. *Historical Capitalism.* London: Verso.

———. 1989. *The Modern World-System III: The Second Era of Great Expansion of the Capitalist World-Economy, 1730–1840s.* New York: Academic Press.

Williams, Samuel C., ed. 1927. *Lieutenant Henry Timberlake's Memoirs.* Johnson City, Tenn.: Watauga.

———. 1928. *Early Travels in the Tennessee Country, 1540–1800.* Johnson City, Tenn.: Watauga.

Wolf, Eric R. 1982. *Europe and the People without History.* Berkeley: University of California Press.

12

Resistance through Healing among American Indian Women

Carol Ward, Elon Stander, and Yodit Solomon

Although recovery and healing from alcohol use and abuse has become a significant dimension of personal and community life for many American Indians, efforts to document and understand this process, pursued largely by anthropologists, have left many questions unanswered. Research has typically focused on the individual needs or problems associated with alcoholism and its treatment and the effective use of indigenous healing methods (Weibel-Orlando 1989). However, this research has seldom addressed questions regarding the link between personal healing and recovery and community-level processes or the effects of the macrolevel changes on American Indian communities. (For exceptions, see Lurie 1971; Brady 1992, 1995.) For example, generally this work has neglected the role of the recent cultural renewal observed in many Indian communities (Nagel 1996) as well as tribal efforts to establish greater control over their economic and natural resources (Cornell and Kalt 1992) and to use sovereignty rights (Vinje 1996) for development purposes, all of which have contributed to the emergence of a new array of cultural, economic, and political opportunities in contemporary Indian communities. The central question for this chapter is, How do the recent experiences of American Indian community members with recovery and healing relate to the cultural, economic, and political changes affecting Indian communities?

An important perspective that enhances our ability to understand these phenomena is world-systems theory as it has recently been applied to the resurgence of indigenous peoples. For example, Dunaway (1996), Friedman (1994, 1998), and Hall (1998b and chap. 13 of the present volume) have suggested that indigenous peoples' recent attempts to assert and protect their cultural, economic, and political interests can be understood not only as the culmination of long-term efforts to resist incorporation into the capitalist world-system, but that such resistance is linked in important ways to the current cycle of hegemonic decline within the modern world-

211

system. Importantly, Kathryn Ward's (1993) work emphasizes that the roles of women in these recent developments are very poorly understood. Thus, a third question for this chapter concerns the experiences and roles of American Indian women participating in recovery and healing processes and their relationship to resistance.

The discussion will begin with a brief review of the social, cultural, and political economic conditions that resulted from the incorporation of American Indians into the modern world-system. This will be followed by a review of the theoretical perspectives typically used for identifying the social forces that produce current problems in Native American communities. The next section will address the benefits of using a world-systems perspective for understanding recent efforts to create changes in American Indian communities. The last half of the essay draws on a case study of the Northern Cheyenne Tribe of southeastern Montana that provides a relevant setting and data for illustrating changes in reservation life and women's roles. The final sections present empirical data on the cultural and political dynamics of recovery and healing from alcohol abuse, particularly as it has involved women. Qualitative data obtained from a focus group of women from several Northern Plains and southwestern tribes reveal important themes concerning how women view and interpret their experiences with healing, and relate them to family, community, and tribal interests.

AMERICAN INDIAN ASSIMILATION INTO U.S. SOCIETY

Over a period of a hundred years, the United States government has employed a variety of approaches designed to "assist" Indian people to assimilate into American society. Some of the earlier forms of incorporation included not only military conquest and removal to reservations but also efforts to systematically undermine tribal sovereignty, through the creation of congressional trusteeship, removal of children from families for education in boarding schools, urban relocation, termination of the special tribal-U.S. relationship, and various attempts to obtain, control, or manage tribal land and natural resources (Snipp 1986, 1988; Ward and Snipp 1996; Cornell and Kalt 1990). The policy of confinement to reservations is of particular importance because American Indians continue to experience the ramifications of this practice. Snipp (1988) describes the history of the changing status of Indian peoples as a movement from "captive nations" to "internal colonies," suggesting that economic exploitation has been added to political dependency in the process of recent "development." Maria Anna Jaimes Guerrero further argues:

> The systematic displacement of native peoples from their land base, which was both their economic source of livelihood as well as their spiritual foundation, set up conditions for colonially induced despair. Benchmarks in U.S. state policy provide a clear picture of these processes of colonization. They are: Allotment, in the 1800s; Termination, in the 1950s; and Relocation, which began in the 1940s and continues today. . . . The ideology of "eminent domain" for the "common good," used in main-

stream population centers, coupled with the designation of certain sites for "national security," mask a massive displacement of native peoples under the guise of pro-development schemes. "Common good" has meant damage to the environment and the ecosystem and the practice of environmental racism so widespread that native peoples and their cultures are now being classified as endangered species in the U.S. (Guerrero 1997, 110)

The impact of reservation life has been the subject of research examining American Indian poverty and the resources needed for improving life chances. Sandefur and Tienda conclude:

Despite substantial increases in educational attainment over the past few decades, many Native Americans remain unprepared to compete in a highly technical and bureaucratized world of work. . . . This situation may improve as larger shares of Native Americans complete post-secondary schooling. However, to the extent that job possibilities on reservations remain limited while larger shares of Native Americans reside on them, the prospects for economic parity with whites will not be realized. Moreover, within the American Indian population, sharp differences in relative economic status between reservation and non-reservation residents will remain. (Sandefur and Tienda 1989, 9)

Ward and Snipp's (1996) analysis of the 1990 census data on a number of human capital and quality-of-life measures reveals limited progress, and in some instances a decline, in the social and economic circumstances of American Indians. With respect to educational attainment, American Indian students had the highest dropout rate (estimated at 35 to 42 percent), and only 26 percent obtained a baccalaureate or higher degree, the lowest rate of any group. In 1990, both American Indian men and women had the highest unemployment rates of any minority group at 15.4 percent and 13.1 percent, respectively, compared to 13.7 percent and 12.1 percent of black men and women, and 5.1 percent and 4.8 percent among white men and women. Family income levels for American Indians declined by 7.3 percent between 1979 to 1989 from $23,440 to $21,750. Additionally, 27 percent of American Indian families live in poverty compared to 7 percent of white families. Despite some improvement over the last four or five decades, significant gaps persist between American Indians and other Americans in educational attainment, participation in the labor force, health conditions, and socioeconomic status (Jorgenson 1978; Talbot 1981; Sandefur and Scott 1983; Snipp and Sandefur 1988; Snipp 1989, 1996; Cornell and Kalt 1990).

ALCOHOL USE: DISEASE, ADAPTATIVE MECHANISM, OR FORM OF RESISTANCE?

One of the most serious conditions affecting reservation populations is substance abuse. Alarming rates of alcoholism have been and are currently reported among American Indian tribes: 80 to 90 percent are not uncommon rates (Milam and

Ketcham 1983). Several approaches have been used to account for or explain alcoholism. The adaptive view, according to Miller and Gold (1990), attributes addiction to moral or psychological weakness. Lack of will power and strength of character allow a person to continue drinking even though it means loss of family, employment, and self-respect. Though decried as a myth by professionals, these notions often inform public images and stereotypes concerning Indian substance abuse and addiction.

The medical model views alcoholism as a disease and attributes the development of the disorder in part to a genetic dysfunction. Heredity is also implicated as a factor in the development of alcoholism: one indication is the 4 to 1 odds of becoming alcoholic if a parent is alcoholic. Research indicates an interplay between heredity and various metabolic, hormonal, and neurological processes to explain susceptibility (Milam and Ketcham 1983).

Another approach supported by an extensive body of literature points to psychological, social, and environmental factors that need to be considered in the etiology of addiction (Miller and Gold 1990; Dovonan 1986; Segal 1987; Tater and Edwards 1988). Using a sociobiology framework, Segal (1987) suggests that alcoholism is not a single disease but a group of syndromes with various etiologies. Each syndrome is a product of multiple forces with differing inputs. For example, social disorganization undermines the protective influence of neighborhood and community. Especially vulnerable to substance abuse, according to Segal, are groups that have lost informal social control. Weakening of old belief systems and hopes as well as disillusionment with religious beliefs predisposes such groups to substance use as an escape, as does loss of a sense of continuity, traditions and habits.

Sorkin (1996) reports that alcohol abuse and addiction have been prevalent coping responses to the circumstances of reservation life that for many include unemployment, poverty, inadequate education, poor health, and low self-esteem. Other causative factors include feelings of lack of control over their lives and the pressures and loss of self-esteem associated with acculturation. Holmes and McPeek (1988) suggest that important underlying causes of alcoholism for many Indians are related to their experience of loss and feelings of grief. These involve the loss of loved ones to accidents, violence, disease, and illness, which occur at higher rates in Indian country than elsewhere. Additionally, Indians grieve for the loss of cultural practices that integrated their communities socially and spiritually. Therefore, Indian drinking is one mechanism for adapting to the social disorganization (loss of culture and identity) that has resulted from acculturation.

Lurie (1971) advances an alternative theory for Indian drinking that challenges the assumptions of these conventional explanations: Indian drinking is an important and valuable form of resistance or protest. In other words, drinking is a useful way of bringing attention to certain issues that may eventually lead to change and, at the very least, serves as a forum for expressing opposition. "Indian drinking is an established means of asserting and validating Indianness and will be either a managed and culturally patterned recreational activity or else not engaged in at all in direct proportion to the availability of other effective means of validating Indianness" (Lurie 1971, 315). Other avenues of validation may include knowledge of cultural traditions, the ability to protect and advance community interests vis-à-vis the dominant society, and the "ability to maintain the basic ideals of dignity, responsibility, resourcefulness, respect for others, and reciprocal generosity" (Lurie 1971, 325). Similar interpretations have been documented regarding the meanings associated with Australian Aborigine drinking patterns. A number of studies have explained Aboriginal drinking as expressing resistance to the dominant culture thereby strengthening group identity (Brady 1992).

OTHER FORMS OF RESISTANCE RELATED TO ALCOHOL USE: MEDICAL TREATMENT AND AA VERSUS TRADITIONAL HEALING

As part of the reservation system, a network of agencies has been created to meet the social and health needs of American Indians. The Indian Health Services and Bureau of Indian Affairs are two of the most prominent ones. Among the most urgent needs addressed by these agencies are health and other conditions related to personal and family well-being. While overall health conditions have improved dramatically over the last several decades, rates of alcohol and drug-related illness have increased or remained at significant levels in many Indian communities. Thus, mental health and the effects of addiction have become increasing concerns.

Efforts to address chemical dependency have typically begun with medical treatment (Sorkin 1996). Additionally, Alcoholics Anonymous (AA), Alanon, and related twelve-step programs are almost always a part of the therapeutic prescription for individual recovery from alcoholism. However, criticisms have been levied against these programs, particularly when dealing with women in recovery. While acknowledging the efficacy of twelve-step programs in helping many people on the road to recovery, Kasl (1992) points out that the AA model derives from a 1930s perspective and was constructed using the experiences of a hundred men and one woman. Additionally, it has not been updated to either include new feminist insights or to be applicable to contemporary life. For example, admonitions to become humble and submissive are helpful for males with inflated egos but are not sensitive to women acculturated in a patriarchal society in which they are more apt to be suffering from low self-esteem. In a society in which many women suffer sexual assault and physical abuse, messages to simply overcome resentments translates into advice to accept the status quo rather than leave a personally damaging situation.

Additional critiques of AA and related twelve-step programs by Holmes and McPeek, for example, include the following view of Holmes:

> From my experience of working with Indians in alcohol treatment centers, I have had a growing conviction that for most Indian people alcoholism is a symptom. It indicates that another problem, usually that of pathological grief, is the cause or root of

the addiction. To treat one without the other is to leave the job half done. The sober, but chronic griever is still not able to grow spiritually, and eventually will slip right back into drinking. If it was pathological grief that triggered the alcoholism in the first place, then we need to deal with the grief, as well as the addiction before that person can return to wholeness. (Holmes and McPeek 1988, 57)

Holmes contends that AA is limited in its ability to deal with the multifaceted problems of oppressed people who have had little opportunity to confront the sources of their loss and grief. Similarly, Brian Maracle, the author of *Crazywater* (1993), asserts that alcoholism symbolizes the genocide experienced by American Indians. The extensive loss associated with the destruction of native lives and lifestyles beginning several generations ago continues to be felt among native people as they cope with the many problems created by conquest and the reservation system. While programs such as AA have been and continue to be vital to the survival of many Indian people, they represent only the beginning of the healing process. Furthermore, these programs lack insights into the social and cultural realities of American Indian life that would allow them to address problems in a more effective and appropriate manner (Beauvais and LaBoueff 1985).

In response to growing criticism of the suppression of traditional forms of healing, non–Native American health providers are now more frequently encouraged to work with traditional healers or at least to respect the client's desires to consult with them. Psychiatrist Carl Hammerschlag (1988, 17) validates the efficiency and wisdom of the native Shaman and asserts that "healing is a powerful, culturally endorsed ritual. There is no doubt that if you trust the practitioner and if you share the same cultural myths, healing is better achieved." Paniagua (1994) also recommends that a medicine man or woman should be consulted for diagnosis and treatment of a Native American client due to the central role these healers often play in the life of indigenous people. In fact, he suggests that clients be encouraged to consult with traditional healers and to discuss ways to coordinate treatment. Suggestions for making treatment efforts more effective and appropriate have included not only involving Indian clients in the planning and application of treatment programs but also integrating traditional healers in the process (Sorkin 1996).

Some American Indian treatment programs have taken this approach even further incorporating traditional social and cultural practices and using native healing methods exclusively ("culture is treatment"). This is based on the idea that substance abuse and addiction can be attributed to the loss of indigenous culture. However, Brady (1995) cautions that this approach runs the risk of reinforcing artificial categories that label Indian people as being "traditional" (having culture) and "nontraditional" (having lost culture). Additionally, it neglects other causes of drinking and the fact that cultures change. In most treatment programs, however, the use of traditional healing and cultural education are one aspect of a multidimensional healing process.

This move toward the inclusion of traditional healing practices in the treatment of substance abuse is not a new phenomenon. In the early 1900s the Peyote religion was used as a means for controlling excessive drinking among the Nebraska Winnebago. "With the adoption of Peyotism, the requirement of avoiding alcohol was no longer merely a personal or secular call, it possessed a religious backing" (Hill 1990, 258). Known as the Native American Church, this group continues to have many followers in reservation communities today. Critical elements of this religion that have contributed to its success in curbing heavy drinking include the following: it provides a social network of support among members, engages members in a variety of activities that would prevent or make it more difficult for them to drink, provides successful role models, takes a holistic approach that helps "resolve underlying problems that in some cases led to heavy drinking as a coping mechanism" (Hill 1990, 261), and, finally, helps strengthen a set of collective ideals or values that are incompatible with drinking activities (Hill 1990). Quintero's (1992) study of alcoholism and recovery among Navajo members of the Native American Church provides further evidence of how the Peyote religion contributes to the healing process. Kunitz and Levy's (1994) longitudinal study of Navajo drinking and recovery patterns also indicates the importance of the Peyote religion and other indigenous healing methods for this population. Similarly, Slagle and Weibel-Orlando's (1986) research on the Indian Shaker Church in northern California identifies certain traditional cultural elements that have contributed to the successful treatment of alcoholism among its members.

In her review of anthropological research supporting the effectiveness of indigenous healing practices related to alcohol treatment and recovery, Weibel-Orlando (1989) asserts that research conclusions may have neglected the complexities of the healing process. For example, in her own work she has identified areas, such as the long-term effectiveness of traditional healing and the role of community dynamics, for which she now believes additional research is needed to fully document the superiority of native healing methods over others. She recommends that new research in this important area should be more comprehensive and fully address the many factors of the social and cultural environment that may affect recovery.

In the contemporary context, ethnic and gender critiques of the twelve-step model and preferences for the use of traditional healing represent a new form of community-level resistance to dominant cultural forms both at the individual and the collective level. They are related to the cultural renewal processes occurring in many Indian communities. Nagel defines this important cultural phenomenon as follows:

> Individual ethnic renewal is the acquisition or assertion of a new ethnic identity by replacing a discarded identity, adding to an existing ethnic identity repertoire, or filling in a personal void. Replacing a discarded identity might entail religious conversion or reinitiation (decision by a Native American to follow traditional spiritual practices). . . . Collective ethnic renewal involves the reconstruction of community: building or rebuilding institutions, culture, traditions, or history, by old or new members. Institution building might involve the creation of new organizations or religions (e.g., formation in the first half of the twentieth century of the Native American Church). Cultural renewal might involve the creation of new or revision of traditional cultural practices. (Nagel 1996, 10)

POSTMODERN VIEWS OF ETHNIC RESISTANCE

One perspective for understanding American Indian cultural renewal and resistance efforts at the macrostructural level is postmodern thought. "For those who have lost the security of the old world and have not found a place in the new there is the sense of limbo; of going nowhere" (Booth 1977, 83). These words of a Native American woman echo the emptiness and despair, the fragmentation of minority lives and of the subjugated in many lands. The recent postmodern discourse tells of transition and contingency, of globalization and consumerism, of fragmentation and lack of criteria for evaluating contemporary circumstances.

Smart (1993) contends that postmodern views may simply reflect efforts to come to terms with the limits and limitations of modernity—that is, its unfulfilled promises. The Enlightenment, embracing positivism and rationality, was thought to be the custodian of universal truths that marked a linear path to progress, prosperity, and security (Bauman 1992). Thus, the quest for power in modern times was rationalized by the obligation to share these truths with those that did not have them. Tolerance for difference, therefore, was incongruous, and variation was antithetical (Smart 1993). The spread of Western knowledge and civilization to "less advanced" societies was a convenient way to legitimize the subjugation of other lands and peoples (Bauman 1993). The legacy of this imperialistic expansion of Western cultures, along with their political and economic organization, military and technological skills, continues to manifest itself in the discontents of today. Acquisition and relinquishment of colonial lands have spawned consequences that are visible in ongoing struggles over boundaries and control. Within America, it is not difficult to see the effects of imperialistic treatment of minority groups. It is in response to this that the Native American poet quoted earlier provides us with a window to the pathos that has been the result for many.

The postmodern signature is plurality and diversity. Challenges to existing national organizations are aimed at both form and identity, and regional groups not only express their differences but demand recognition and autonomy (Smart 1993; Nagel and Snipp 1993). The plethora of cultural forms emerging from complex and unpredictable accommodations and articulations transform the topography into a multicultural maze.

Such racial and ethnic expressions signal that the assimilationist version of the American "melting pot" ideology is no longer appropriate. In its place some propose the more pluralistic goal of "celebrating diversity." Nevertheless, forces of the postmodern world continue to create ethnic conflict. As the world shrinks through globalization processes, the potential for friction between peoples multiplies. Global communication networks provide more opportunities for crossing cultural boundaries and increasing the risk of offense, fragmentation, and aggression (Smart 1993).

Giddens (1991) contends that the renewal of ethnic and religious beliefs and convictions represents a resurgence of the oppressed. Minority or status cultures were never truly abandoned or discarded since they certainly failed to disappear as predicted by Marx, Durkheim, and Weber. In fact, contemporary religion is faced with the creation of new cultural expressions of spirituality. This period marks the beginning of greater tolerance and acceptance of diversity among ethical and religious beliefs. Consequently, where once the religious practices of minority groups such as American Indians were suppressed, they are now protected by law.

WORLD-SYSTEMS THEORY AND ETHNIC RESISTANCE

Albert Bergesen (1995; chap. 10 of this volume) asserts that the resurgence of ethnic minority cultures addressed by postmodern thought and discourse can be best understood from the perspective of world-systems theory, which originated in the mid-1970s in part as a response to the criticisms of the dependency paradigm. Dependency theory identifies the causes of underdevelopment to be the exploitative and uneven nature of trade among nations and the continued flow of surplus from the periphery to the core areas. The world-systems approach focuses on long-term historical dynamics. An important contribution of this work has been to trace the origins of the capitalist world-economy to sixteenth-century Europe. The processes of commercialization of agriculture and industrialization eventually developed into a single system organized on the basis of a trilevel international division of labor: a developed core, an underdeveloped periphery, and a partially developed semiperiphery (Ward 1993).

The continued expansion of the capitalist world-system has involved the incorporation of regions or areas existing outside the system. Hall (1988, 24) conceptualizes incorporation as "movement along a continuum from initial contact to full absorption." The pattern of incorporation varies by setting: in the premodern systems access to trade and resources was as important as political factors for incorporation. In this instance, there was a certain degree of acceptance or tolerance of ethnic differences provided that the objectives of the state were being fulfilled. In contrast, in the modern capitalist world-system, much more emphasis has been placed on cultural assimilation. Ethnic or national unity has been considered an important aspect of the development of nation-states. Outcomes of this type of incorporation process are exemplified by the loss of cultural autonomy among the colonized native peoples of North America, and the loss and subsequent replacement of old identities as in the case of Africans brought as slaves (Hall 1998b; chap. 13). In the contemporary context, however, there have been significant transformations in the nature of ethnicity (1984, 1998a). "Ethnicity, and identity in general, has become increasingly politicized" (Hall 1998b; chap. 13).

Bergesen (1995) suggests that the world-systems perspective provides important tools for understanding the development of ethnic cultural renewal and attention to multiculturalism today. As the hegemonic power of the United States has declined over the last several decades, space has opened up for minority and other previously suppressed groups to reinitiate or assert their cultural and political interests. Thus, a return to particularism and ethnic concerns is consistent with a specific phase of the cyclical processes of world-systems changes. Friedman (1998, 249) continues, "The world of decline is a complex world, one that combines balkanization and globalization of cultural and social identities, in which the multicultural invades the center and the global and central state hierarchy disintegrates at the same time as new cosmopolitan elites identify with the larger world instead of with the hegemon itself." A central form of this cultural and political process is that indigenous populations reinstate their traditions and claim their indigenous rights. Demographic consequences can be significant; for example, through both natural increase and reidentification, the American Indian population more than doubled from 1970 to 1990 (Friedman 1998, 243; Snipp 1989, 1996).

Several scholars have noted the importance of these political and economic shifts for understanding recent social changes in reservation life. Hall (1986, 1987, 1988, 1989) suggests that status measures indicating reservation population characteristics represent an ethnic group's power relative to other groups and the success of their economic and cultural resistance or adaptations to specific historical circumstances. Similarly, Cornell's (1988) analysis of recent American Indian political resurgence emphasizes that changes in the political and economic arenas at the societal level have increased opportunities for American Indians to assert their political interests. Thus, resistance earlier in the century began to include political organization at multiple levels—tribal, regional, and national—as new demands and opportunities developed for political activism (Cornell 1988). Dunaway's (1996) work shows well how Cherokee tribal efforts to resist colonial cultural pressures as well as economic and political domination have evolved in relation to new needs and circumstances.

Nagel (1996) concludes that the cultural renewal seen in many American Indian communities today was fueled and given momentum by the Red Power movement, which occurred primarily in the 1960s and 1970s. Thus, indigenous efforts to resist domination—both political and cultural—have not only persisted but have expanded in recent years. However, as Hall points out (chap. 13; 1998b), these efforts are played out on Euro-American turf; for example, battles for control over resources have utilized the concept of sovereignty and the tenets of American law. To fight battles on this turf necessarily has meant the acceptance "of some of the premises of that turf" (Hall 1998b; chap. 13). Thus, the losses and disruption of native societies continue, but the battles today now often include ideological, cultural, religious, and educational issues as well as resources and land.

This discussion suggests the relevance of both cultural renewal and political resurgence to the lives of American Indians today. However, the experiences and roles of women in recent efforts to develop and utilize traditional forms of healing and, therefore, contribute to resistance must also be considered.

AMERICAN INDIAN WOMEN AND RECOVERY

Among those most likely to be represented at the new battle lines of resistance to incorporation are women. A number of scholars have documented over the last fifteen years that stereotypical ideas about the passive cultural roles of American Indian women are not only overstated but often erroneous (Albers 1983). Albers found that, among Lakota women, participation in the economic sector allows them to support and contribute to important social and religious ceremonies, such as give-aways, organized by family groups and others in the community. She states:

> Understanding the influence and control Sioux women exercise in the social networks organized for provisioning and ceremonial purposes helps to explain why women are now achieving greater recognition within a community-wide political realm. . . . Their influence in mobilizing and monitoring political support is inseparable from their contribution in ceremonial and interhouse provisioning. (Albers 1983, 216)

She attributes the increased influence of women, both socially and politically, not only to their stable and independent economic positions but also to the renewal of their traditional egalitarian relation to men. Within the context of a reemerging cultural system, the significance of women's economic roles in the family and community is once again gaining legitimacy.

Recent work by historians (e.g., Shoemaker 1995b) and anthropologists (e.g., Klein and Ackerman 1995b) also shows that American Indian women have often been the strongholds of traditional cultural resources and have provided for the survival of their people through their ability to make needed cultural adaptations. And yet, the role of indigenous women in resistance activities—both cultural and political—has often been overlooked. Ward (1993) suggests that resistance to incorporation by the capitalist world-system often includes greater participation in the informal economy and other adaptations in which women play dominant roles. However, world-systems theorists have devoted little effort to fully understanding the meaning and significance of women's activities, either in the local economies in which they live or in the larger system.

This case study is an effort to begin to understand more clearly the nature of several dimensions of the resistance activities of American Indian women who have drawn on their experiences with healing from substance abuse to create new lives for themselves, their families, and communities. The study uses data from several focus groups held in the summer of 1994 on the Northern Cheyenne reservation. These groups included women who either worked closely with or were from the Northern Cheyenne reservation, the Ft. Peck reservation, the Ft. Berthold reservation, the White Mountain Apache reservation, and the Navajo Nation. Thus, these groups represented experience and knowledge of an array of Plains Indian communities as well as southwestern tribal groups. The subject of the focus groups was the recovery experiences of American Indian women and its implications for their lives, families and communities.[1]

The Northern Cheyenne case will serve as a reference point for illustrating the circumstances faced by many reservation populations since the beginning of the reservation system. The following discussion will provide a brief history of the Northern Cheyenne reservation, particularly intergenerational changes in the experiences of women, and then will identify and explore the process and outcomes of women's recovery from substance abuse and addiction. The presentation of focus group data that follows demonstrates important dimensions of the women's experiences with recovery and illustrates the common themes.

THE NORTHERN CHEYENNE CASE HISTORY

The conquest and colonization experienced by American Indians exemplify significant aspects of the modernization process that oppressed numerous indigenous peoples. The Northern Cheyenne provide a specific case for examining the historical process that has produced many forms of resistance. These include a contemporary American Indian recovery community that offers the possibility for collective as well as individual healing and development. In this case study, the experiences of Cheyenne women with recovery will be described and discussed as they relate to both an individual process of healing and community-level process of cultural recovery that includes greater support for the practice of traditional medicine.

After a difficult struggle to maintain their traditional nomadic lifestyle, the Northern Cheyenne Indians were sent by the U.S. government to Oklahoma to join a related band, the Southern Cheyenne, in the 1870s (Weist 1977). Unaccustomed to the hot, humid climate and subjected to poor rations and provisions, the Northern Cheyenne experienced illness and death until, in desperation, a sizeable group left Indian Territory and started the long trek back to their traditional "home" near the northern Rockies. Once again pursued by the military, starving, and fragmented, this group of Cheyenne people fought against extinction. Finally, with their numbers depleted, the Northern Cheyenne won the right to select the site of their reservation in southeastern Montana; in 1884, the Tongue River Reservation was established. There the struggle went on as the Northern Cheyenne faced the challenges of supervision by government agents, developing new relations with their non-Indian neighbors, and coping with a world in which the old ways were no longer seen as relevant. The forced assimilation the Cheyennes experienced in the ensuing years included the denigration of and legislation against their language, customs, and religious practices (Weist 1977). Traditional subsistence activities were replaced eventually with farming and ranching, but many found these new activities unacceptable. The social, cultural, and personal changes and loss evoked many responses among the Cheyenne, including the use of alcohol.

The trajectories of Northern Cheyenne men's and women's lives, like those of other American Indians, were greatly affected by the dramatic shift from seminomadic life on the Plains to reservation life. While many believe that men's lives changed the most dramatically because hunting roles were replaced with agricultural production and wage work, and warrior societies and the Council of Forty-four were replaced by federal agents, in fact, women's lives and roles in the home and community were also transformed in important ways (Sanday 1981; Bonvillain 1989). As boarding schools were established and children were removed from their families and sent away for schooling, prominent roles in the home such as the socialization of children were replaced or subordinated to wage work or work alongside their husbands in the fields (Szasz 1974). Both men and women experienced the suppression of their traditional roles in religious ceremonies and activities (Weist 1977). Thus, the oldest genera-tion of women and men alive on the reservation today experienced the effects of the first wave of changes associated with reservation life that removed their parents and grandparents from their traditional roles in the family and community (Hinckley 1994). As these men and women had children, more changes were experienced: in the aftermath of the Depression and the world wars, federal relief programs created new opportunities to develop agriculture and other types of jobs, and new schools, services, and communities were established on the reservation. These led to improve-ments of the quality of life for many people, including greater opportunities for education, new work roles (especially for women), and more resources (Sawyer 1993).

With the development of programs supported by the War on Poverty and the Civil Rights activities in the 1960s and 1970s, additional resources flowed into reserva-tions along with new opportunities for active involvement in the solutions of reser-vation problems faced by the members of the second generation, the children of the early reservation residents discussed earlier. However, the sharp departures from the old ways, increased interactions with the larger society (especially by men in the wars and in political activities), and the pressures of continuing scarcity produced by uneven development led to continuing poverty and instability for a large segment of this generation.

Finally, for the third generation, young adults today, there have been still more changes, some positive and other negative. For example, while they have benefited from the expansion of schools, social services, and economic opportunities, this gen-eration has also experienced the problems associated with the previous generations' involvement with alcohol and drug abuse, a persistent element in the social context of many reservation communities (Sawyer 1993). Thus, despite a general increase in the opportunity structure and advancement in the areas of education and work, there is also increasing fragmentation in the lives of many Cheyennes who have faced family problems, obstacles to accessing schooling and employment, as well as diffi-cult social adjustments, particularly among Vietnam veterans. In addition, recent federal funding cutbacks have reversed earlier gains by decreasing both services and work opportunities in reservation organizations. Nevertheless, most young Cheyenne adults continue to have the same goals as their parents and grandparents to live and work on the reservation, support their families, and participate in reservation community life (Ward 1992; Ward et al. 1995).

WOMEN'S EXPERIENCES WITH RECOVERY

In many American Indian communities, the losses and suppression of cultural traditions have developed in relation to the assimilationist policies and practices that outlawed traditional religious activities and replaced education in the home with formal schooling. Adults in the oldest generation living on the reservation today were forced as children to learn English and were punished for speaking their native language, and they were separated from their families for long periods of time while they attended boarding schools. This separation from their families contributed to the loss of important community and cultural experiences as well as development of a sense of shame about being Indian. It also had important results for their children. Because they did not have the full range of early experiences with their parents that would socialize them for traditional parenting, this generation often faced problems as they raised their own children. Additionally, many did not have much knowledge of their cultural heritage to share with their children.

The ideas and views expressed in the following quotes reflect the experiences of American Indian women from the different reservation settings involved in the focus groups. They address the important themes of cultural loss, shame, and identity:

And it isn't even a matter of self-esteem. It's a confusion of identity. No sense of self. People don't know who they're supposed to be. . . . The culture aspect comes in. "Well, I'm a traditionalist," but yet their behavior doesn't fit it. So it's a confusion of identity. It's not even a sense of self-esteem. It's no sense of self. And that's why we have to work with the spirit of the person. How's the spirit been damaged?

. . . [Y]ou have to look at each generation and what was going on, and what their mothers brought into that, the whole perspective. And at the core of everything is shame, shame for who we are, shame for what we can do, what we think, how we look. At the core of all of that is the shame, and the grieving for the self. . . .

[S]ome women talked about the fact that they felt judged by people who practiced tradition because they didn't, and they didn't grow up with it. And the one thing I hear repeatedly around here is a lot of people didn't grow up with traditions, they don't know what tradition is, and so when they learn it they've got to . . . learn it from someplace else. . . .

And I see my uncles' and my aunts' and my grandparents' teaching being [lost], and it's really sad because there's so much that we can learn from them. And I've talked to all my younger cousins and they said that they felt that my uncle was a dying breed. And I said, "Well, who's going to carry on after he's gone? Who's he going to take it to the grave." It was like it didn't really matter.

The development of a particular type of drinking behavior, much of which was excessive rather than social drinking, developed in a setting in which cultural suppression and shame were prevalent. Influences on the development of this drinking pattern include early trading practices with trappers who often drank excessively and traded alcohol for hides, and later the need to drink quickly before the reservation agent caught Indian people with alcohol. Regardless of the causes that helped establish and perpetuate the excessive drinking pattern, social drinking is much less prevalent in many reservation communities. However, not everyone drinks irresponsibly.

> Most of the people that I know that have done drinking on the reservation and now are in recovery have been [drunk] where they were gone for months. They didn't know what they were drinking or how many or anything, so it's a real different experience for them to be around a social drinker . . . because there's no concept on the reservation of social drinking.
>
> It's such a dichotomy, an either/or thing, that there's completely down-and-out drinking, [or no drinking]. And I've heard more and more people, women talking about the drinking and the nondrinking crowd. And the drinking crowd could include those women who just go down on Friday night and drink, [but who] take care of their families. The AA people lock them in with the winos, the pitiful drinkers; they are no good drinking people. It's this real Puritanical no drinking . . . [but] if they live long, . . . they will stop drinking and be sober, or turn into responsible drinking people which, of course, doesn't fit the AA model at all. . . . And I think it's all tied back to the spirituality issues and finding yourself and working through the pain or whatever it might be.

For many people, the impetus to get sober or to recover from alcohol abuse or addiction is related to their need to take responsibility for their family obligations. This seems to be particularly true for women.

> For me, that's what got me sober. What created the need for me to get sober is I had twin daughters who were my youngest and I think they were about twelve when I got sober, and for two years I had physical custody but they were living with their father. . . . There were a lot of reasons for many years that I should quit drinking, but the part of that whole denial system, it just didn't puncture, didn't get through. But what did it was they were coming to visit me for a month, their month visit to me, and I forgot to pick them up at the bus station. . . . And for two days they were trying to find me to come pick my daughters up. And that got through to me. Something inside of me, and I also knew that there was no way that I could drink for the rest of that month . . . and so I called the hotline and went to AA.
>
> So we have this way of kind of thinking that we are taking care of our children; it's the delusion, it's the denial, it's all part of that. To me, I think it was. I really was surprised as I got sober, the more and more sober I got, and the more I looked back, the more I could see how abusive, neglectful; I don't know how many times I abandoned my children. That was just . . . I was overwhelmed with the shame in that. And the thought of redoing that was really what kept me sober for a long time.

For many women, AA was the source of help that made a difference in their being able to stop drinking. For example, on the Northern Cheyenne reservation, counselors at the Recovery Center report that the majority of the people they help to begin the recovery process are women. The treatment programs offered on many reservations involve both medical supervision of detoxification and individual and group counseling that is largely based on the twelve-step model associated with the AA program. Following treatment, recovering alcoholics are advised to continue to attend AA meetings to support their recovery. Women who have experienced recovery through the AA program frequently report that it helped to dramatically change the way they lived their lives.

> Recovery to me when I first got in was learning how to get sober and learning how to live sober. And that's what it was for probably two to three years. . . . I think initially it's a feeling of being encouraged to be myself for the first time in my life.
>
> Yeah, when I first started, it [recovery] meant getting sober. And then after I got sober, it seemed like there was still something missing. And then after I sobered up, I wanted to find out who I was and where I fit in this community. And so I started talking to people, mostly to my relatives. And then I started realizing there was a lot of things I didn't know about who I was. And then I started searching and talking to my elders. And then that's when I started learning about my culture and being raised by white people and not have that closeness that I felt with my real family. And then having to find out how I fit in this community and what I could do to help other people find what I had found. And then I think just mainly working at pulling the things that I felt comfortable with out of my culture and the white man culture, and putting them together to where I was comfortable and my family was comfortable.

After a period of sobriety, some women began to see recovery differently and to have different expectations of AA meetings and groups: not only to help maintain their sobriety but also to help them reconnect to their family and community. In fact, some women voice strong criticisms of AA.

> The feeling I came out with was I was supposed to begin acting like a recovery robot, . . . or little dolls, recovery dolls. . . .
>
> I believe that part of the reason that AA doesn't work is it's an oppressive program to cultures, to minorities. It puts us all in a box, labels us. And the treatment is all the same. And it robs me of my individuality. Everyone of us in this room is absolutely unique. We do share sameness, but we share differences. And if I really want to love you unconditionally, I have got to understand the differences so that I can do that. . . . I think it's wonderful initially to help people kind of get on their feet. The most loving thing to do is to help people find other avenues of expression, other avenues of treatment and healing, developing their sense of self.

These women also commented on the significance of spirituality or spiritual growth in relation to recovery. However, AA groups often did not address women's needs in this important dimension of their healing.

It's that [spirituality] that keeps me from wanting to drink. It isn't the meetings. The meetings make me want to drink! So that's the whole [thing] in a nutshell. It's a spiritual disease, and it's spiritual discovery, recovery. If you miss that, then you have to cling to those groups. They're your lifeline. And you have to surround yourself with people that talk like you do, think like you do, dress like you do. Because if you go away from that, you have nothing to tend to. It's an inside-outside. They will tell you that, but they won't help you find it. Because they haven't got it either. And that's why you hear people that are twenty years sober, and they go out and get drunk. They never did develop that inner sources of power, I think. I think that's the boat we miss because we just say, go to AA . . . go to Sexual Abuse Anonymous. That's not it.

I think the greatest damage that is done in the twelve-step groups is the stifling of the spirit.

I think when I began getting disenchanted . . . with the twelve-step recovery process was in my attempts to begin finding for myself, what does it mean, what is religion, what is spirituality, what is God? And getting the door closed on me, getting the "Well, just read your big book, do the fourth step, you're not working the program right; trust." And also another factor that was involved in all of this is for the first seven years of my sobriety, there was not one day in that seven years that I did not think of killing myself, not once. And nobody would hear it. . . . And it was like I cried for help; help me, nobody hears me, nobody's helping me. I didn't want to drink again, but what I was fighting was the need to stay on this planet. I need to stay here because I really don't want to go, but nothing's changing inside.

Today's spirituality is the essence of who I am. I do not align myself as being Christian . . . [but] I try to be Christ-like . . . And I am very religious in my spiritual practices and beliefs.

As these quotes show, spiritual growth and expression were central to many women's recovery and to their new life. When they could not adequately address these needs in AA groups, women typically have gone outside the AA program and found meaningful spiritual experiences within traditional ceremonies, the Native American Church, or their own programs of spirituality.

Well, definitely me practicing the Native American Church religion and sweat lodge ceremony made a tremendous impact on my recovery. Spiritually, I grew each time I went to a ceremony or sweat because each time . . . each ceremony made me look at . . . the reality of myself and my recovery. . . . And so I guess that what helped me to grow was the more I went, the more I realized about life and the importance of a drug and alcohol-free lifestyle which the Native American Church really pretty much emphasizes.

I went to recovery and I followed the twelve steps religiously . . . and [my friend] invited me to come help her at a sun dance. . . . the ceremony and being a part gave me connectedness with myself that I never had, with the earth, and this sounds so . . . with the universe I was part of. I had never felt a part of, never.

I see more of my recovery coming from learning from my elders, learning from my uncles, my grandparents, learning to live back off the land and for the land rather than to live against it.

Another part of recovery has often involved returning to school to obtain additional credentials needed for work. As women have begun the healing process, they often have had to face problematic situations at home, including the need to support their children as single parents or to make more of a contribution to their family's support. Many women have seen education as an important resource for increasing their access to jobs. Other women who already held jobs have sought new educational credentials that would enhance their contributions to their work as well as increase their earnings through promotions or new positions. In both cases, education and work have been as important to women as their sobriety for achieving the roles they wanted and expected to play in their homes, work, and community. In school, either at adult education programs or tribal colleges, women found the support of new friends, many also involved in the recovery process. At Northern Cheyenne, since women have been the majority of the tribal college graduates in recent years (Dull Knife Memorial College graduation programs, 1987–94), they have found support for both their recovery and their educational goals. Thus, the recovery process has been closely linked to educational experiences, especially for those women entering the counselor training or human services programs at the tribal college. However, recovery has been linked just as closely to the expanded roles women have increasingly played in the workforce over the last three generations. As women have entered the work place, gained experience, and moved into higher-level positions of local agencies, businesses, and schools, they have realized greater economic power and increased their participation in a variety of community activities. These experiences have lent support to the greater expectations that women have for the recovery process.

The developments involving women's expanded roles in work, community activities and recovery suggest that empowerment is an important dimension of women's experience.

> [The term] empowerment maybe has been used loosely in the last few years. But I see it as the key. That self and that sense of spirit that I talked about. Empowerment is what I needed. And empowerment is what was discouraged. I see that in men and women here on the reservation. Indian people are disempowered and have been. . . .
>
> I would like to see for women a sense of self-empowerment, the right to define what it is. And that's self-efficacy to me . . . it's the ability to feel that I can do it, and when you're so totally overwhelmed in grief, grave circumstances, that you have not found support that allows you—to help you—define what it is. . . . And that is what some of the traditional women have. They have more respect for your self-autonomy. I think of it as a sense of being capable of doing things that are not unrealistic. . . . In recovery you can say I'm in a tough situation, but this would challenge anybody. The reason I'm not doing well isn't because I'm incapable, it's just because it's a challenging situation, and anyone would have to struggle to get out of it. But it doesn't reflect on me; I'm capable.

As recovery, spiritual growth, and personal empowerment have brought American Indian women a greater sense of their cultural identity and efficacy in their lives, many have become more active in the cultural traditions and spiritual ceremonies of their communities. However, as women have begun to participate more, they sometimes have encountered opposition to their participation.

> [T]his Cheyenne man that I was talking about that had traditional parents said women just aren't supposed to sing. They [the women] can support the men, but they can't sing. So I took that and asked another man who I respect . . . , so I began questioning him and he said there are a lot of Northern Cheyenne men who are just male chauvinists. . . . He gave me some history and he said that the whole ceremony was brought by a woman. There are many tribes that the women do the whole ceremony. The women do the drumming, the women sing, they drum, the whole thing. And so I have to look at that and say, "We need to begin challenging some of this." We cannot continue taking this at face value. . . . [We need to say], "Wait a minute. I think you might be wrong." But there's so much fear behind a woman stepping out and doing something.

> . . . And I find that a lot of men still don't want the women to be a part of anything, even though they (I've heard this a lot from men) that all of our ceremonies come from women . . . they're the givers of life, you know. And to me, it just looks sort of [bad] because they come around and they continue to abuse the women. They [the women] are more used for cooking, like maids.

Women who have experienced such opposition to their participation in cultural activities interpret it in two ways. One is related to the personality or personal relationships between specific actors while the other is more sociological. In the latter interpretation, the women see Indian men's responses as representing the influence of American society and culture. Specifically, men have adopted the more negative views of women found in American society generally and have used these views to rationalize the exclusion of women. However, as the quotes reveal, Indian women are actively resisting this opposition.

DISCUSSION AND CONCLUSION

In this chapter we have presented qualitative evidence concerning elements of the process of recovery from substance abuse and addiction that have been identified by women of Northern Cheyenne and other Plains and southwestern reservation communities. The experiences described by the women in these focus groups suggest common themes and patterns of resistance across reservation settings. These themes begin with some of the reasons for excessive drinking which can be found in the ways that Indian people adapted to changes or coped with the difficulties associated with the last 150 years of their histories, including conquest, forced assimilation, suppression of important integrative cultural traditions and spiritual activities, and political economic oppression. While excessive drinking became the norm for increasing proportions of people across several generations, the recovery movement has grown as well, particularly over the last several decades. Interestingly, Indian women often have been among the first to begin the recovery process and typically comprise the majority of those recovering in many reservation communities today. The explanation for this pattern has been identified to include, at least in part, the responsibilities of women for care and socialization of children. At the end of the boarding school period, as women have resumed their central roles in family life, this responsibility has provided the impetus for many women to begin recovery. The need to work outside the home to help support their families has provided incentive for recovery as well.

Once they begin the process of achieving sobriety, many women have responded to the treatment and twelve-step support programs by engaging in a process of self-discovery and initiation of improvements in their way of life. For many women, this includes a return to school to obtain GEDs or college credentials that can mean not only better access to jobs or promotions but increased earnings, participation, and roles in the workplace. It also often includes a quest for new perspectives on women's identity and cultural roles, an important dimension of which is related to a discovery of spirituality, which may include traditional spiritual practices. This quest takes some women to older members of their own families for help while others go to tribal elders or healers, or to Native American Church leaders. The spiritual growth gained through this recovery and healing process also leads many women to want to clarify their positions and roles in the workplace and the reservation communities in which they live. Thus, through educational and work experience as well as recovery, many women expand their roles, improve their skills, and increase their voice within their personal lives, home, work, and community.

The new insights and experiences women have in recovery, educational, and work contexts often result in enhanced self-esteem and increased efficacy in the whole range of activities in which they participate. Increased efficacy, in turn, has prompted some women to seek a greater voice in the traditional cultural activities about which they are learning and in which they participate. Although they may draw on cultural traditions to support their desire for increased involvement in religious and community activities, many of these women encounter opposition from those (men) who have assumed greater control over these areas since the reservation period began. Women's responses to such opposition often include greater efforts to include men in the recovery and spiritual growth that they are pursuing.

Although there is variety in the experiences of the American Indian women with recovery and healing, spiritual growth, and the improvements they are making in their lives as well as the lives of their families and communities, the patterns identified here represent significant dimensions of a number of women's experiences. Importantly, this process of personal recovery represents a challenge to the dominant group's forms of treatment that can be understood from the perspective of world-systems researchers (e.g., Kathryn Ward) concerned with the roles of women. In this view, these women's experiences provide evidence of a significant new dimension of resistance in the area of culture: they are microlevel examples of how Indian women have used new opportunities created at the macrolevel (by the decline of American political and cultural hegemony) to expand American Indian resistance to political-economic incorporation, to include as well challenges to the cultural oppression that has characterized so much of their history. In other words, the increased attention to self-determination and greater acknowledgment of indigenous rights over the last several decades, as suggested by Hall, Bergeson, and Friedman, have created the possibilities for Indian women to move beyond the political and economic arenas to engage in cultural resistance.

As they have undergone personally trying and difficult experiences with recovery, Indian women have responded to the Western forms of treatment provided on the reservation with critiques and challenges. They sought answers to the most basic questions about their cultural identities and address the need to restore the self-esteem their oppressive circumstances have robbed from them. The coincidence of the end of boarding schools with new support for bilingual and multicultural education as well as new efforts to protect and create greater religious freedom and the use of native healing has given women currently participating in recovery programs support for their resistance. Thus, these women not only have challenged the loss of their culture but also have made their voices heard as they have found alternative paths of healing that, for many, have direct links to their traditional heritage. Women also have found support for their recovery in the new and expanded roles they have assumed in education, work, and community activities.

The new ways in which American Indian women are experiencing recovery and healing are not occurring in isolated situations. Evidence from the diverse group of women in these focus groups suggests instead that this new dimension of resistance is gaining momentum. The data indicate that as American Indian women engage in recovery and healing, they have new critical responses to the recovery process, new attention to their personal identity, cultural roles, and spiritual growth. Importantly, they have been empowered to pursue new forms of healing and to renew important and meaningful traditions as well as forge new cultural forms in their communities.

SUGGESTED READINGS

Sandefur et al. (1996) pull together information from fifteen contributors on current issues related to demography, health, social, and economic conditions of American Indians. Weist (1977) is an anthropological and historical account of the Northern Cheyenne Tribe of southeastern Montana up to the late 1970s. Moore (1996) provides a comprehensive examination of the Cheyenne people, culture, and society from their prehistoric origins to their present life on the reservations in Oklahoma and Montana. Shoemaker (1995a) covers the history of ten tribal groups throughout North America. The book contests the idea that European colonization led to a loss of Native American women's power and instead presents a more complex picture of the adaptation to and subversion of the economic changes introduced by Europeans. Klein and Ackerman (1995a) dispel the myths and correct the misunderstandings of the roles and power of Native women within their societies. The chapters deal with the history of tribes from ten culture areas since their colonization.

Bachman (1992) uses quantitative and qualitative research to present an overview of historical and contemporary perspectives on American Indians related to violence and its causes. The latter include the destructive effects of alcohol and other addictive substances. Kunitz and Levy (1994) present data from a longitudinal demographic and ethnographic study of drinking and recovery patterns of three groups of Navajos.

NOTE

1. Focus groups are small groups in which individuals who are knowledgeable about a research topic freely discuss the topic drawing on their personal experiences, views, and so forth. The group discussion is intended to include dialogue that helps illuminate multiple dimensions of the topic and to represent the experiences of the participants from their viewpoints. This type of group interview is part of an approach, often called "participatory research" (Park et al. 1996), which emphasizes the integrity of the participants in the research by including them as collaborators who help to shape the research questions, collect the data, interpret the data, and decide how the research results will be used. In the research process, researchers provide information and assistance to the participants regarding completion of the research tasks but also act as "learners," not experts, about the research topic. In this way, they help empower the research participants for full participation in and use of the research.

The impetus for this project came from research on the changing roles and experiences of Northern Cheyenne women across generations (Ward et al. 1995) as well as research concerning the experiences of women on the Ft. Peck reservation with survival of domestic violence (Baird-Olson 1993).

REFERENCES

Albers, Patricia. 1983. "Sioux Women in Transition: A Study of Their Changing Status in a Domestic and Capitalist Sector of Production." Pp. 175–236 in *The Hidden Half: Studies in Plains Indian Women,* edited by Patricia Albers and Beatrice Medicine. New York: University Press of America.

Baird-Olson, Karren. 1993. "Survival Roles of Plains Indian Reservation Women." *Family Perspective* 27:4:445–470.

Bachman, Ronet. 1992. *Death and Violence on the Reservation*. Westport, Conn.: Auburn House.

Bauman, Zigmaunt. 1992. *Intimations of Postmodernity*. London: Routledge.

———. 1993. *Postmodern Ethics*. Cambridge: Blackwell.

Beauvais, Fred, and Steve LaBoueff. 1985. "Drug and Alcohol Abuse Intervention in American Indian Communities." *International Journal of the Addictions* 20:1:139–171.

Bergesen, Albert. 1995. "Post-modernism: A World-System Explanation." *Protosoziologie* 7:54–59, 304–305.

Bonvillain, Nancy. 1989. "Gender Relations in Native North America." *American Indian Culture and Research Journal* 13:2:1–28.

Booth, Barbara Wolf. 1977. "In the Shallows." P. 83 in *I Am the Fire of Time,* edited by Jane B. Katz. New York: Dutton.

Brady, Maggie. 1992. "Ethnography and Understandings of Aboriginal Drinking." *Journal of Drug Issues* 22:3(Summer):699–712.

———. 1995. "Culture in Treatment, Culture as Treatment: A Critical Appraisal of Developments in Addictions Programs for Indigenous North Americans and Australians." *Social Science and Medicine* 41:11:1487–1498.

Cornell, Stephen. 1988. *The Return of the Native.* New York: Oxford University Press.

Cornell, Stephen, and Joseph P. Kalt. 1990. "Pathways from Poverty: Economic Development and Institution-building on American Indian Reservations." *American Indian Culture and Research Journal* 14:1:89–125.

———. 1992. "Reloading the Dice: Improving the Chances for Economic Development on American Indian Reservations." Pp. 1–60 in *What Can Tribes Do? Strategies and Institutions in American Indian Economic Development.* Los Angeles: University of California Press.

Donovan, James M. 1986. "An Etiologic Model of Alcoholism." *American Journal of Psychiatry* 143:1(Jan.):1–11.

Dunaway, Wilma A. 1996. "Incorporation as an Interactive Process: Cherokee Resistance to Expansion of the Capitalistic World-Systems, 1560–1763." *Sociological Inquiry* 66:4(Fall):455–470.

Friedman, Jonathan. 1994. *Culture Identity and Global Process.* Thousand Oaks, Calif.: Sage.

———. 1998. "Transnationalization, Socio-political Disorder and Ethnification as Expressions of Declining Global Hegemony." *International Political Science Review* 19:3(July):233–250.

Giddens, Anthony. 1991. *Modernity and Self-Identity.* Stanford, Calif.: Stanford University Press.

Guerrero, Marie Anna Jaimes. 1997. "Civil Rights versus Sovereignty: Native American Women in Life and Land Struggles." Pp. 101–121 in *Feminist Genealogies, Colonial Legacies, Democratic Futures,* edited by M. Jacqui Alexander and Chandra Talpade Mohanty. New York: Routledge.

Hall, Thomas. 1984. "Lessons of Long-term Social Change for Comparative and Historical Study of Ethnicity." *Current Perspectives on Social Theory* 5:121–144.

———. 1986. "Incorporation in the World-System: Toward a Critique." *American Sociological Review* 51:3(June):390–402.

———. 1987. "Native Americans and Incorporation: Patterns and Problems." *American Indian Culture and Research Journal* 11:2:1–30.

———. 1988. "Patterns of Native American Incorporation." Pp. 23–38 in *Public Policy Impacts on American Indian Economic Development,* edited by C. Matthew Snipp. Albuquerque: Native American Studies, University of New Mexico.

———. 1989. *Social Change in the Southwest, 1350–1880.* Lawrence: University Press of Kansas.

———. 1998a. "The Effects of Incorporation into World-Systems on Ethnic Processes: Lessons from the Ancient World for the Contemporary World." *International Political Science Review* 19:3(July):251–267.

——. 1998b. "Civilizational Incorporation of Indigenes: Toward a Comparative Perspective." *Comparative Civilizations Review* 39(Fall):10–37.

Hammerschlag, Carl. 1988. *The Dancing Healers.* San Francisco: Harper & Row.

Hill, Thomas W. 1990. "Peyotism and the Control of Heavy Drinking: The Nebraska Winnebago in the Early 1900s." *Human Organization* 49:3(Fall):255–265.

Hinckley, Gregory S. 1994. "Coming Full Circle: Changes in Northern Cheyenne Women's Sex Roles." Unpublished master's thesis, Sociology Department, Brigham Young University, Provo, Utah.

Holmes, Arthur, and G. McPeek. 1988. *The Grieving Indian: An Ojibwe Elder Shares His Discovery of Help and Hope.* Winnipeg: Intertribal Christian Communications.

Jorgensen, Joseph G. 1978. "A Century of Political Economic Effects on American Indian Society, 1880–1980." *Journal of Ethnic Studies* 6:Fall:1–82.

Kasl, Charlotte. 1992. *Many Roads, One Journey.* New York: Harper Perennial.

Klein, Laura F., and Lillian A. Ackerman, eds. 1995a. *Women and Power in Native North America.* Norman: University of Oklahoma Press.

——. 1995b. "Introduction." Pp. 3–16 in *Women and Power in Native North America,* edited by Laura Klein and Lillian Ackerman. Norman: University of Oklahoma Press.

Kunitz, Stephen J., and Jerrold E. Levy. 1994. *Drinking Careers: A Twenty-five Year Study of Three Navajo Populations.* New Haven, Conn.: Yale University Press.

Lurie, Nancy Oestreich. 1971. "The World's Oldest On-Going Protest Demonstration: North American Indian Drinking Patterns." *Pacific Historical Review* 40:3(Aug.):311–332.

Maracle, Brian. 1993. *Crazywater: Native Voices on Addiction and Recovery.* Toronto: Penguin.

Milam, James R., and Katherine Ketcham. 1983. *Under the Influence.* New York: Bantam.

Miller, Norman S., and Mark S. Gold. 1990. "The Disease and Adaptive Models of Addiction: A Re-evaluation." *Journal of Drug Issues* 20:1(Winter):29–35.

Moore, John H. 1996. *The Cheyenne.* Cambridge, Mass.: Blackwell.

Nagel, Joane. 1996. *American Indian Ethnic Renewal: Red Power and the Resurgence of Identity and Culture.* New York: Oxford University Press.

Nagel, Joane, and C. Matthew Snipp. 1993. "Ethnic Reorganization: American Indian Social, Economic, Political, and Cultural Strategies for Survival." *Ethnic and Racial Studies* 16:2(April):203–235.

Paniagua, Freddy. 1994. *Assessing and Treating Clinically Diverse Clients.* Thousand Oaks, Calif.: Sage.

Park, Peter, Mary Brydon-Miller, Budd Hall, and Ted Jackson. 1996. *Voices of Change: Participatory Research in the United States and Canada.* Westport, Conn.: Bergin & Garvey.

Quintero, Gilbert. 1992. "Walking the Good Road: Alcoholism and Recovery in the Native American Church." Unpublished master's thesis, Department of Anthropology, Northern Arizona University, Flagstaff.

Sanday, Peggy. 1981. *Male Dominance and Female Autonomy.* New York: Cambridge University Press.

Sandefur, Gary D., Ronald R. Rindfuss, and Barney Cohen, eds. 1996. *Changing Numbers, Changing Needs: American Indian Demography and Public Health.* Washington, D.C.: National Academy Press.

Sandefur, Gary D., and Wilbur Scott. 1983. "Minority Group Status and the Wages of Indian and Black Males." *Social Science Research* 12:1(March):44–68.

Sandefur, Gary D., and Marta Tienda, eds. 1988. "Introduction: Social Policy and the Minority Experience." Pp. 1–22 in *Divided Opportunities: Minorities, Poverty and Social Policy.* New York: Plenum.

Sawyer, Kae. 1993. "Women's Choices about Work and Family: Three Generations of Northern Cheyenne Women." Unpublished master's thesis, Sociology Department, Brigham Young University, Provo, Utah.

Segal, Boris M. 1987. "Drinking Motivation and the Cause of Alcoholism: An Overview of the Problem and a Multidisciplinary Model." *Alcohol and Alcoholism* 22:3:301–311.

Shoemaker, Nancy, ed. 1995a. *Negotiators of Change: Historical Perspectives on Native American Women.* New York: Routledge.

———. 1995b. "Introduction." Pp. 1–25 in *Negotiators of Change: Historical Perspectives on Native American Women,* edited by Nancy Shoemaker. New York: Routledge.

Slagle, A. Logan, and Joan Weibel-Orlando. 1986. "The Indian Shaker Church and Alcoholics Anonymous: Revitalistic Cults." *Human Organization* 54:4(Winter):310–319.

Smart, Barry. 1993. *Postmodernity.* New York: Routledge.

Snipp, C. Matthew. 1986. "The Changing Political and Economic Status of the American Indians: From Captive Nations to Internal Colonies." *American Journal of Economics and Sociology* 45:2(April):145–157.

———. 1988. *Public Policy Impacts on American Indian Economic Development.* Albuquerque: Native American Studies, University of New Mexico.

———. 1989. *American Indians: The First of This Land.* New York: Russell Sage Foundation.

———. 1992. "Sociological Perspectives on American Indians." *Annual Review of Sociology* 18:351–371.

———. 1996. "The Size and Distribution of the American Indian Population: Fertility, Mortality, Residence and Migration." Pp. 17–52 in *Changing Numbers, Changing Needs: American Indian Demography and Public Health,* edited by Gary D. Sandefur, Ronald R. Rindfuss, and Barney Cohen. Washington, D.C.: National Academy Press.

Snipp, C. Matthew, and Gary D. Sandefur. 1988. "Earnings of American Indians and Alaskan Natives: The Effects of Residence and Migration." *Social Forces* 66:4(June):994–1008.

Sorkin, Alan. 1996. "The Changing Health Status of Native Americans Residing on or near Reservations." Pp. 97–130 in *Human Capital and Development,* edited by Carol Ward and C. Matthew Snipp. Greenwich, Conn.: JAI.

Szasz, Margaret C. 1974. *Education and the American Indian.* Albuquerque: University of New Mexico Press.

Talbot, Steve. 1981. *Roots of Oppression: The American Indian Question.* New York: International Publishers.

Tater, Ralph E., and K. Edwards. 1988. "Psychological Factors Associated with the Risk for Alcoholism." *Alcoholism Clinical and Experimental Research* 12:4(Aug.):471–480.

Vinje, David. 1996. "Economic Development and Indigenous Peoples of the United States, Canada, and Australia: A Comparative Analysis." Pp. 69–98 in *Human Capital and Development,* edited by Carol Ward and C. Matthew Snipp. Greenwich, Conn.: JAI.

Ward, Carol. 1992. "The Social and Cultural Influences on the Schooling of Northern Cheyenne Youth." Unpublished Ph.D. dissertation, Sociology Department, University of Chicago.

Ward, Carol, G. Hinckley, and K. Sawyer. 1995. "The Intersection of Ethnic and Gender Identities: Northern Cheyenne Women's Roles in Cultural Recovery." Pp. 201–227 in *American Families: Issues in Race and Ethnicity,* edited by Cardell K. Jacobson. New York: Garland.

Ward, Carol, and C. Matthew Snipp. 1996. "An Introduction to American Indian Human Capital and Development." Pp. 1–16 in *Human Capital and Development,* edited by Carol Ward and C. Matthew Snipp. Greenwich, Conn.: JAI.

Ward, Kathryn B. 1993. "Reconceptualizing World-System Theory to Include Women." Pp. 43–68 in *Theory on Gender/Feminism on Theory,* edited by Paula England. New York: Aldine de Gruyter.

Weibel-Orlando, Joan. 1989. "Hooked on Healing: Anthropologists, Alcohol and Intervention." *Human Organization* 48:2(Summer):148–155.

Weist, Thomas. 1977. *A History of the Northern Cheyenne People.* Billings: Montana Council for Indian Education.

13

Frontiers, Ethnogenesis, and World-Systems: Rethinking the Theories

Thomas D. Hall

As the European world-system expanded, it absorbed new areas and peoples. This expansion created and transformed frontiers between itself and the external world. The incorporation of new areas and peoples has not been limited to the expanding European system. Writers who have pushed world-systems analysis into the distant past all agree that the tendency to expand and incorporate new areas and peoples is common to *all* world-systems (Abu-Lughod 1989, 1993; Chase-Dunn and Hall 1991, 1997; Frank and Gills 1993; Peregrine and Feinman 1996). This is why world-systemic processes have shaped the formation, transformation, fossilization, and obliteration of frontiers since ancient times. These processes remain relatively understudied and undertheorized. This chapter is one attempt to remedy this lack.

Our understanding of past patterns has been hampered by relative concentration on core-generated processes and neglect of frontiers. The sampling of our explanatory universe has been biased by this lack of attention to processes and events that have occurred on far peripheries in both recent and ancient times. If we seek to understand the evolutionary dynamics of the system itself, and to understand what is truly new at the turn of the second millennium, we must study world-system structure and dynamics in all places and all times. That is, far peripheries are as relevant as the core. This is not a matter of political correctness or humane inclusiveness—though they are important—but a matter of valid theory building.

I have argued many times (especially 1989a, 1989b) that some processes can be best, or often only, studied on the far peripheries—because that is where they occur most often and where they are most visible. Peter Sahlins begins his masterful study of the French-Spanish "fossilized" boundary with an epigraph from Pierre Vilar which is singularly appropriate here: "The history of the world is best observed from the frontier" (1989, xv). Hence the study of frontiers is indispensable to understanding the modern world.

My goal in this chapter is to begin to develop a world-systems analysis of frontiers. I summarize much of the work done from this perspective, illustrate the points with various examples, and suggest many issues and topics for further discussion. I seek to convince readers that a comparative world-systems perspective on the formation, transformation, fossilization, and obliteration of frontiers is vital to our understanding of them. Furthermore, I hope to persuade readers that it is interesting, exciting, and "doable." This approach cannot explain everything. A comparative, historical world-systems perspective is a *necessary* but *not sufficient* requirement to understand the dynamics of frontiers.

I begin with a thumbnail sketch of world-systems analysis and its extension into the ancient world. I elaborate the analysis of world-system incorporation, focusing on frontier dynamics, illustrated with various examples. I conclude with a discussion of theoretical and empirical problems that need further study.

COMPARATIVE WORLD-SYSTEMS ANALYSIS

A world-system is an intersocietal system that has a self-contained division of labor with some degree of internal coherence. It is a "world," but not necessarily global. It is a key unit of analysis within which all other social structures and processes should be analyzed (see chaps. 1 and 2). Thus, a world-system must be studied as a whole. The study of social, political, economic, or cultural change in any component of the system must begin by understanding the role of that component role within the system, whether it be a nation, state, region, ethnic group, class, gender role, household, or nonstate society. Conversely, changes in any of its components affect the entire system (Bach 1980).

Three components compose a world-system:

- a *core* that employs advanced industrial production and distribution and has strong states, a strong bourgeoisie, and a large working class;
- a *periphery* that specializes in raw materials production and has weak states, a small bourgeoisie, and many peasants; and
- a *semiperiphery* that is intermediate between core and periphery, in its economic, social, and political roles and its own internal social structure.

Core capitalists use various sorts of coercion to promote unequal exchange and accumulate capital,[1] which leads simultaneously to core development and peripheral impoverishment, or underdevelopment.

By the 1980s archaeologists found that world-systems analysis, while suggestive, could not be used without major modifications (see chap. 3; Hall and Chase-Dunn 1993). This prompted Christopher Chase-Dunn and me (1991, 1997, 1998) to extend world-systems analysis by transforming many of its assumptions into empirical questions. Several changes are germane to the discussion of frontiers:

1. World-systems, or core/periphery structures, date back at least to the Neolithic revolution (approximately ten thousand to twelve thousand years ago).
2. Core/periphery structures are a major locus of social change.
3. Not all change can be explained from the world-system level, but system processes are a crucial part of all social change.
4. World-systems, themselves, have evolved.
5. World-systems have several types of dynamic cycles.
6. World-systems have multiple, seldom coinciding, boundaries and bounding mechanisms that change through time.

The last two points are most relevant to the study of frontiers. Various dynamic cycles drive the rates of expansion, and occasionally contraction:

a. All world-systems pulsate—that is, expand and contract, or expand rapidly, then more slowly.
b. All state-based world-systems (i.e., since about 3,000 B.C.E.) have cycles of rise and fall of core states. A typical, but not universal process is the displacement or conquest of the dominant core state by a semiperipheral marcher state (Chase-Dunn and Hall 1997, chap. 5).
c. In the capitalist world-system, this becomes the hegemonic cycle wherein one core power displaces others in succession: the Dutch, followed by the British, followed by United States.
d. State-based systems also seem to oscillate between public and private dominant forms of capital accumulation (Arrighi 1994).

Precapitalist tributary systems range from very private forms of accumulation, feudal systems, to very centralized forms, a centralized empire. The modern capitalist system ranges from accumulation sponsored or fostered by states to accumulation concentrated in private holdings. We are currently in the more private phase of this cycle.

World-systems typically have four non-coterminous boundaries and bounding mechanisms (Chase-Dunn and Hall 1997, chap. 3). From the narrowest to the widest they are (see fig. 13.1) as follows:

1. the bulk goods exchange network (BGN) is narrowest;
2. somewhat larger is the network of political/military interactions (PMN);
3. still larger is the network of prestige or luxury goods exchange (PGN);
4. of comparable size, but not coincident, is an information exchange network (IN).

The ways in which these networks are nested and overlap are a matter of continuing empirical and theoretical research. These four boundaries only coincide on isolated island systems or in the late twentieth century. The recent convergence of these four boundaries at the limits of planet Earth is what many people call "globalization."

Figure 13.1   Spacial Boundaries of World-Systems

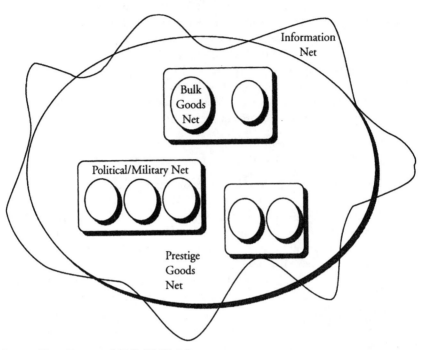

Source: Chase-Dunn and Hall (1997, 54).

The steady growth in size of all four networks, and their gradual coincidence, has been going on since the formation of the earliest world-systems. Recently, however, the process has become both planetary and rapid, and hence visible to even the most casual observer (see Chase-Dunn and Hall 1997 for a fuller account).

GROUP FORMATION, TRANSFORMATION, AND INCORPORATION INTO A WORLD-SYSTEM

When new areas are absorbed into an expanding world-system, frontiers are formed and transformed (Wallerstein 1974; Hall 1996b; Hopkins et al. 1987; Markoff 1994). Thinking of a frontier as a membrane is helpful (Slatta 1990, 1997, 1998). From a global perspective, a frontier is relatively narrow and sharp, but from nearby it is a broad zone with considerable internal spatial and temporal differentiation. Its permeability varies with the direction of flow and the things moving through it—types of goods, groups, and individuals. A frontier is, as Sahlins (1989) argues, the result of an often long, complex, and highly politicized process of negotiation.

A frontier, in the history of the western United States, typically refers to areas with population densities less than two persons per square mile. Guy and Sheridan (1998b, 10) define frontiers "as zones of historical interaction where, in the brutally direct phrase of Baretta and Markoff (1978:590), 'no one has an enduring monopoly on violence.'" Guy and Sheridan continue, "Frontiers were, in a most basic sense, contested ground." My definition is "a region or zone where two or more distinct cultures, societies, ethnic groups, or modes of production come into contact" (Hall 1997, 208). Furthermore, I note here that contact often leads to conflict. Thus, borders are the political markings around which frontiers form. Following Sahlins (1989), a border marks a politically negotiated limit of state sovereignty. Typically it is surrounded by a zone of interaction that persists long after the specific boundary line has been drawn.

Incorporated areas and peoples often experience profound, and occasionally devastating, effects from incorporation even when incorporation is relatively weak. They also resist and react against these effects to the degree possible. Such resistance, in turn, often shapes policies and actions of the incorporating system. These interactions also reshape the ethnic landscape. Frontiers are zones where ethnogenesis, ethnocide, culturicide, and genocide are common.[2] Thus, the study of incorporation entails close attention to local conditions, actors, and actions. To explore this more fully it is useful to summarize the analysis of world-system incorporation.

INCORPORATION IN WORLD-SYSTEMS ANALYSIS

Because my studies of the region that became the southwestern United States demonstrated that the effects of incorporation have important, interactive effects long before colonization is complete, I extended Wallerstein's dichotomous concept of incorporation to a continuum that ranges from weak to strong (Hall 1986, 1987, 1989a). I also argued that changes in *the degree of* incorporation affect those incorporated, and conversely their reactions shape the incorporation process.[3] At the weakest extreme of incorporation are areas external to a world-system. With even slight contact, *external arenas* become *contact peripheries*. With somewhat stronger, partial incorporation these become *marginal peripheries*. Marginal peripheries are analogous to what Beltran (1979) calls a "region of refuge." Marginal incorporation can freeze social change within such areas. This "preserves" the area in two ways. First, it sets it aside for future development by the state. Second, it "preserves" older, "traditional" social forms and hence is a "refuge" for social forms that have been destroyed elsewhere in the system. At the strongest extreme of incorporation such areas become *full-blown* or *dependent peripheries*. Most studies of the modern world-system focus primarily only on dependent peripheries. Calling this entire range "incorporation" hides variations and makes the study of change in degree of incorporation more difficult to understand (see fig. 13.2).

Figure 13.2   Continuum of Incorporation

Strength of Incorporation	None	Weak	Moderate	Strong

Impact of Core on Periphery	None	Strong	Stronger	Strongest
Impact of Periphery on Core	None	Low	Moderate	Strong

Names for Types of Peripheries

Hall and Chase-Dunn and Hall	External Arena	Contact	Marginal or Region of Refuge	Full-Blown/ Dependent
Wallerstein	None	External Arena	Incorporation	Peripheralization
Arrighi	None	N o m i n a l o r F o r m a l		Effective or Real
Sherratt		M a r g i n		Periphery or Structural Interdependence
Frank and Gills		H i n t e r l a n d		Periphery

Source: Chase-Dunn and Hall (1997, 63).

Some important aspects of incorporation are the amount of goods exchanged, the type of goods (e.g., the amount of labor involved in their production and whether they are raw products, manufactured goods, or prestige goods), degree of centralization of the exchange process, and relative importance of the transfer to each economy. At the weak pole of the continuum the primary influence is from core areas to peripheral areas.

For instance, furs from northern North America were not vital to European economies, yet the trade produced major social and economic changes among indigenous groups (Wolf 1982, chap. 6; Kardulias 1990; Abler 1992). Difference in size of economies is important. Often trade with one peripheral region has no major impact on the core, whereas trade with all peripheral areas combined may have significant impact.

The fur trade brought several dramatic changes to indigenous social organization. Most dramatic was a dependency on European goods, as metal pots and cutting implements replaced more breakable pots, baskets, and stone tools. As the desire for

European goods turned into a need, more furs were required to acquire European goods. This led to overharvesting of game and a chain reaction of expansion in search of new hunting grounds, which in turn heightened warfare and depleted the environment.

Heightened competition for hunting grounds transformed warfare from line fighting to what we would today call guerrilla fighting, and it transformed guns from luxuries to necessities. The intensified harvesting of furs led families to scatter in the winter to pursue fur bearing animals (Abler 1992; Kardulias 1990; Meyer 1994). This left women in charge of home camps while men were absent for long periods. Intensified conflict encouraged stronger fortifications. Finally, increased travel, contact, and migration helped spread diseases, often long before direct contact with Europeans. The impacts of disease often were devastating to the maintenance of local social structures (Reff 1991; Thornton 1987; Thornton et al. 1991).[4]

Wallerstein (1989, chap. 3) argues that plunder by a core power is not necessarily incorporation unless local production processes have become integrally linked into the commodity chains of the larger world-system. Rather, plundering typically has profound consequences for local groups. Thus, I include regularized plunder as a form of incorporation. This is not to deny that peripheralized production is more stable since it typically is not as disruptive to a peripheral region as plundering, nor does it exhaust resources as rapidly. Furthermore, regularized plundering often is a first step toward stronger incorporation and subsequent control of local productive processes.

Under this conceptualization, West Africa was incorporated into the European world-system when slave raiding became regularized, rather than when colonial agriculture began. And, if trade in bullion is the criterion for incorporation, then West Africa has been linked to the Mediterranean and Near Eastern core regions since at least the ninth century C.E. when coins minted in Byzantium and Egypt began using West African gold (Curtin 1984, 32). These coins played an important part in the Eurasian world-system (Abu-Lughod 1989; Moseley 1992; and Willard 1993).

Similarly, Peru, Mexico, and what is now the southeastern United States became incorporated—albeit weakly—into the European world-system with first contact by Spanish explorers.

Another important aspect of incorporation is its reversibility. Changes in the world-system, or in the specific incorporating state, or in local resistance to incorporation, or any combination of these often change the degree of incorporation. Although weakening of degree of incorporation is not rare historically, the tendency is for regions and peoples to become more strongly incorporated as world-systems expand. Strengthening incorporation typically is easier than weakening incorporation. Thus, incorporation has a grain or direction to it like wood or corduroy.

This is why when degree of incorporation weakens for a region, it does not return to the status quo ante automatically, contrary to early claims by Arrighi (1979). Indeed, a weakened degree of incorporation can produce nearly opposite effects in different settings. On the one hand, capturing slaves from a region typically undermines local prosperity. If weakened incorporation decreases slave raiding, local prosperity may recover. On the other hand, if the core supplies goods that have no local substitutes, prosperity that depended on those goods would decrease if the flow of goods declined along with weakened incorporation. The loss of trade connections to Mesoamerica undermined the entire Chaco Canyon economy in the twelfth and thirteenth centuries in what is now northwestern New Mexico (Mathien and McGuire 1986). According to So (1984), weakening incorporation of Canton in the 1840s gave rise to unemployment, peasant uprisings, and rebellions.

The volatility of local changes is, if anything, greater in weakly incorporated areas than in more strongly incorporated areas. Thus, weakly incorporated areas tend to vary more widely through time and hence are vital objects of study to discern the detailed mechanisms of incorporation-induced social changes (Hall 1989b). This is why frontiers *seem* similar but on closer examination always show considerable differences. (For detailed examples, see Guy and Sheridan 1998a; Hall 1989a, 1996a, 1998a, 1998c; Slatta 1997; Weber and Rausch 1994.)

Finally, these variations in the incorporation process are a major reason that frontiers tend to shift location and change through time. A specific frontier often follows a complex path, or trajectory, of incorporation with successive strengthening and weakening phases. In fact, I argue that entire trajectories of incorporation, not just its degree at any one point in time, must be studied comparatively.

INCORPORATION AND ETHNOGENESIS, TRANSFORMATION, AND ETHNOCIDE

Incorporation also has major effects on individuals and groups. Many American Indian groups we know today were socially constructed from an aboriginal base of loosely connected bands during the process of incorporation—for example, the Diné (Navajo) (Hall 1989a). While language, customs, and a vague sense of being the same "people" predate the arrival of Europeans, Diné-wide institutions such as the Navajo Tribal Council were developed much later, among other things, to deal with incorporation into American society.

These effects are widespread. Ferguson and Whitehead (1992a, 1992b) have collected studies of the "tribal zone," a region along state borders where contact with nonstate groups is common. They purposely use the vague term *tribe* to refer to a range of nonstate societies from the simplest band groups to sprawling chiefdoms. They are well aware of the vagueness and misuses of the term *tribe* (Ferguson 1997, 475–476), which are discussed later in this chapter. They argue that four fallacies underpin the common Hobbesian image of "tribal" peoples:

- postcontact conditions and relations are a continuation of precontact conditions and relations;
- ethnic divisions are survivals of precontact divisions;
- "tribal" warfare is unreasoned hostility;

• ethnographers, ethnohistorians, and historians typically have an adequate understanding of the relevant contexts of contact.

Studies of many regions show that both state-nonstate and internonstate warfare increased substantially after state contact. Contact and even weak incorporation *increase* regional violence. Ferguson and Whitehead do *not* claim that all was idyllic prior to initial incorporation. Only that state contact intensified violence and often transformed it into more virulent forms. Incorporation can either amalgamate or fragment previously existing groups. The transformation from an autonomous nonstate society to a subordinate ethnic group is a complex process in which identities, cultures, and social organizations are transformed. Thus, the Navajo example cited earlier is typical, not unusual.

Ferguson and Whitehead argue that "tribal" warfare is not "unreasoned hostility" but typically conflict over access to state supplied goods and the uses of local resources controlled or produced by the contacting state. These goods and resources may be used for very different political ends by leaders in nonstate societies, often to enhance prestige and to gain followers. Thus, local leaders not only need access to trade goods and local resources, but also must block the access of their rivals. These needs are a source of inter-, and sometimes intra-, group warfare.

Because leaders in kin-ordered prestige hierarchies used goods differently, state officials often did not understand this origin of conflict, and called it "unreasoned hostility." Such conclusions were all too easy for them to draw, given the ethnocentric assumption by state officials that nonstate peoples are "barbarians," "backward," and "inferior." Interestingly enough, officials who actually understood the customs of nonstate societies were not rare. Typically these officials had been born in the contact zone or they had spent a long time there. Equally typically, central state officials did not listen to their advice. (For detailed examples, see Barfield 1989; Hall 1989a.)

Reconstruction of precontact social conditions is frequently very difficult because all these effects occur nearly simultaneously with contact. That a representative of a literate society is present to record events and conditions means that change already may be extensive. To avoid such ethnographic upstreaming (reading the present into the past; Sheridan 1992), critical and cautious combinations of archaeological data with reports of the earliest observers must serve to temper these distortions. Anderson's (1994) reconstruction of Savannah River chiefdoms is an excellent example of such work. He combines archaeological data with careful, critical reading of the reports of the De Soto expedition. He shows how conditions can change very rapidly. Sometimes, however, there is little change and aboriginal conditions and organization persist. The point is neither change nor persistence should be assumed but should be determined empirically. Chase-Dunn and I (1998) have sketched some of the precontact world-systems in North America, and Chase-Dunn and Mann (1998) provide an exemplary analysis of the Wintu in what is now northern California on the eve of European contact.

Incorporation can also have divisive effects. For the White Earth Anishinaabeg (Chippewa or Ojibwa), increasing incorporation fractured old clan and band distinctions and created a new division between more, and less, assimilated Anishinaabeg, or in local parlance, between full- and mixed-bloods (Meyer 1994). Sandra Faiman-Siva (1997) finds much the same processes among the Mississippi Choctaw. Indeed, the full-blood/mixed-blood distinction is itself an important consequence of incorporation into the European world-system with far-reaching legal consequences.

Gender roles and gender relations are also reshaped by incorporation. Women are often harmed by incorporation even while men may benefit. There are gender and class differentials in contraception (Bradley 1997), fertility (Ward 1984), labor force participation (Ward 1990; Ward and Pyle 1995; Bose and Acosta-Belen 1995), and household structure and function (Smith et al. 1988). The key process here seems to be that new resources are differentially accessible by gender, giving increased power to men, and decreasing social power and changing the social roles of women. Both Dunaway (chap. 11 in this volume; 1996a, 1997) and Faiman-Siva (1997) find this holds for Cherokee and Choctaw. It remains unclear, however, whether the direction of this differential is a direct result of incorporation into a capitalist system, or an indirect result of the strong patriarchy of the British, Spanish, and French states. Mies (1986) and Shiva (1993) address the general issue of patriarchy and capitalism in considerable detail. Still, as Ward (1993) argues, the effect of incorporation on gender differentials and gender roles is an area in need of further empirical and theoretical development. Chapters 11 and 12 by Wilma Dunaway and Carol Ward et al. are important contributions to such development.

These examples make a compelling case for why special care is needed in dealing with the labels we apply to groups, especially nonstate groups that have experienced some degree of incorporation. Sometimes a shared identity of a group is the product of incorporation. Conversely, unified groups can become divided internally, or even shattered by incorporation. These examples also make a compelling case for viewing frontiers as regions where ethnogenesis, ethnic transformation, and transmutation, culturicide, ethnocide, and genocide are common (Chappell 1993a, 1993b; Felon 1997, 1998). This is one reason that they are so important to study: this is where these processes occur with the greatest variety.

INCORPORATION IN PRECAPITALIST CONTEXTS

The extension of world-systems analysis into precapitalist settings suggests refinements to the analysis of incorporation:

- First, incorporation is not unidimensional but multidimensional along the four types of world-system boundaries. Incorporation can be economic (for either bulk goods or luxury goods), political/military, or cultural, which includes all types of information and symbols.

- Second, incorporation creates multiple frontiers, corresponding to each of the boundaries.
- Third, *ceteris paribus,* incorporation will begin at the furthest boundaries, cultural, symbolic, informational, or luxury goods and proceed to narrower, more intense forms along the political/military boundary and finally along the bulk goods dimension.
- Fourth, relations among the dimensions of incorporation and the resulting frontiers is complex theoretically and empirically.

Sherratt's (1993a, 1993b, 1993c, 1993d, 1995) studies of the interactions between the Near Eastern and Mediterranean urbanized core region and peripheral regions in Europe during the Bronze Age support this analysis. He argues that beyond what he calls the periphery is a *margin* whose characteristics include "that it is dominated by time-lag phenomena—'escapes'—rather than structural interdependence with the core" (1993a, 43). According to Sherratt, goods followed long chains from the Near Eastern core through the periphery and then into northern Europe. He notes that the nodal points of connection to Europe in the Bronze Age eventually became important centers in their own right (shown in his fig. 12, 1993, 44). He uses these distinctions to divide the Bronze Age world-system into zones of core, periphery, and margin (his fig. 13, 1993, 44 and fig. 14, 1993, 45).

Sherratt's "margin" approximates the range of incorporation from contact through marginal peripheries. The alternative label for this range, regions of refuge, highlights that his "escapes" are regions where older social forms may be preserved. In short, they "escape" pressures for change exerted in more strongly incorporated areas. His "structural interdependence" and "periphery" correspond to full-blown peripheral and semiperipheral regions. Similarly, the concept of "hinterland" used by Frank and Gills (1993), Gills and Frank (1991), and Collins (1978, 1981) also corresponds roughly to the weaker range of incorporation (see fig. 13.2).

FURTHER ANALYSIS OF FRONTIERS

I begin with a return to the fur trade. In the southeast Cherokees became extensively involved in what Dunaway (1994, 1996a, 1996b) calls "a putting-out system financed by foreign merchant-entrepreneurs" (1994, 237). Incorporation via the fur trade transformed both techniques of production and associated culture:

> As export production was entrenched, there emerged new hunting and warfare techniques, an altered division of labor within the household and within the village, and a reformed relationship with precolonial society toward secular and national governance, eventually creating the "tribal half-government" that permitted the Europeans to treat the Cherokees as a unified corporate entity. (Dunaway 1994, 237)

The export of about one hundred thousand deer hides transformed a luxury trade into a bulk goods trade and constituted full-blown or dependent incorporation. This trade played an important role in European economies and vastly disrupted local relations, changing gender relations and roles, local political organization, and cultural values. Increased incorporation promoted changes along all world-system boundary criteria: information/culture, luxury trade, political/military interaction, and bulk goods trade.

The spread of horses from colonies in New Mexico to various foraging groups illustrates a much weaker form of incorporation. Horses dramatically transformed production techniques, political and social organization, and the cultures of all groups who acquired them (Secoy 1953; Mishkin 1940). Horses differ from most other European goods because they can reproduce without human intervention. Horses transformed the buffalo hunt from a rare and dangerous task into a far more successful enterprise. It allowed both concentration and dispersal of populations and transformed erstwhile sedentary (or sometimes semisedentary) horticulturalists into full-time nomadic hunters. Groups who continued occasional gardening could now congregate in larger base camps because they could hunt over larger territories. Conversely, groups who were, or became, entirely nomadic could use a greater range of territory (Secoy 1953). The access to a high-quality protein[5] source allowed population efflorescence and attracted migrants to the plains, especially those from eastern forested areas who were losing territory to better armed fur gatherers. Competition and warfare increased. West (1998, 33–93) summarizes the literature on this quite cogently, providing a detailed account of interactions with immigrating Americans in the nineteenth century.

As horses were spreading from northwestern New Spain toward the north and east, guns were spreading from the northeast, primarily through French fur traders, toward the southwest. These two frontiers crossed in the mid- to late eighteenth century, giving the groups in the cross-over region nearly equal access to guns and horses. This coincidence gave rise to the celebrated Plains cultures and gave Comanche bands the tools to dominate the South Plains (Secoy 1953; Hall 1989a). Indeed, Comanches actually sold guns to Spanish peasants in New Mexico where the government tried to maintain a monopoly on weapons.

Elsewhere in Latin America the spread of feral horses and cattle also vastly disrupted indigenous world-systems (Baretta and Markoff 1978), and in the Pampas region of southern South America it gave rise to a whole new type of society: the gauchos (Slatta 1983, 1997). Access to horses produced some of the most dramatic social transformations that came from the mildest degree of incorporation into the European world-economy.

Both examples illustrate why incorporation can be so hard to reverse. Short of removing horses, Plains Indians groups could not have returned to the status quo ante. Indeed, their incorporation was so weak that many would consider them to have remained outside the world-system, in the external arena, especially after horses became feral in the west. This highlights the inherent fuzziness of the lower limits of incorporation. The point is not to draw some arbitrary line but rather to emphasize that the lower limits of incorporation are empirically and theoretically problematic.

What if the fur trade had ceased? The source of metal goods would have been lost. In time, as animal and human populations might have recovered from epidemics and excessive predation, social relations and organization might have returned to previous conditions. But this would have been unlikely because epidemics frequently undermine the very basis of culture and social structure, especially if too many of those who possess special knowledge die. These effects are especially devastating in oral cultures, where knowledge dies with its holder if it has not already been transmitted to the next generation. Finally, if population did recover, people might have discovered new solutions to old problems. Thus, new social structures and organization typically would not have been identical to those predating initial incorporation.

Once the source of useful goods (e.g., metals) was lost, alternative sources or substitutes would be sought. Here, we should not lose sight of the impact of basic knowledge. Once a group has seen a technology, they at least know that it exists. Depending on resources and fundamental skills they might, or might not, have been able to reconstruct or obtain it. Even where they could not reconstruct or obtain it, knowledge of its existence might precipitate quests to acquire it that would not otherwise have occurred. Thus, even minimal contact can lead to vast disruptions through the spread of new knowledge.

The history of what is now the Southwest of the United States suggests another example. In the late eighteenth century, toward the end of the era of Spanish control, incorporation increased and frontier warfare declined. When incorporation decreased during the unrest which accompanied the rebellion of Mexico from Spain (1810–21) fighting with nomads increased on the frontier as state resources used for rations to guarantee the peace became scarce (Griffen 1988a, 1988b; Hall 1989a). This illustrates how even relatively small shifts in degree of incorporation can cause significant social, political, and cultural changes.

As we have seen, when goods obtained from state peoples significantly affect survival, as with metal tools in North America, or guns nearly everywhere, these goods become necessities. For the recipients, luxury trade becomes bulk goods trade, and incorporation has increased. In West Africa, members of a group without guns often found themselves on a one-way voyage to the Americas. When there was no legitimate way to acquire state-controlled goods—whether because of a ban on trade or a lack of suitable goods to trade—raiding could become an alternative means of acquisition. Thus, trading and raiding frequently alternate in frontier settings. This was one source of the nearly constant warfare between nomadic and sedentary peoples on the frontiers of the Spanish empire (Guy and Sheridan 1998a; Hall 1989a; Hall 1998a; Slatta 1997, 1998; Jones 1998). Similar processes occurred almost everywhere nomadic peoples confronted sedentary peoples (Lattimore 1951, 1962a, 1962b, 1962c; Barfield 1989, 1990, 1991; Hall 1991a, 1991b; Chase-Dunn and Hall 1997, especially chap. 8). All these processes have spatial dimensions.

SPATIAL ASPECTS OF INCORPORATION AND FRONTIER (TRANS)FORMATION

As Slatta's metaphor of the frontier as membrane suggests, some of the complexity of the effects of incorporation on frontier formation and transformation stems from problems of scale of analysis. Incorporation occurs at multiple geographical scales—local, regional, global—simultaneously and recursively. The world-system serves as the largest, though not fully determinative, context for broad regional processes. In turn, regional (or state) processes serve as a context for subregional (or substate) processes. These all serve as contexts for the most local processes. Furthermore, all these levels of change interact, simultaneously and recursively, with class, ethnic, gender, economic, and political relations. Seeking a key factor for these processes is a fool's errand. Rather, *interactions* among all these factors shape social relations and processes on frontiers.

At times the different levels and different factors may counteract each other. Sometimes the counterforces may be so well balanced that no changes occur. In this case situations that appear "static" may in fact be dynamic but temporarily balanced. One of the key insights of Sahlins's analysis in *Boundaries* is that the French-Spanish border bisecting Cerdanya was not fossilized but in dynamic balance.

PHYSICAL AND SOCIAL GEOGRAPHY

The effects of interaction between peoples living in different ecological zones are also shaped by world-systemic processes. Many observers (Barfield 1991, forthcoming; Barth 1969; Lattimore 1940, 1962a; McNeill 1964) have highlighted the salience of the boundary between the steppe and the sown—that is, between nonarable grasslands and arable lands. This boundary is actually a zone where regular rainfall drops below the limit for agriculture. Nomads live on the steppe side, whereas farmers live on the sown side. Small changes in the drought resistance of crops or slight changes in climate may cause the boundary zone to shift somewhat, but it is relatively fixed. While the farmer-nomad difference typically is seen as racial, ethnic, or cultural, it is most often a matter of adaptation. People on opposite sides of the zone differed culturally and ethnically, but not physiologically. Barth and Haaland (1969) report many cases in which individuals or families switch modes of adaptation. Usually, they change identities at the same time. Ecological borders often produce interesting cross-border interactions that alter the social structure and organization of groups on both sides.

There are analogous differences between plains (or valleys) and mountainous regions. Because of their low level of differentiation plains facilitate gradients, continua, and general similarity in adaptations and cultures. Highly differentiated mountainous areas, like the Basin and Range Province or Great Basin in the western United States (D'Azevedo 1986) or mountainous regions of California (Heizer 1978), encourage sharper boundaries and close grained local differentiation. On the one hand, American Indian Plains cultures on the Great Plains in the United States exhibited considerable cultural convergence, especially after adoption of horses (Lowie 1954; Hoebel 1982; West 1998, 33–93). On the other hand, Apachean groups, many of whom dwelt in the Basin and Range Province, were highly differentiated (see Hall 1989a, 1996a, 1998c; Melody 1977; Ortiz 1979, 1983).

Clearly, the natural environment and society interact. That interaction defines the land usage and limits the social organization of those who exploit it. For instance, Lindner (1981, 1983) argues that the Huns abandoned the use of horses after moving west of the Carpathians. Forested areas made horses a liability, hampered a nomadic lifestyle and promoted sedentary living.

The interaction between valley people and hill people is perhaps the earliest form of world-system formation (see Chase-Dunn and Hall 1997, chap. 7; Chase-Dunn and Mann 1998). Ecological borders (a point stressed by Lattimore) are fertile locations to study intergroup interactions. Such interactions typically transform the social organization and structure of the participating groups and precipitate the formation of systems composed of very different kinds of peoples. Thus, frontiers frequently coincide with geographical differences.

GEOPOLITICS

The geopolitical role of incorporated groups also shapes frontier dynamics. Their location relative to world-system expansion is a major factor in determining whether they function as buffers or barriers. A group that occupies a region in the path of expansion is typically treated as a barrier to be pacified, displaced, or destroyed. A group that occupies a region between the expanding system and its other adversaries is typically treated as a geopolitical buffer. Spanish officials in northern New Spain incessantly sought to settle or remove Apache bands because they blocked internal communication, whereas they made alliances with Comanche bands (after 1786) to buffer intrusion by European adversaries. After the war with Mexico (1846–48) the situation reversed, as Comanches became an internal barrier and Apaches at most a border nuisance. This reversal accounts for the relatively strong survival of Apachean groups and the near annihilation of Comanches, whose population reached a low of approximately one thousand at the turn of the twentieth century (Hall 1986, 1989a, 1998a).

The distinction between buffer and barrier is close to, but distinct from, the distinction between internal and external frontiers. This distinction is very close to Mikesell's (1960) distinction between frontiers of inclusion and frontiers of exclusion. An *external frontier* exists along the edges of an expanding world-system. A group that lives on an external frontier is more likely to serve as a buffer than a barrier. In contrast, an *internal frontier* is encapsulated within an expanding system. For example, the interior of North America, which Europeans settled after settling the coastal regions. This internal frontier was the setting of most of the last battles with Native Americans.

These distinctions clarify the varied responses of nonstate peoples to world-systems encroachment. If an expanding world-system needs territory, conquest and removal of indigenous populations are highly likely. Conversely, if the world-system needs people, often for slaves or other coerced workers, retreat to an external arena by indigenous populations is likely and constitutes defeat for the system and escape from incorporation for those people. This is why the same action—leaving traditionally occupied territory—can be either a defeat or a victory, depending on the goals of the expanding world-system (see Gunawardana 1992 for further examples).

Geopolitical considerations also illuminate processes of marginal incorporation. If incorporation is weak, the incorporating agency's concern with frontier groups will diminish if these groups retreat beyond the frontier zone into less desirable environments. The retreat affords these "marginal" groups a respite and time and space to adapt to new circumstances. This may determine whether they survive or are destroyed by the encounter. Thus, the degree of incorporation shapes the probability of survival for indigenous groups as much as their tenacity and skills of resistance. Inversely, a group with great cultural tenacity and strong resistance might survive weak incorporation, while a less tenacious or weaker group might not.

There are two other types of buffer zones: an "empty zone" (Upham 1992) and contested peripheries (Allen 1997) and/or semiperipheries (Berquist 1995). Also there are two types of empty frontiers. First are *emptied* frontiers, which are frontiers from which indigenous populations have been driven by invasion, warfare, disease, or other disaster (Dunaway 1996a, 1996c). Empty frontiers might be confused with neutral frontiers that develop between competing groups and buffer unwanted contacts. The difference between the two can sometimes blur. At the extreme, emptied frontiers are due to conquest, whereas neutral frontiers form by more or less mutual consent. In some areas of California, ethnographers have reported the existence of "neutral territories" that no group claimed. These areas were available for exploitation by any group that chose to do so (Heizer and Treganza 1956, 356). Johnston (1978) surmises that there was a "no man's land" along the east side of the Sacramento River between Central Wintun (Nomlaki) and Yana settlements and that this served as a buffer zone between the two groups (Chase-Dunn and Hall 1997, chap. 7; Chase-Dunn and Mann 1998). Meyer (1994) reports a similar buffer between the Anishinaabe and Lakota peoples in what is now the Dakotas and Minnesota.

Anderson (1994) reports the existence of such zones in precontact southeastern United States. Indeed, these zones were so effective that when De Soto crossed them, he actually "surprised" the people on the other side. That is, these zones so effectively cut communications, people on the far side had no advanced warning of De Soto's impending arrival, in contrast to his experiences in populated areas, where runners spread word of his presence long in advance of his arrival.

As noted, a neutral frontier implies a more pacific and cooperative system. The existence of such empty buffer zones or neutral grounds provides an unoccupied territory between competing groups that reduces the likelihood of destructive encounters. These zones might also serve as barriers to incorporation by obviating contact between groups. They have also served as a zone for ecological recuperation where game and vegetative matter that had been overharvested might recover.

Contested peripheries (Allen 1997) and semiperipheries (Berquist 1995) are special types of buffer zones or buffer frontiers. Both are found in the interstices of expanding and competing, and typically conflicting, world-systems. These have been relatively rare in the last five hundred years or so but were more common in ancient times when there were many isolated, or nearly isolated, world-systems. Until now the theoretical discussion of them has been rooted in ancient Middle Eastern examples—with an obvious exception implicit in the title, *Contested Ground* (Guy and Sheridan 1998a). Clearly, there have been more. Southeast Asia is another such frontier between India and China. It was both shaped by, and shaped, the patterns of interaction of cores of these erstwhile separate world-systems (see Chase-Dunn et al. in press for more details). Central Asian nomad confederations and states are other prominent examples of contested semiperipheries (Barfield 1989; Chase-Dunn and Hall 1997a, chap. 8; Frank 1992).

The key relationships are the complex interactions between the contested frontier and the two (or more) competing world-systems neighboring the region. Residents of contested frontiers are able to play one world-system against the other(s) to maintain, and occasionally increase, local autonomy. For their part, the world-systems have an interest in maintaining some autonomy so as to create a buffer between themselves and avoid costly wars. Such wars are typically unwinnable because of the large intervening distances and the need to cross through the territory of competing, if subsidiary states. Such wars violate what Collins (1978, 1981) has called "the no intervening heartland rule." The cost of such geopolitical overextension is loss of legitimacy at home, and often collapse of the core state.

Local states in contested frontiers appear to develop peculiar state structures that depend on continued interaction with the larger world-systems. They depend on them for goods, wealth, and to some extent protection, if only via implied threat, from other world-systems. These frontiers differ from other types of buffer frontiers in that they exist for much longer times, often centuries, and are a consequence of a prolonged dynamic balance. If, or when, the balance tips toward one world-system that incorporates other world-systems, they become more fully incorporated. Clearly, there is need for many more empirical studies of such contested frontiers to further develop a theoretical understanding of them (see Chase-Dunn and Hall 1997, chap. 4, especially 61–63 and 76–77 for further discussion of world-systems mergers).

Geographical location, technology, and environment significantly shape the process of incorporation. Incorporation is likely to remain weak, in the marginal range, if the frontier area contains few valuable resources and if it does not block expansion. When this is the case, indigenous groups have a higher likelihood of survival with some degree of social and cultural integrity. This accounts, in part, for the con-

tinued survival of Native American groups in Arizona, New Mexico, and much of
the interior western United States (Hall 1989a). Conversely, it also explains much
of the drive for their removal in eastern United States.

Changes in definitions of what constitutes resources by expanding systems or
changes in dominant technologies typically change the degree of incorporation. After
World War II when water, uranium, oil, and coal increased in value, more serious
attempts were made to incorporate energy resource owning groups more fully into
the United States political economy to exploit those resources. It is important to note
that this move toward dependent peripheralization did little to benefit those groups
(Snipp 1988).

Much of this discussion has emphasized economic, or at least material, conditions
and early phases of incorporation. It is worthwhile to focus on cultural and political aspects of incorporation, especially in more recent times.

POLITICAL AND CULTURAL INCORPORATION

Several writers have treated political aspects of incorporation, most notably Cornell
(1988) and Champagne (1989, 1992) (see, too, Meyer 1994; Dunaway 1996a;
Faiman-Silva 1997). Cornell and Champagne emphasize the ways in which political organization has changed and adapted to inclusion within the American political system. Cornell in particular has emphasized changing relations of tribal, fac-
tional, and national or pan-Indian identities as the political context has changed.
Meyer, through a detailed examination of the White Earth Anishinaabeg, has examined how increasing involvement in farming and timber industries has led to a
split between full-bloods and mixed-bloods. I should note here that the full-/
mixed-blood distinction is more of a cultural metaphor than a true indication of
biological heritage. Like race in the Caribbean or Brazil, perceptions are strongly
shaped by social characteristics. Meyer explicitly critiques my earlier concept of incorporation for failure to address such issues. Dunaway, as we have seen, describes
political and cultural changes that emanated from involvement in the fur trade.
Faiman-Silva (1997) has extended and synthesized Cornell's approach to political
incorporation with my work in her analysis of Choctaw history. She examines in
considerable detail the effects of involvement with Weyerhauser corporation and the
timber industry in eastern Oklahoma in the last several decades.

Faiman-Silva's synthesis is rich and complex. She sees the Choctaw moving from
"nation" to "tribe" to "ethnic minority" (1997, 23). She also notes that these relationships "connote asymmetrical political realities and cultural dominance relationships" (24) in addition to economic relationships. Specifically, Americans sought to
undermine traditional clan relations in favor of more centralized chiefly roles. By
the 1820s most Choctaws remained small-scale subsistence agriculturists, but a small
group of capitalist-oriented planters was emerging (30). New factions emerged
around these class relationships, often expressed in the idiom of full- and mixed-blood

differences. "Progressives drafted a constitution in 1826 that dispensed with many ancient Choctaw practices, including hereditary chiefs, traditional burial practices, infanticide, polygyny, and matrilineal inheritance" (33). Elected leaders were increasingly mixed-bloods. By the late twentieth-century Choctaws were much more heavily involved in the capitalist economy: directly through participation in capitalist enterprises, notably Weyerhauser, and indirectly by administering their own affairs, albeit with federal money.

Faiman-Silva emphasizes that tribal economic development "remains foremost a problem that is simultaneously structural and political-economic, rooted in historic and contemporary indigenous sovereignty issues and class exploitation" (214). She argues that three obstacles to full Choctaw sovereignty remain formidable. First, the asymmetry between Native American ethnicities and the U.S. political economy. Second is the volatile nature of high-stakes bingo. Local opposition centers on the moral issue of gambling, the regressive nature of the business, and low skill and low pay in the service employment it generates. Third are the continuing high rates of unemployment. She asks, "As the twenty-first century looms, will Choctaw culture survive as more than a distant memory played out at the annual Labor Day tribal gathering?" (218). Much the same question arises with respect to the Miamis in Indiana (Rafert 1996). Indeed, they have been unable to convince the federal government that they remain Indians.

One may ask along with Choctaw Chief Roberts whether tribalism is becoming extinct, "as Choctaws transform into a rural ethnic minority community." Indeed, one may turn Faiman-Silva's closing sentence into a key question: Is it true that "[T]ribal culture persists as a superstructural bas relief over a base driven by world economic forces in a global market place" (Faiman-Silva 1997, 224)? I should note here, that several Choctaws disagree with Faiman-Silva's analysis, but it is clear that even they see Choctaw identity as problematic, albeit for different reasons.

What Faiman-Silva describes as transformation from "nation" to "tribe" to "ethnic minority" might also be described as a transformation from autonomous society, to marginal to dependent periphery. These transformations are accompanied by, or are processes of, culturicide (Fenelon 1997, 1998), in which the internal culture of the group is transformed, yet some generic identity remains. This is assimilation, but with maintenance of a cultural boundary. If the assimilation were complete, and the identity lost, then the process would be ethnocide. Clearly, these are processes that are part and parcel of world-system incorporation. They *cannot* be understood solely locally. Rather, they must be understood in a larger, world-systemic, context. What remains unclear is whether culturicide and ethnocide are the cultural analogs of full-blown or dependent economic incorporation and the complex relations among the two.

Here Nagel's (1996) discussion of symbolic ethnicity is singularly germane. Symbolic ethnicity is "a source of personal meaning, and the benefits of ethnic identity involve mainly emotional fulfillment, social connectedness, or sometimes recreational pleasure" (25). For Native Americans, this is much more problematic because of the legacy of a biological, racialist, rooting of American concepts of ethnicity. This, in turn, is overlain by the competing levels of Native American identity, as a member of a clan and/or tribal faction, as a member of a native nation, and as an American Indian. Since acceptance by the federal government often carries important economic consequences, the drive for recognition further politicizes ethnic identity. Fenelon (1999) discusses some of these issues in his analysis of the politicization of Native American "mascots" (see, too, Churchill 1994, 65–113).

These issues become prominent in discussions of who is an authentic "Indian" and who can produce authentic American Indian art (Nagel 1996, chap. 2; Churchill 1994, 89–113, 1996, 483–499; Durham 1992). Again, these "symbolic issues" are tightly intertwined with important economic issues: making a living as an artist and/ or the right to collect fees from tourists. Given the circumstances of high unemployment rates on all Indian reservations, these are far from trivial issues. They are vital issues.

Thus, we may ask whether one of the late twentieth-century aspects of incorporation is a transformation of identity from something that "just is" into a resource around which political and cultural mobilization can crystallize? If so, can we still speak of incorporation? Finally, drawing on the work of Jonathan Friedman (1994, 1998) (and Bergesen in chap. 10), I note that these changes coincide with decline in world-system hegemony in the late twentieth century. This is a context that encourages cultural proliferation, and especially contests over and about ethnic identities. In short, ethnic conflict and ethnic violence can be expected to continue, and perhaps increase, as long as world-system hegemony is low. As Bergesen suggests, this will foster numerous cultural changes. This situation raises, in turn, four other, closely intertwined issues: gender roles, spirituality, culture, and law.

GENDER ROLES, SPIRITUALITY, CULTURE, AND LAW

As we have seen, gender roles have been transformed time and again during incorporation. Yet maintenance of gender roles can itself be a form of resistance to incorporation. This may be in the form of persisting matrilineal traditions, or in different allocations of gender roles in political processes, or in gender differentials in political participation (Jaimes and Halsey 1992), or in the use of indigenous spirituality to resist alcohol and drug abuse (chap. 12). Following the prescriptions of Ward (1993), this question might best be inverted to ask, What can we learn about incorporation by examining gender role changes, rather than what does degree of incorporation tell us about gender roles?

In early colonial La Plata and Northern New Spain, captives taken from indigenous populations by Spaniards seem to have been predominantly women and children (Hall 1989a; Socolow 1992; Jones 1994, 1998). Adult males were seen as problematic captives. First, the obligation to participate in warfare, defense, and resistance made them particularly intractable and hazardous. Second, they were much harder to resocialize than children—although this applies to adult females as well. Third, captives taken from foraging societies were not very useful as coerced laborers in agricultural settings. Spaniards all over the Western Hemisphere complained of this, as did American farmers (Knack 1987; Hurtado 1988). Conversely, adult males taken from sedentary societies may not be useful in nomadic societies because the cost of supervision surpassed the value of any work they could be forced to do.

The situation is different for women and children. When infant mortality rates are high, as was true for *all* societies before the mid–nineteenth century, the reproductive capacity of women is especially valuable. In cultural settings where polygyny was accepted, captive women could become secondary wives who could do menial labor under the supervision of the primary wife. At times, this labor was a major motive for capturing women. Ties of affection for children born in captivity may have tempered any desires to flee. According to Socolow (1992), this is one reason that many European women often chose to stay with Indian captors rather than return to Spanish society.

Captured Indian women presented a significantly different problem to European societies. They could be forced to become servants and/or concubines, but they had no acceptable role. Their children took on a lower caste status, even on extreme frontiers where such distinctions were sometimes not so tightly drawn (Gutiérrez 1991). Again, attachment to children could temper desires to flee.

Children of captured Spanish women were accepted more readily into the Indian society. This differential acceptance of their children also encouraged them to remain in Indian societies. The same differential acceptance was accorded captured children. Among most nomadic foragers a child raised in captivity often was adopted and became a full fledged member of the group. The best a captured Indian child raised in Spanish society could hope for (in New Mexico at least) was to be considered *genízaro* (Horvath 1977, 1979; Hall 1989a, 1989b).

Children raised in captivity could not return to their native group. They would have been socialized for life in the capturing group, typically inappropriate for their native group. Furthermore, their emotional ties would have been to their captors who had raised them. The general acceptance of such children within Indian societies obviated the pressure to form distinctive communities. These differences are relative; they are rules of thumb, not absolutes.

Indian women continue to play different roles in modern times (Jaimes and Halsey 1992). Ward et al.'s analysis in chapter 12 is instructive here. They suggest that Native women, drawing on indigenous traditions, are developing ways to resist some of the most debilitating consequences of conquest. In doing so, they build on traditional female roles but simultaneously reconstruct Native ethnicities and identities. That is, their resistance to cultural incorporation, or cultural dominance, is, in itself, a form of ethnogenesis.

Similarly, Native spirituality has become highly politicized. Many writers have vociferously opposed the (mis)use of Indian culture and spirituality. Rose (1992) and Churchill (1992, 1994, 1996) have protested mightily against white shamanism, calling it a "new hucksterism." Their complaint is that Euroamericans, who long since have given up their own premodern religious practices and beliefs, are trying to regain them through (mis)appropriating practices of various Native American groups. They both note the irony of Euroamericans try to co-opt what they spent so many years attempting to destroy during the allotment era (Hoxie 1984). Deloria (1995) and Cook-Lynn (1996) have echoed this same complaint.

All these differences carry important legal implications. As noted earlier, Native American groups, and indigenous peoples throughout the world, have had considerable success in manipulating European notions of sovereignty to their own benefit (Wilmer 1993). By putting European derived states in the position of denying their own bases of sovereignty when they deny the claims of Native groups, indigenous peoples have been able to retain some degree of sovereignty. European states have found the ideological havoc wreaked by outright denial of Native sovereignty too dear a price to pay for the few resources they have sought from Native Peoples. Again, with important exceptions.

This, of course, has not stopped them from trying, as both Ward (1997) and Wilson (1997) argue (see, too, Gedicks, 1993). This has meant that Europeans have had to resort to different sources of justifications—with promises of economic development, jobs, increased access to industrial goods and services, education and health care chief among them—to justify claims on Native resources.

Although Native peoples have met with some success in this arena, they have had to fight on European grounds—within European law (for detailed examples from northern New Spain, see Cutter 1995a, 1995b, 1995c). Recently, one of the more outstanding successes has been to use the doctrine of sovereignty to set up various gaming operations. By exploiting the contradictory desires for access to gambling and desire to forbid it, Indians have begun to turn considerable profits. But as noted for the Choctaw, this success is fragile and volatile and subject to redefinition by the federal Congress.

The question remains of how much they have had to give up to win these victories. By fighting European civilization on its own turf, they had to accept some of the premises of that turf. Biolsi (1995, 543) persuasively argues that the law is "a fundamental constituting axis of modern social life—not just a political resource or an institution but a constituent of all social relations of domination" (see also Biolsi 1992, n.d.). That is, political incorporation has entailed major changes driven by European legal traditions. In other words, incorporation has disrupted, and continues to disrupt, massively Native American traditions. This is shown most dramatically and forcefully in the works of contemporary Native American writers (Alexie 1993, 1996; Erdrich 1984, 1988; Power 1994; Seals 1979, 1992; Silko 1977, 1991; and many more).

CONCLUSON

This type of exploratory and explicative chapter calls not for conclusions but rather a concluding discussion that relates to the general themes of the chapter and the entire collection. If each of the four boundaries of a world-system generates its own zone of incorporation, and if only two broad types of world-system are considered, tributary and capitalist, and if the range of nonstate societies is lumped into three broad categories of bands, tribes, and chiefdoms (see Hall 1989, chap. 3, for a discussion of these distinctions), there are at least twenty-four potential types of zones of incorporation, frontiers, or border regions. When the buffer-barrier and internal-external frontier distinctions are added, the types of frontiers proliferate enormously (Hall 1998b). This point in itself is revealing. It helps explain why so many scholars have found frontiers fascinating objects of study yet tremendously intractable in terms of developing parsimonious theories of their processes and changes.

Though complex, the preceding analysis provides conceptual tools for much broader, multitiered comparisons among frontiers. It also provides a means of rendering more intelligible comparisons of, say, the Chinese frontier, the Roman frontier, and the Spanish frontiers in northern New Spain and southern La Plata. By attending to such factors as the type of world-system doing the incorporating, the motives behind the incorporation, including especially the types of resources sought, the roles of the incorporated regions and peoples in the larger system, and the preincorporation social organization of the incorporated groups, analysts cannot only understand the incorporation process better but also highlight the complex ways in which incorporated peoples have resisted, and continue to resist, incorporation.

As Dunaway (1996b, 1997) observes, the resistance may not always be effective, but it often gives incorporated peoples a marginal degree of autonomy. In this light, many of the actions of Native American groups—whether they were attempts to flee a jurisdiction, like that of Chief Joseph; the myriad cases of armed resistance; legal attempts to preserve and enhance tribal autonomy; the pursuit of some new economic opportunities such as gambling facilities; or the rejection of others like strip mining of coal or insisting on writing their own histories—become readily understandable. More importantly, these are clearly not haphazard or irrational resistance to change but rather intelligent, often well-chosen attempts to control and resist the processes of incorporation.

As was seen earlier, the same action, flight, may have precisely opposite consequences depending on the specific constellation of conditions under which it occurs. The lens of incorporation also helps highlight and clarify the differences between the incorporation of indigenous peoples and the "assimilation" of migrants. Although both groups come to be seen as ethnic minorities, their origins and goals are often quite different. Incorporated indigenous peoples most often are concerned with preservation of autonomy and distinctiveness. Sovereignty is a vital issue, as Wilmer (1993) has argued. For migrants, fair treatment and economic opportunities are more often the most salient issues. Often the two interact and compete, as Chappell (1993b) has shown to be the case in contemporary New Caldonia and Mulroy (1993) has shown for the Seminole maroons. These differences, like so many discussed in this chapter, are matters of degree, not absolutes.

Most indigenous groups have experienced several waves of incorporation. The results of incorporation range from genocide, through culturicide (Fenelon 1997, 1998), to assimilation, or to transformation into a minority group. Some of these ethnic minorities may, like some Native American groups, retain a degree of political and cultural autonomy within the larger system. The interplay of world-systemic, national, regional, and local factors that shape these consequences remains complex and poorly understood. However, it is not only the degree of incorporation, but entire history of the incorporation process, the trajectory of incorporation, that shapes the final result. How and the degree to which this is so are far from clear because entire trajectories have seldom been studied, and even less frequently been compared (Hall 1989a is an elementary attempt).

But incorporation into a world-system itself is changing. By the late twentieth century there were virtually no nonstate societies that had not already experienced at least some degree of incorporation. One of the important emergent constraints on the modern world-system is that there are no more new territories or peoples to incorporate. On the one hand, repeated incorporations have reduced human cultural diversity. On the other hand, frontiers have long been regions of human creativity, and zones of ethnogenesis. Often in adapting to volatile conditions and repeated interactions, new forms of social organization and identities have grown with frontiers. But incorporation and the consequent formation of frontiers have not stopped, only changed forms. Now internal frontiers are far more common than external frontiers, especially frontiers between zones of the world-system itself, as where the core meets the periphery (or semiperiphery in the view of some) along the United States–Mexico border.

As the discussion of political and cultural incorporation indicates, new frontiers are forming along the edges of, and within, the modern world-system. As a world-system penetrates deeper into social processes, as more relations are commodified, new frontiers of change are created. What is highly problematic is how these will manifest themselves spatially. Robertson (1992, 1995) has observed that globalization (King 1997) carries with it glocalization, the localized flavoring of global processes. This too, however, is an ancient process, long predating the modern world-system. Incorporated groups have always resisted, and often modified the system that incorporated them (e.g., Miller 1993).

World-systems are simultaneously homogenizing and heterogenizing forces. Only those seduced by modernization theory, or the "end of history," have been surprised by the continuous creation and maintenance of ethnic and cultural differences. How and why this occurs varies with the type of world-system, the type of region being incorporated, and the specific context of the incorporation process. While the specific mechanisms and the particular raw materials on which these forces impinge are new at the end of the second millennium, the underlying dialectic is ancient indeed.

I have argued elsewhere (Hall 1998b), as have Chappell (1993a), McNeill (1986), and Smith (1986, 1991), that ethnic diversity within states is the modal or normal (in the statistical sense) condition of states, not ethnic homogeneity. Why anyone thought otherwise is, itself, a complex problem that, while not discussed here, warrants investigation. A major conclusion of those arguments, the one that is most relevant here, is that there is much to be learned about our contemporary, "modern" world from the study of ancient worlds. Similarly, we can learn much about modern frontiers and borderlines by studying how older, even ancient, frontiers and borderlines were formed, transformed, and transcended.

Clearly, a world-systems approach to incorporation of new regions and peoples as a fundamental creator and transformer of frontiers cannot answer all our questions. But I hope it is equally clear that most of our questions cannot be answered without such an approach. To paraphrase Vilar, to understand the center, go to the frontier.

NOTES

I have presented many parts and versions of this essay in a variety of venues: Political Economy of the World-System Roundtables at the American Sociological Association meetings in 1992, 1997, 1998; Association of American Geographers in 1994; Social Science History meetings in 1995 and 1996; International Society for the Comparative Study of Civilizations in 1997; International Studies Association in 1997 and 1998; American History Association in 1998; and Political Economy of the World-System meeting in 1998. I thank the many commentators, discussants, and listeners for useful comments. All the contributors to this volume have influenced this chapter. Much of it reflects the benefits of long conversations with Christopher Chase-Dunn, Wilma A. Dunaway, James Fenelon, Carol Ward, and David Wilson. As always, none of these colleagues is to be held accountable for my failures to heed their often sage advice. I would also like to thank the Faculty Development Committee and the John and Janice Fisher Fund for faculty development at DePauw University for support to attend these meetings. I wish to thank Westview Press for permission to reproduce figures 13.1 and 13.2 © Westview Press.

1. Capital accumulation refers to amassing wealth in any form; capitalist accumulation to "the amassing of wealth by means of the making of profits from commodity production" (Chase-Dunn and Hall 1997, 271).

2. These terms are often confused. *Ethnogenesis,* literally, means the creation of an ethnic identity and/or group through complex social actions. *Ethnocide* refers to the destruction, the "killing" of an ethnic identity, without necessarily actually killing individuals. This is assimilation from the assimilated's point of view. *Genocide* does refer to the actual killing of all or nearly all members of a group. *Ethnic cleansing,* in contrast, is a "limited" form of genocide: not killing all members of a group, only those who live in the area being "cleansed." Finally, Fenelon (1997, 1998) has developed a concept of *culturicide,* by which he means the destruction of a culture, without necessarily destroying the group identity or the individuals. The distinctions among these terms are *not* sharp but rather indicate relative strengths of cultural, social, and physical processes. They are discussed in more detail later in this chapter.

3. For fuller explication of the conventional world-system analysis of incorporation, see Chase-Dunn and Hall (1997, chap. 4) or Dunaway (1996a, chaps. 1 and 2).

4. The end of the film *The Black Robe* portrays this in a dramatic way when a French priest arrives at an Algonquin village to find it nearly destroyed by disease.

5. Meat from the American bison has higher protein content than beef or chicken but lower fat, cholesterol, and calories than beef, chicken, or pork.

REFERENCES

Abler, Thomas S. 1992. "Beavers and Muskets: Iroquois Military Fortunes in the Face of European Colonization." Pp. 151–174 in *War in the Tribal Zone,* edited by R. Brian Ferguson and Neil L. Whitehead. Santa Fe, N.M.: School of American Research Press.

Abu-Lughod, Janet. 1989. *Before European Hegemony: The World System A.D. 1250–1350.* New York: Oxford University Press.

———. 1993. "Discontinuities and Persistence: One World System or A Succession of World Systems?" Pp. 278–291 in *The World System: Five Hundred Years or Five Thousand?* edited by Andre Gunder Frank and Barry K. Gills. London: Routledge.

Aguirre Beltran, Gonzalo. 1979. *Regions of Refuge.* Monograph No. 12. Washington, D.C.: Society for Applied Anthropology. (Originally, *Regiones de Refugio.* Mexico City: Ediciones Especiales, No. 46, Instituto Indigenista Interamericano, 1967.)

Alexie, Sherman. 1993. *The Lone Ranger and Tonto Fistfight in Heaven.* New York: Atlantic Monthly Press.

———. 1996. *Indian Killer.* New York: Atlantic Monthly Press.

Allen, Mitchell. 1997. "Contested Peripheries: Philistia in the Neo-Assyrian World-System." Unpublished Ph.D. dissertation, interdepartmental archaeology program, University of California, Los Angeles.

Anderson, David G. 1994. *The Savannah River Chiefdoms: Political Change in the Late Prehistoric Southeast.* Tuscaloosa: University of Alabama Press.

Arrighi, Giovanni. 1979. "Peripheralization of Southern Africa. I: Changes in Production Processes." *Review* 3:2(Fall):161–191.

———. 1994. *The Long Twentieth Century: Money, Power and the Origins of Our Times.* New York: Verso.

Bach, Robert L. 1980. "On the Holism of a World-System Perspective." Pp. 289–318 in *Processes of the World-System,* edited by Terence K. Hopkins and Immanuel Wallerstein. Beverly Hills: Sage.

Baretta, Silvio R. D., and John Markoff. 1978. "Civilization and Barbarism: Cattle Frontiers in Latin America." *Comparative Studies in Society and History* 20:4(Oct.):587–620.

Barfield, Thomas J. 1989. *The Perilous Frontier: Nomadic Empires and China.* London: Blackwell.

———. 1990. "Tribe and State Relations: The Inner Asian Perspective." Pp. 153–182 in *Tribe and State Formation in the Middle East,* edited by Philip S. Khoury and Joseph Kostiner. Berkeley: University of California Press.

———. 1991. "Inner Asia and Cycles of Power in China's Imperial Dynastic History." Pp. 21–62 in *Rulers from the Steppe: State Formation on the Eurasian Periphery,* edited by Gary Seaman and Daniel Marks. Los Angeles: Ethnographics Press, Center for Visual Anthropology, University of Southern California.

———. Forthcoming. "Empires: Imperial State Formation along the Chinese-Nomad

Frontier." In *Empires,* edited by Carla Sinopoli, Terence D'Altroy, Kathleen Morrison, and Susan Alcock. Cambridge: Cambridge University Press.

Barth, Frederick. 1969. *Ethnic Groups and Boundaries.* Boston: Little, Brown. (See especially editor's "Introduction.")

Berquist, Jon. 1995. "The Shifting Frontier: The Achaemenid Empire's Treatment of Western Colonies." *Journal of World-Systems Research* 1:17 (electronic journal: http://csf.colorado.edu/wsystems/jwsr.html).

Biolsi, Thomas. 1992. *Organizing the Lakota: The Political Economy of the New Deal on the Pine Ridge and Rosebud Reservations.* Tucson: University of Arizona Press.

———. 1995. "Bringing the Law Back In: Legal Rights and the Regulation of Indian-White Relations on the Rosebud Reservation." *Current Anthropology* 36:4(Aug.–Oct.):543–571.

———. n.d. "Law and the Production of Race Relations: Indian and White on and off Rosebud Reservation." Unpublished manuscript, Portland State University, Oregon.

Bose, Christine E., and Edna Acosta-Belen, eds. 1995. *Women in the Latin American Development Process.* Philadelphia: Temple University Press.

Bradley, Candice. 1997. "Fertility in Maragoli: The Global and the Local." Pp. 121–138 in *Economic Analysis Beyond the Local System,* edited by Richard Blanton, Peter Peregrine, Deborah Winslow, and Thomas D. Hall. Lanham, Md.: University Press of America.

Champagne, Duane. 1989. *Indian Societies: Strategies and Conditions of Political and Cultural Survival.* Cambridge, Mass.: Cultural Survival.

———. 1992. *Social Order and Political Change: Constitutional Governments among the Cherokee, the Choctaw, the Chickasaw, and the Creek.* Stanford, Calif.: Stanford University Press.

Chappell, David A. 1993a. "Ethnogenesis and Frontiers." *Journal of World History* 4:2(Fall):267–275.

———. 1993b. "Frontier Ethnogenesis: The Case of New Caldonia." *Journal of World History* 4:2(Fall):307–324.

Chase-Dunn, Christopher, and Thomas D. Hall, eds. 1991. *Core/Periphery Relations in Precapitalist Worlds.* Boulder, Colo.: Westview.

———. 1997. *Rise and Demise: Comparing World-Systems.* Boulder, Colo.: Westview.

———. 1998. "World-Systems in North America: Networks, Rise and Fall and Pulsations of Trade in Stateless Systems." *American Indian Culture and Research Journal* 2:1:23–72.

Chase-Dunn, Christopher, Thomas D. Hall, and Susan Manning. In press. "Pulsations in the Afro-Eurasian System: Indic City and Empire Growth and Decline." *Social Science History.*

Chase-Dunn, Christopher, and Kelly M. Mann. 1998. *The Wintu and Their Neighbors: A Very Small World-System in Northern California.* Tucson: University of Arizona Press.

Churchill, Ward. 1992. *Fantasies of the Master Race: Literature, Cinema and the Colonization of American Indians,* edited by M. Annette Jaimes. Monroe, Maine: Common Courage.

———. 1994. *Indians R Us: Culture and Genocide in Native North America.* Monroe, Maine: Common Courage.

———. 1996. *From a Native Son: Selected Essays on Indigenism, 1985–1995.* Boston: South End.

Collins, Randall. 1978. "Some Principles of Long-Term Social Change: The Territorial Power of States." Pp. 1–34 in *Research in Social Movements, Conflicts, and Change,* edited by Louis Kriesberg. Greenwich, Conn.: JAI.

————. 1981. "Long-Term Social Change and The Territorial Power of States." Pp. 71–106 in *Sociology since Midcentury* by Randall Collins. New York: Academic Press.

Cook-Lynn, Elizabeth. 1996. *Why I Can't Read Wallace Stegner and Other Essays: A Tribal Voice.* Madison: University of Wisconsin Press.

Cornell, Stephen. 1988. *The Return of the Native: American Indian Political Resurgence.* New York: Oxford University Press.

Curtin, Philip D. 1984. *Cross-Cultural Trade in World History.* Cambridge: Cambridge University Press.

Cutter, Charles R. 1995a. "Judicial Punishment in Colonial New Mexico." *Western Legal History* 8:115–129.

————. 1995b. "Indians as Litigants in Colonial Mexico." Paper presented at Social Science History Association meeting, Chicago, November.

————. 1995c. *The Legal Culture of Northern New Spain, 1700–1810.* Albuquerque: University of New Mexico Press.

D'Azevedo, Warren L. 1986. *Handbook of North American Indians,* vol. 11: *Great Basin.* Washington, D.C.: Smithsonian Institution.

Deloria, Vine. 1995. *Red Earth, White Lies: Native Americans and the Myth of Scientific Fact.* New York: Scribner's.

Dunaway, Wilma A. 1994. "The Southern Fur Trade and the Incorporation of Southern Appalachia into the World-Economy, 1690–1763." *Review* 18:2(Spring):215–242.

————. 1996a. *The First American Frontier: Transition to Capitalism in Southern Appalachia, 1700–1860.* Chapel Hill: University of North Carolina Press.

————. 1996b. "Incorporation as an Interactive Process: Cherokee Resistance to Expansion of the Capitalist World-System, 1560–1763." *Sociological Inquiry* 66:4(Fall):455–470.

————. 1996c. "The Incorporation of Mountain Ecosystems into the Capitalist World-System." *Review* 19:4(Fall):355–381.

————. 1997. "Rethinking Cherokee Acculturation: Women's Resistance to Agrarian Capitalism and Cultural Change, 1800–1838." *American Indian Culture and Research Journal* 21:1:231–268.

Durham, Jimmie. 1992. "Cowboys and . . . Notes on Art, Literature, and American Indians in the Modern American Mind." Pp. 423–438 in *The State of Native America: Genocide, Colonization, and Resistance,* edited by M. Annette Jaimes. Boston: South End.

Erdrich, Louise. 1984. *Love Medicine: A Novel.* New York: Holt, Rinehart & Winston.

————. 1984. *Tracks: A Novel.* New York: Holt.

Faiman-Silva, Sandra L. 1997. *Choctaws at the Crossroads: The Political Economy of Class and Culture in the Oklahoma Timber Region.* Lincoln: University of Nebraska Press.

Fenelon, James V. 1997. "From Peripheral Domination to Internal Colonialism: Socio-Political Change of the Lakota on Standing Rock." *Journal of World-Systems Research* 3:259–320 (electronic journal: http://csf.colorado.edu/wsystems/jwsr.html).

————. 1998. *Culturicide, Resistance, and Survival of the Lakota ("Sioux Nation").* New York: Garland.

————. 1999. "Indian Icons in the World Series of Racism: Institutionalization of the Racial Symbols Wahoos and Indians." Pp. 24–45 in *The Global Color Line: Racial and Ethnic Inequality and Struggle from a Global Perspective,* edited by Pinar Batur-Vanderlippe and Joseph Feagin. Series 6, *Research in Politics and Society.* Stamford, Conn.: JAI.

Ferguson, R. Brian. 1997. "Tribes, Tribal Organization." Pp. 475–476 in *Dictionary of Anthropology,* edited by Thomas Barfield. London: Blackwell.

Ferguson, R. Brian, and Neil L. Whitehead, eds. 1992a. *War in the Tribal Zone: Expanding States and Indigenous Warfare.* Santa Fe, N.M.: School of American Research Press.

———. 1992b. "The Violent Edge of Empire." Pp. 1–30 in *War in the Tribal Zone,* edited by R. Brian Ferguson and Neil L. Whitehead. Santa Fe, N.M.: School of American Research Press.

Frank, Andre Gunder. 1992. *The Centrality of Central Asia.* Comparative Asian Studies No. 8. Amsterdam: VU University Press for Center for Asian Studies Amsterdam (CASA).

Frank, Andre Gunder, and Barry K. Gills, eds. 1993. *The World System: Five Hundred Years or Five Thousand?* London: Routledge.

Friedman, Jonathan. 1994. *Cultural Identity & Global Process.* Thousand Oaks, Calif.: Sage.

———. 1998. "Transnationalization, Socio-Political Disorder and Ethnification as Expressions of Declining Global Hegemony." *International Political Science Review* 19:3(July):233–250.

Gedicks, Al. 1993. *The New Resource Wars: Native and Environmental Struggles Against Multinational Corporations.* London: South End.

Gills, Barry K., and Andre Gunder Frank. 1991. "5000 Years of World System History: The Cumulation of Accumulation." Pp. 67–112 in *Core/Periphery Relations in Precapitalist Worlds,* edited by Christopher Chase-Dunn and Thomas D. Hall. Boulder, Colo.: Westview.

Griffen, William. 1988a. *Apaches at War and Peace: The Janos Presidio, 1750–1858.* Albuquerque: University of New Mexico Press.

———. 1988b. *Utmost Good Faith: Patterns of Apache-Mexican Hostilities in Northern Chihuahua Border Warfare, 1821–1848.* Albuquerque: University of New Mexico Press.

Gunawardana, R. A. L. H. 1992. "Conquest and Resistance: Pre-state and State Exapansionism in Early Sri Lankan History." Pp. 61–82 in *War in the Tribal Zone,* edited by R. Brian Ferguson and Neil L. Whitehead. Santa Fe, N.M.: School of American Research Press.

Gutiérrez, Ramón A. 1991. *When Jesus Came, the Corn Mothers Went Away: Marriage, Sexuality, and Power in New Mexico, 1500–1846.* Stanford, Calif.: Stanford University Press.

Guy, Donna J., and Thomas E. Sheridan, eds. 1998a. *Contested Ground: Comparative Frontiers on the Northern and Southern Edges of the Spanish Empire.* Tucson: University of Arizona Press.

———. 1998b. "On Frontiers: The Northern and Southern Edges of the Spanish Empire in America." Pp. 3–15 in *Contested Ground: Comparative Frontiers on the Northern and Southern Edges of the Spanish Empire,* edited by Donna J. Guy and Thomas E. Sheridan. Tucson: University of Arizona Press.

Haaland, Gunnar. 1969. "Economic Determinants in Ethnic Processes." Pp. 53–73 in *Ethnic Groups and Boundaries,* edited by Frederick Barth. Boston: Little Brown.

Hall, Thomas D. 1986. "Incorporation in the World-System: Toward A Critique." *American Sociological Review* 51:3(June):390-402.

———. 1987. "Native Americans and Incorporation: Patterns and Problems." *American Indian Culture and Research Journal* 11:2:1–30.

———. 1989a. *Social Change in the Southwest, 1350–1880.* Lawrence: University Press of Kansas.

———. 1989b. "Is Historical Sociology of Peripheral Regions Peripheral?" Pp. 349–372 in *Studies of Development and Change in the Modern World,* edited by Michael T. Martin and Terry R. Kandal. New York: Oxford University Press.

————. 1991a. "Civilizational Change: The Role of Nomads." *Comparative Civilizations Review* 24:34–57.

————. 1991b. "The Role of Nomads in Core/Periphery Relations." Pp. 212–239 in *Core/Periphery Relations in Precapitalist Worlds,* edited by Christopher Chase-Dunn and Thomas D. Hall. Boulder, Colo.: Westview.

————. 1996a. "World-Systems and Evolution: An Appraisal." *Journal of World-System Research* 2:1–109 (electronic journal, http://csf.colorado.edu/wsystems/jwsr.html).

————. 1996b. "The World-System Perspective: A Small Sample from a Large Universe." *Sociological Inquiry* 66:4(Fall):440–454.

————. 1997. "Frontier." Pp. 208–209 in *Dictionary of Anthropology,* edited by Thomas Barfield. London: Blackwell.

————. 1998a. "The Río de La Plata and the Greater Southwest: A View from World-System Theory." Pp. 150–166 in *Contested Ground: Comparative Frontiers on the Northern and Southern Edges of the Spanish Empire,* edited by Donna J. Guy and Thomas E. Sheridan. Tucson: University of Arizona Press.

————. 1998b. "The Effects of Incorporation into World-Systems on Ethnic Processes: Lessons from the Ancient World for the Modern World." *International Political Science Review* 19:3(July):251–267.

————. 1998c. "World-Systems and Evolution: An Appraisal." Pp. 1–25 in *Leadership, Production, and Exchange: World-Systems Theory in Practice,* edited by P. Nick Kardulias. Boulder, Colo.: Rowman & Littlefield.

Hall, Thomas D., and Christopher Chase-Dunn. 1993. "The World-Systems Perspective and Archaeology: Forward into the Past." *Journal of Archaeological Research* 1:121–143.

Heizer, Robert, F., ed. 1978. *Handbook of North American Indians,* vol. 8: *California.* Washington, D.C.: Smithsonian Institution.

Heizer Robert F., and A. E. Treganza. 1971. "Mines and Quarries of the Indians of California." Pp. 346–359 in *The California Indians: A Sourcebook,* 2d ed., edited by Robert F. Heizer and Mary A. Whipple. Berkeley: University of California Press.

Hoebel, E. Adamson. 1982. *The Plains Indians.* Bloomington: Indiana University Press.

Hopkins, Terence K., I. Wallerstein, Resat Kasaba, William G. Martin, and Peter D. Phillips. 1987. *Incorporation into the World-Economy: How the World System Expands,* special issue of *Review* 10:5/6(Summer/Fall):761–902.

Horvath, Steven. 1977. "The *Genízaro* of Eighteenth-Century New Mexico: A Reexamination." *Discovery,* School of American Research: 25–40.

————. 1979. "The Social and Political Organization of the *Genízaros* of Plaza de Nuestra Señora de los Dolores de Belén, New Mexico, 1740–1812." Unpublished Ph.D. dissertation, Brown University, Providence, R.I.

Hoxie, Frederick E. 1984. *A Final Promise: The Campaign to Assimilate the Indians, 1880–1920.* Lincoln: University of Nebraska Press.

Hurtado, Albert L. 1988. *Indian Survival on the California Frontier.* New Haven, Conn.: Yale University Press.

Jaimes, M. Annette, with Theresa Halsey. 1992. "American Indian Women: At the Center of Indigenous North America." Pp. 311–344 in *The State of Native America: Genocide, Colonization, and Resistance,* edited by M. Annette Jaimes. Boston: South End.

Johnston, Jim, 1978. "The Wintu and Yana Territorial Boundary." Paper presented to the annual meetings of the Society for California Archaeology, Yosemite National Park, March.

Jones, Kristine L. 1994. "Comparative Ethnohistory and the Southern Cone." *Latin American Research Review* 29:1:107–118.

————. 1998. "Comparative Raiding Economies: North and South." Pp. 97–114 in *Contested Ground: Comparative Frontiers on the Northern and Southern Edges of the Spanish Empire*, edited by Donna J. Guy and Thomas E. Sheridan. Tucson: University of Arizona Press.

Kardulias, P. Nick. 1990. "Fur Production as a Specialized Activity in a World System: Indians in the North American Fur Trade." *American Indian Culture and Research Journal* 14:1:25–60.

King, Anthony D., ed. 1997. *Culture, Globalization and the World-System: Contemporary Conditions for the Representation of Identity*. Minneapolis: University of Minnesota Press.

Knack, Martha C. 1987. "The Role of Credit in Native Adaptation to the Great Basin Ranching Economy." *American Indian Culture and Research Journal* 11:1:43–65.

Lattimore, Owen. 1951. *Inner Asian Frontiers*. 2d ed. Boston: Beacon. (Originally 1940, New York: American Geographical Society).

————. 1962a. "The Frontier in History." Pp. 469–491 in *Studies in Frontier History: Collected Papers, 1928–58*, edited by Owen Lattimore. London: Oxford University Press.

————. 1962b. "Inner Asian Frontiers: Defensive Empires and Conquest Empires." Pp. 501–513 in *Studies in Frontier History: Collected Papers, 1928–58*, edited by Owen Lattimore. London: Oxford University Press.

————, ed. 1962c. *Studies in Frontier History: Collected Papers, 1928–58*. London: Oxford University Press.

Lindner, Rudi Paul. 1981. "Nomadism, Horses and Huns." *Past & Present* 92:(Aug.):3–19.

————. 1983. *Nomads and Ottomans in Medieval Anatolia*. Vol. 144. Indiana University Uralic and Altaic Series. Bloomington, Ind.: Research Institute for Inner Asian Studies.

Lowie, Robert H. 1954. *Indians of the Plains*. New York: Natural History Press.

Markoff, John. 1994. "Frontier Societies." Pp. 289–291 in *Encyclopedia of Social History*, edited by Peter N. Stearns. New York: Garland.

Mathien, Frances Joan, and Randall McGuire, eds. 1986. *Ripples in the Chichimec Sea: Consideration of Southwestern-Mesoamerican Interactions*. Carbondale: Southern Illinois University Press.

McNeill, William H. 1964. *Europe's Steppe Frontier, 1500–1800*. Chicago: University of Chicago Press.

————. 1986. *Polyethnicity and National Unity in World History*. Toronto: University Toronto Press.

Melody, Michael E. 1977. *The Apaches*. Bloomington: Indiana University Press.

Meyer, Melissa L. 1994. *The White Earth Tragedy: Ethnicity and Dispossession at a Minnesota Anishinaabe Reservation, 1889–1920*. Lincoln: University of Nebraska.

Mies, Maria. 1986. *Patriarchy and Accumulation on a World-Scale*. London: Zed.

Mies, Maria, and Vandana Shiva. 1993. *Ecofeminism*. London: Zed.

Mikesell, Marvin W. 1960. "Comparative Studies in Frontier History." *Annals of American Association of Geographers* 50:1(March):62–74. (Reprinted pp. 152–171 in *Turner and the Sociology of the Frontier*, edited by Richard Hofstadter and Seymour Martin Lipset. Boston: Basic Books.)

Miller, David Harry. 1993. "Ethnogenesis and Religious Revitalization beyond the Roman Frontier: The Case of Frankish Origins." *Journal of World History* 4:2(Fall):277–285.

Mishkin, Bernard. 1940. *Rank and Warfare among the Plains Indians*. Monograph no. 3 of the American Ethnological Society. Seattle: University of Washington Press. (Reprinted 1992, Lincoln: University of Nebraska Press.)

Moseley, Katherine P. 1992. "Caravel and Caravan: West Africa and the World-Economies ca. 900–1900 AD." *Review* 15:3(Summer):523–555.

Mulroy, Kevin. 1993. "Ethnogenesis and Ethnohistory of the Seminole Maroons." *Journal of World History* 4:2(Fall):287–305.

Nagel, Joane. 1996. *American Indian Ethnic Renewal: Red Power and the Resurgence of Identity and Culture*. Oxford: Oxford University Press.

Ortiz, Alfonso, ed. 1979. *Handbook of North American Indians*, vol. 9: *Southwest*. Washington, D. C.: Smithsonian Institution.

———, ed. 1983. *Handbook of North American Indians*, vol. 10: *Southwest*. Washington, D.C.: Smithsonian Institution.

Peregrine, Peter N., and Gary M. Feinman, eds. 1996. *Pre-Columbian World-Systems*. Monographs in World Archaeology No. 26. Madison, Wisc.: Prehistory Press.

Power, Susan. 1994. *The Grass Dancer*. New York: Putnam.

Rafert, Stewart. 1996. *The Miami Indians of Indiana: A Persistent People, 1654–1994*. Indianapolis: Indiana Historical Society.

Reff, Daniel T. 1991. *Diseases, Depopulation, and Culture Change in Northwestern New Spain, 1518–1764*. Salt Lake City: University of Utah Press.

Robertson, Roland. 1992. *Globalization: Social Theory and Global Culture*. Newbury Park, Calif.: Sage.

———. 1995. "Globalization: Time-Space and Homogeneity-Heterogeneity." Pp. 25–44 in *Global Modernities*, edited by Mike Featherstone, Scott Lash, and Roland Robertson. Newbury Park, Calif.: Sage.

Rose, Wendy. 1992. "The Great Pretenders: Further Reflections on Whiteshamism." Pp. 403–421 in *The State of Native America: Genocide, Colonization, and Resistance*, edited by M. Annette Jaimes. Boston: South End.

Sahlins, Peter. 1989. *Boundaries: The Making of France and Spain in the Pyrenees*. Berkeley: University of California Press.

Seals, David. 1979. *The Powwow Highway: A Novel*. New York: Plume.

———. 1992. *Sweet Medicine*. New York: Orion.

Secoy, Frank R. 1953. *Changing Military Patterns of the Great Plains Indians (17th Century through Early 19th Century)*. Monographs of the American Ethnological Society 21. Locust Valley, N.Y.: Augustin. (Reprinted 1992, Lincoln: University of Nebraska Press.)

Sheridan, Thomas E. 1992. "The Limits of Power: The Political Ecology of the Spanish Empire in the Greater Southwest." *Antiquity* 66:153–171.

Sherratt, Andrew G. 1993a. "What Would a Bronze-Age World System Look Like? Relations between Temperate Europe and the Mediterranean in Later Prehistory." *Journal of European Archaeology* 1:2:1–57.

———. 1993b. "Core, Periphery and Margin: Perspectives on the Bronze Age." Pp. 335–345 in *Development and Decline in the Mediterranean Bronze Age*, edited by Clay Mathers and Simon Stoddart. Sheffield: Sheffield Academic Press.

———. 1993c. "Who Are You Calling Peripheral? Dependence and Independence in European Prehistory." Pp. 245–255 in *Trade and Exchange in Prehistoric Europe*, edited by Chris Scarre and Frances Healy. Prehistoric Society Monograph. Oxford: Oxbow.

————. 1993d. "The Growth of the Mediterranean Economy in the Early First Millennium B.C." *World Archaeology* 24:3:361–378.

————. 1995. "Reviving the Grand Narrative: Archaeology and Long-Term Change." The Second David L. Clarke Memorial Lecture. *Journal of European Archaeology* 3:1:1–32.

Silko, Leslie Marmon. 1977. *Ceremony.* New York: Viking.

————. 1991. *Almanac of the Dead: A Novel.* New York: Simon & Schuster.

Slatta, Richard W. 1983. *Gauchos and the Vanishing Frontier.* Lincoln: University of Nebraska Press.

————. 1990. *Cowboys of the Americas.* New Haven, Conn.: Yale University Press.

————. 1997. *Comparing Cowboys and Frontiers.* Norman: University of Oklahoma Press.

————. 1998. "Spanish Colonial Military Strategy and Ideology." Pp. 83–96 in *Contested Ground: Comparative Frontiers on the Northern and Southern Edges of the Spanish Empire,* edited by Donna J. Guy and Thomas E. Sheridan. Tucson: University of Arizona Press.

Smith, Anthony D. 1986. *The Ethnic Origins of Nations.* Oxford: Blackwell.

————. 1991. *National Identity.* Reno: University of Nevada Press.

Smith, Joan, Jane Collins, Terence K. Hopkins, and Akbar Muhammad, eds. 1988. *Racism, Sexism and the World-System.* New York: Greenwood.

Snipp, C. Matthew. 1988. "Public Policy Impacts and American Indian Economic Development." Pp. 1–22 in *Public Policy Impacts on American Indian Economic Development,* edited by C. Matthew Snipp. Albuquerque: Native American Studies, University of New Mexico.

So, Alvin Y. 1984. "The Process of Incorporation into the Capitalist World-System: The Case of China in the Nineteenth Century." *Review* 8:1(Summer):91–116.

Socolow, Susan Migden. 1992. "Spanish Captives in Indian Societies: Cultural Contact Along the Argentine Frontier, 1600–1835." *Hispanic American Historical Review* 72:1(Feb.):73–99.

Thornton, Russell. 1987. *American Indian Holocaust and Survival.* Norman: University of Oklahoma Press.

Thornton, Russell, Tim Miller, and Jonathan Warren. 1991. "American Indian Population Recovery Following Smallpox Epidemics." *American Anthropologist* 93:1(March):28–45.

Upham, Steadman. 1992. "Interaction and Isolation: The Empty Spaces in Panregional Political and Economic Systems." Pp. 139–152 in *Resources, Power, and Interregional Interaction,* edited by Ed Schortman and Patricia Urban. New York: Plenum.

Wallerstein, Immanuel. 1974. *The Modern World-System: Capitalist Agriculture and the Origins of European World-Economy in the Sixteenth Century.* New York: Academic Press.

————. 1989. *The Modern World-System. III: The Second Era of Great Expansion of the Capitalist World-Economy, 1730–1840s.* New York: Academic Press.

Ward, Carol. 1997. "American Indian Women and Resistance to Incorporation: The Northern Cheyenne Case." Paper presented at the International Society for the Comparative Study of Civilization meeting, Provo, Utah, May.

Ward, Kathryn B. 1984. *Women in the World-System: Its Impact on Status and Fertility.* New York: Praeger.

————, ed. 1990. *Women Workers and Global Restructuring.* Ithaca, N.Y.: ILR.

————. 1993. "Reconceptualizing World-System Theory to Include Women." Pp. 43–68 in *Theory on Gender/Feminism on Theory,* edited by Paula England. New York: Aldine.

Ward, Kathryn B., and Jean Larson Pyle. 1995. "Gender, Industrialization, Transnational

Corporations, and Development: An Overview of Trends and Patterns." Pp. 37–63 in *Women in the Latin American Development Process*, edited by Christine E. Bose and Edna Acosta-Belen. Philadelphia: Temple University Press.

Weber, David J., and Jane M. Rausch. 1994. *Where Cultures Meet: Frontiers in Latin American History*. Wilmington, Del.: Scholarly Resources.

West, Elliott. 1998. *The Contested Plains: Indians, Goldseekers, and the Rush to Colorado.* Lawrence: University Press of Kansas.

Willard, Alice. 1993. "Gold, Islam and Camels: The Transformative Effects of Trade and Ideology." *Comparative Civilizations Review* 28(Spring):80–105.

Wilmer, Franke. 1993. *The Indigenous Voice in World Politics: Since Time Immemorial.* Newbury Park, Calif.: Sage.

Wilson, David R. 1997. "From Isolation to the Iron Age: the Northern Cheyenne and Energy Development." Paper presented at the International Society for the Comparative Study of Civilization meeting, Provo, May.

Wolf, Eric R. 1982. *Europe and the People without History*. Berkeley: University of California.

14

Modern East Asia in World-Systems Analysis

Alvin Y. So and Stephen W. K. Chiu

Compared with that of other regions, East Asia's development has been marked by the following distinctive features from the region's incorporation into the capitalist world-economy in the nineteenth century. First, East Asia is the only region to produce a non-European nation (Japan) that not only escaped colonization but also became an economic power challenging U.S. hegemony. Second, the largest state in East Asia (China) not only successfully transformed from an Asiatic empire into a communist state but also attained rapid industrialization in the late twentieth century. Third, East Asia is the only region where nation-states (China, Taiwan, Hong Kong, and South and North Korea) still divide along old cold war lines. Finally, in just half a century, the entire East Asian region attained national upward mobility. Not only are there now no peripheral states, but the entire region is an economic powerhouse for the world-economy.

CURRENT PERSPECTIVES ON EAST ASIAN DEVELOPMENT

This peculiar pattern of East Asian development has been largely discussed in terms of the following four perspectives. First of all, from a *neoclassical* perspective, Balassa (1988) argues that the growth of exports in the East Asian newly industrializing economies (NIEs) accounted for their GDP growth rates, which were among the highest for developing countries. Exports contributed to resource allocation according to comparative advantage, helped these nations overcome the limitations of their small domestic markets, and provided the "carrot and stick" of competition. The East Asian NIEs adopted such export policies because of their stable incentive system, limited government intervention, and reliance on private capital.

Second, there is the *cultural* explanation. The culturalists (e.g., Rozman 1992) argue that what the successful East Asian economies have in common is their Confucian traditions. Confucianism placed the family as the paramount institution within society, which led to the emergence of family entrepreneurship in East Asia. Furthermore, Confucianism shaped a new pattern of personalistic corporate management that is different from the West's rational, bureaucratic management. Finally, Confucianism glorified the established authority of the better-educated and rationalized their claims of superiority on the basis of their possessing specialized wisdom.

Third, there is the *statist* perspective, which contends that East Asian states play a strategic role in taming domestic and international market forces and harnessing them to national ends (Amsden 1989). Onis (1991) points out that East Asian industrializing states benefited from the unusual combination of both bureaucratic autonomy and public-private cooperation. As a result, strong autonomous states—Japan, South Korea, and Taiwan—emerged in East Asia, capable of not only formulating strategic developmental goals, but also translating these broad national goals into effective policies to promote rapid industrialization.

Finally, there is the *dependency* perspective. Bello and Rosenfeld (1990) contend that the NIEs were highly dependent upon the U.S. political hegemony after World War II. Serving the United States as its front lines against communism, South Korea and Taiwan received massive U.S. aid, loans, contracts, and advice, which helped them recover from World War II damages. In addition, military tension in East Asia justified the authoritarian NIE states in building up the military, banning labor unions and strikes, and suspending democratic elections. The states also used anticommunism as a hegemonic ideology to control the civil society (Koo 1993). After the end of the cold war in the 1980s, however, the NIEs experienced profound developmental problems, such as the threat of global protectionism, growing labor unrest and environmental disasters, and intense competition from other developing countries.

Although each perspective makes significant contributions to our understanding of the East Asian phenomenon, they suffer from the following problems. First, except for the dependency perspective, the other perspectives are unidisciplinary, focusing only on factors stressed by their own disciplines. Second, even when scholars in these traditions use the term *East Asia*, it becomes clear that they exclude socialist China and North Korea from it, supposedly because these countries possess "a different political and social system from the rest" of East Asia (Balassa 1988, 289). Third, each perspective restricts the unit of analysis, confining itself to the study of nations. Except for the dependency perspective, the literature also tends to ignore the regional and global dimensions of East Asian development. Fourth, these perspectives often adopt a short historical time span, mostly focusing on post–World War II development.

What is needed, then, is a comprehensive perspective that goes beyond disciplinary boundaries, adopts a larger unit of analysis to analyze socialist-capitalist state interactions and Asian regional dynamics, and traces a longer historical time span to examine the origins and transformation of East Asia.

TOWARD A WORLD-SYSTEMS ANALYSIS

In Wallerstein's (1991) formulation, world-systems analysis protests against the ways in which social science inquiry itself has been structured since its inception in the nineteenth century. First, Wallerstein questions whether academic disciplines can be separated from one another in the first place. Can the economy, the polity, and the society have, even hypothetically, autonomous activity? For instance, as markets are sociopolitical creations, can a true economic price somehow be stripped of its political and social bases? As an alternative, Wallerstein proposes a new historical social science that encourages researchers to examine the interactions among economics, politics, and society.

Second, Wallerstein's world-systems analysis fights a war on two fronts regarding the concept of time: against the traditional historians who would provide a narrative "episodic history" unique to the epoch described; and against the social scientists who search for "eternal laws" applicable across time. To go beyond these nineteenth-century conceptions of time, Wallerstein calls for a shift of focus from the historian's "episodic time" and the social scientist's "eternal time" to the historical social scientist's study of "structural time (the *longue duree*) and cyclical time (*conjonctures*)."

Third, Wallerstein questions the treatment of the state/society as the unit of analysis, asking where and when do the entities within which social life occurs exist? Thus, Wallerstein insists on taking the large-scale "historical world-system" rather than the state/society as the unit of scientific inquiry. For Wallerstein this is more than a mere semantic substitution, because the term *historical world-system* rids us of the central connotation that "society" is linked to the "state," and that the nation-state represents a relatively autonomous society that develops over time.

Wallerstein's (1979) task, therefore, is to apply such analysis to the modern capitalist world-economy (hereafter abbreviated as "CWE"). The CWE possesses a trimodal structure consisting of core, semiperiphery, and periphery, and dynamics of incorporation (the expansion of the CWE's outer boundaries to other parts of the world), deepening (processes through which production in one zone is geared into production in other zones of the world-economy), hegemony and rivalry (among the core states), regionalization (of the semiperipheral states), peripheralization, and cyclical rhythms. Furthermore, as the accumulators of capital lose their ability to resolve the CWE's contradictions, a worldwide wave of antisystemic movements emerges to transform the CWE towards a more democratic, egalitarian direction (So 1990).

To conclude, world-systems analysis provides a new mode of thinking that stresses large-scale, long-term, and holistic social change (the instantaneous interactions among politics, economics, and culture). In addition, it formulates many innovative concepts for examining the CWE's structure, dynamics, and crises.

A number of researchers close to the world-systems school now examine how the CWE's dynamics transforms the geopolitics and regional development in East Asia (Arrighi, Ikeda, and Irwan 1993; Cumings 1987; Gereffi 1992; McMichael 1987; Moulder 1977; Palat 1993). Following this line of thinking, this article argues that an examination of the historical processes of incorporation, regionalization, ascent, and centrality in the CWE sheds new light on the contours of East Asian development.

INCORPORATION

In the mid–nineteenth century, during the CWE's upward phase, Great Britain enjoyed the status of hegemon, with a virtual monopoly in industrial production and military technology. This industrial and military strength spurred the British state to promote the doctrine of "free trade" and "market incorporation" abroad—the informal extension of British interests through a vast market network rather than by direct colonial rule (Dixon 1991). But how did this market mode of incorporation into the CWE affect mid-nineteenth-century East Asian development and why did such incorporation result in the rise of Japan on the one hand but the fall of the Chinese empire on the other?

To a certain extent, Japan's path of development echoed that of China. Both states involved themselves extensively with maritime activity, world trade, and large-scale commercialization in the fourteenth and fifteenth centuries (Fitzpatrick 1992; Sanderson 1994); withdrew from world trade in the sixteenth century and grew totally isolated from the CWE from the seventeenth to the early nineteenth century; forcibly became incorporated into the CWE under "unequal treaties"; and, thanks largely to the market mode of incorporation, still possessed a high enough capacity to initiate domestic reforms in response to the challenges from the CWE by the late nineteenth century.

But Japan and China also differed greatly with regard to their processes of incorporation into the CWE. Hall (1986) suggests that the nature of incorporation is itself problematic, depending on a variety of factors that can affect the process. First, with regard to the countries' preincorporated legacies, China was an Asian empire, while Japan was not. As a "Middle Kingdom," China established a Sinocentric world order and extended its Confucian civilization to tributary states. China's bureaucrats and landed upper class were naturally reluctant to borrow techniques and values from outside "barbarians." By contrast, beginning with the Taika Reforms in the seventh century, Japan exhibited a historical legacy of importing new institutions from China and other countries.

Moreover, the structure of the governing elite in the two countries prior to their incorporation also differed. In Qing China, the close alliance between the central state and the landed upper class protected the Chinese empire from completely falling apart in the midst of foreign aggression and domestic antisystemic movements. In Tokugawa Japan, however, the deep cleavage between the Tokugawa government and the lords of outer feudal domains energized the samurai from the outer domains to promote the Meiji Revolution (Moulder 1977).

Second, with regard to the agents of incorporation, China was attacked by Britain, while Japan negotiated with the United States. As a hegemonic super-power in the CWE, Britain forcibly opened China up for trade and investment. On the other hand, the United States, a Western power yet to exercise its military domination in faraway East Asia, approached Japan. Moreover, as Japan was the Asian nation farthest removed from the reach of the great European naval powers, and as the core states perceived China as possessing far more attractive resources and markets, strategic geopolitical location enabled Japan to experience relatively less foreign domination and resource drain than China in the latter half of the nineteenth century (Norman 1975).

Third, with regard to incorporation responses, the Chinese were confronted with the Taiping Rebellion, while the Japanese achieved the Meiji Restoration. In China, the regional shift of foreign trade from Canton to Shanghai resulted in large-scale peasant uprisings; in Japan, the young samurai, seeing what the Westerners did to China, quickly evoked the emperor symbol, merged Confucianism with Shintoism, and promoted nationalistic sentiments during the Meiji Restoration (Gibney 1992).

Finally, with regard to state strength, the Chinese state possessed relatively weaker capacity to promote semiperipheral ascent than the Japanese state. Weakened by both foreign domination and peasant rebellions, the Chinese state could at best put forward a mediocre Self-Strengthening Movement. By contrast, the Japanese state, avoiding both strong foreign domination and large-scale peasant rebellions, and spurred by the imminent threat of colonialism and the urgency of national survival, carried out the far-ranging Meiji Reforms within a short period. Here the historical irony is that in China, the "weak" Qing state was strong enough to suppress the earlier challenges of peasant rebellions, thus precluding a fundamental restructuring of state organization and power, whereas in Japan, the weak Tokugawa government quickly crumbled in face of the nationalistic movement from the outer feudal domains, allowing a strong Meiji state to emerge and institute a variety of reforms (Moulder 1977).

Obviously, this different mode of incorporation greatly impacted the subsequent development of Japan and China. Whereas the Chinese empire started to crumble by the 1890s, Japan was on the road to a robust semiperipheral ascent. By then, the Japanese state began to feel the constraints of limited territory, tiny internal markets, and poor resources. What new policy, then, did the Japanese state adopt in order to sustain its drive toward industrialization?

REGIONALIZATION

By the late nineteenth century, the CWE had reached a downward phase and Britain had become a declining hegemon. In order to overcome stagnation, core states competed with one another for investment rights in the periphery and replaced their previous policy of market incorporation with territorial annexation and colonialism.

Observing this global trend towards colonialism, the Japanese state developed a regionalization strategy of empire building in East Asia. The turning point was Japan's defeat of China in the 1894 war, after which the sudden inflow of Chinese indemnity money paid for the cost of the war, helped finance Japan's development of heavy industry, and enabled Japan to shift to the gold standard. The opening of China's market provided an additional stimulus for Japan's textile industry. Furthermore, Japan emerged from the war as a nascent empire, possessing Taiwan and becoming involved in intrigues in Korea (Duus 1988).

Geopolitics in East Asia shaped Japan's regionalization strategy of continental expansion. First, Japan could afford to invade only her weaker neighbors, since she had yet to develop a strong military to exert her influence beyond East Asia. Second, as to timing, Japan was a "late" imperialist, finding herself already surrounded by Western imperialist powers and their Asian possessions. As a result, Japan's imperialist expansion in East Asia inevitably brought Japan into open conflict with the Western core states. Third, as to security, Japan as a semiperiphery still worried about core domination. In this respect, Korea was of critical geopolitical concern to Japan; if controlled by a hostile power, it could be a potent threat. Finally, in regard to imperialist ideology, Japan's semiperipheral status in East Asia amid Western imperialist powers led Japan's leaders to adopt what Duus (1988, 7) calls "a curious form of anti-imperialist imperialism," characterized by both anti-imperialist, pan-Asian rhetoric and by the imperialist rhetoric of continental expansion.

On the other hand, the rise of Japan led to the fall of China, whose defeat in the Sino-Japanese War in 1894 led to a scramble for concessions over her territory. In order to meet indemnity payments, the Qing government was forced to take out foreign loans, grant railway and mining concessions to foreigners, and raise internal taxes (Thomas 1984). The reforms of the Qing government, such as new schools and a new army, further eroded its traditional bases of social support from gentry-landlords and provincial governors, resulting in its final collapse in 1911.

Japan's regional dominance, however, was soon challenged by the United States—the rising hegemon of the CWE. During the Washington Conference of 1921–1922, the major Western powers managed to establish a basis for military equilibrium in East Asia to avoid an excessive buildup of the Japanese navy. The U.S. global democratic project even led to a brief era of Taisho democracy in Japan. In 1925 universal manhood suffrage was established, further increasing the democratic elements of the Japanese government. In response to liberal domestic reforms, Japan instituted a decade and a half of civilian administration for her Korean and Taiwanese colonies. For instance, she granted Korea a number of modest reforms designed to permit greater self-expression for Koreans, to abolish abuses in the judicial system, and to equalize educational and economic opportunities (Peattie 1988).

The world depression in the 1930s and subsequent dislocations in Japan's economy, nevertheless, encouraged Japan to move towards ultranationalism. Inspired by a romantic view of selfless devotion to the state and the emperor, the ultranationalists resorted to tactics such as terrorist assaults on key political leaders. With the support of these ultranationalists, the military's influence within the government expanded significantly, and Japan renewed her expansion into continental Asia. A "Japan-Manchukuo Economic Bloc" came into being in the 1930s, following the worldwide trend of a "sterling bloc" centered on Great Britain and the Commonwealth and a "dollar bloc" centered on the United States (Nakamura 1988).

As Japan plunged into the Pacific War, her central concerns were consolidation of the empire and the integration of colonial economies to meet the wartime requirements of Japan proper. Accordingly, the colonial governments of Korea and Taiwan drew up plans for major industrialization programs and downgraded agricultural production as an economic priority. In both Taiwan and Korea, therefore, companies created industrial facilities to produce the raw materials—petrochemicals, ores, and metals—needed by Japanese heavy industry (Borthwick 1992).

In China, on the other hand, the Sino-Japanese War stripped bare the contradictions of the Guomindang (GMD) regime. In the early 1930s, nationalist sentiments and boycotts against Japanese goods swept through China. But Chiang Kai-shek of the GMD ignored this surging nationalism because his strategy was first to defeat the communists and then resist against Japanese invasion. The war also sparked China's greatest inflationary rates of all time. The loss of major sources of revenues (the Japanese by then collected customs duties), and the need to finance increased wartime expenditures with resources from a few poor provinces in the interior, laid the roots of this inflation. This inflation not only produced widespread financial speculation and corruption within the GMD, but also affected intellectuals and government workers badly because they depended on a fixed salary (Bianco 1971). In addition, the Sino-Japanese War empowered the Communists by enlarging the area in which the Chinese Communist Party (CCP) expanded its political and military activities. Swallowed up in the vastness of China, the Japanese forces did not have the manpower to effectively control the countryside—where Communist guerrilla bases multiplied rapidly during the war years. Moreover, the war enabled the Communists to present themselves to the Chinese populace as nationalists; posted as the leaders of patriotic resistance against the Japanese, the Communists became saviors of the Chinese in the areas liberated from Japan. Thousands of students went to Yenan to work with the Communists for patriotic resistance; patriotic propaganda, with its program of reduction of land rent, succeeded in winning over more peasants than the agrarian revolution several years previously (Johnson 1962). In this respect, Japan's regionalization strategy set the timing for and acted as a powerful catalyst in the triumph of the Chinese Communist Revolution.

ASCENT

After World War II, the United States emerged as the new hegemonic power in the CWE, advocating free investment and promoting the twin ideologies of democracy and modernization. Under U.S. leadership, the postwar CWE became much more liberal, multilateral, and interdependent, leading to an unprecedented expansion of the capitalist world-economy. The main concern of the United States about East Asia in the late 1940s was containing the spread of communism from China to other parts of the region. Thus, the United States sent warships to protect the defeated Guomindang in Taiwan, dispatched soldiers to fight against the communists in Korea, and imposed an economic embargo on China.

The geopolitics of U.S. anti-communism profoundly impacted China's post-war development. First, due to intensive hostility from the United States, China adopted the Leninist model for state building to strengthen her political order and confront "imperialist enemies." Second, the cold war climate motivated the nascent socialist state to accelerate processes of collectivization, nationalization, and heavy-industry-led growth in order to channel national resources into the defense industry, allowing China to become a strong state capable of defending its territory as soon as possible (Selden 1992). Third, in order to withstand U.S. aggression, China developed a mass mobilization policy to prepare for war. When the United States intensified its military involvement in Vietnam in the mid-1960s, Mao Tse-tung, the chairman of the CCP, started the Cultural Revolution. The Cultural Revolution was aimed at consolidating support from poor peasants and unskilled workers—the disadvantaged masses in the class hierarchy. Thus, entitlement programs for job security, housing, child care, and pensions were granted to the urban working class, while social programs in education, health care, and welfare became increasingly available to the peasantry. Finally, China put forward an inward-looking, "self-reliance" model of development, stressing national autonomy, pride in being a poor country, and developing heartland rather than coastal provinces. By relocating key industries in heartland provinces, the CCP hoped that it could avoid any economic ruin that would be caused by a U.S. invasion (Naughton 1991).

The U.S. anti-communist project also affected Japan's postwar development. In the mid-1940s, Japan fell under the control of the U.S. occupation policies, including demilitarization and democratization in the political sphere, demands for wartime damages, antitrust programs, and land reforms in the economic sphere. However, by the late 1940s, the United States reversed the above occupation policies so as to build Japan up as its junior partner in East Asia to fight communism. Special concessions by the United States to Japan—such as the Korean War procurement, U.S. aid and open markets, and the relief of Japan's defense burden—were instrumental in promoting Japan's economic success. In addition, with U.S. hegemony enforcing relatively "free" transactions in the world economy, Japan obtained the requisite supply of critical inputs from its former colonies without resorting to its prewar strategy of empire building. Under U.S. patronage, Japan promoted industrial and financial policies, incorporated labor into enterprise unionism, installed an ideology of consensus, paternalism, and sexism, and engaged in transborder expansion so as to transform Japan into a core state as soon as possible (Arrighi 1994).

Cold war geopolitics benefited Hong Kong, South Korea, and Taiwan as well. The "windfall profit" from the Chinese Communist Revolution enabled Hong Kong to take advantage of refugee capital and laborers and start its industrial revolution in the early 1950s. In addition, U.S. aid and loans to South Korea and Taiwan carried great weight in alleviating huge government budget deficits, financing investment, paying for imports, and helping their authoritarian states to stay in power. Nearly all U.S. aid before the 1960s was provided on a grant basis, thus making it possible for South Korea and Taiwan to begin their export-led growth in the 1960s without a backlog of debt. In order to provide more incentives for the East Asian NIEs to adopt export-led industrialization, the United States also opened her own market to them. This explains why despite the notorious closed domestic markets of South Korea and Taiwan, their exports still enjoyed unrestricted access to the U.S. market for so long. The U.S. market was critical to the NIEs' economic growth because it was their largest single market throughout the 1960s and 1970s (Chiu 1994).

In the 1970s, a further "organic division of labor" between Japan and the NIEs married Japanese capital and technology to the cheap, relatively docile labor in Hong Kong, South Korea, and Taiwan. The East Asian NIEs became a base for the relocation of Japan's labor-intensive industries as Japan moved upward through each product cycle and progressively toward high-tech production (Cumings 1987).

The economic success of Japan and the NIEs and the decline of the cold war in the late 1970s lured China back to a mercantilist strategy in order to achieve upward mobility in the CWE. During this period, China's top priority was to develop productive forces to catch up with the capitalist core states as rapidly as possible. Thus, China opened four Special Economic Zones and fourteen coastal cities to attract direct foreign investment. In the countryside, China dismantled her communes, gave peasant families plots of land to cultivate, and made them responsible for their own gains and losses. State and collective enterprises were asked to contract out their unprofitable operations to small enterprises or temporary workers. Such a return to mercantilism necessitated a revision in ideology. The new ideology attacked Mao and Maoism; labeled the Cultural Revolution as ten years of disaster; and condemned the values of egalitarianism and collectivism as excessive and unrealistic. Replacing these were a materialistic ideology that emphasized productivity, expertise, technology, modernization, and consumerism (So 1992).

CENTRALITY

In the 1970s, the golden era of postwar economic expansion came to an end and the United States became a declining hegemon. Just like the previous period of declining hegemony, core states formed regional blocs to enlarge their trading networks and protect themselves from competitors in order to overcome economic stagnation. It was through these global dynamics that East Asia became a new epicenter of capital accumulation in the world-economy. One by one, East Asia states came out from under the U.S. shadow and began to assert their own economic, political, and cultural initiatives.

Japan, in particular, grew from a regional to a global economic power, challenging the hegemony of the United States. In one product area after another, Japanese manufacturers outsold their U.S. competitors in such areas as TVs, steel, and automobiles. With their technological and productivity level, Japanese manufacturers controlled a sizable share of the U.S. market in many critical products. In addition, Japan emerged as a superpower in the global financial system. The share of Japanese banks in international assets (36 percent) far exceeded that of the United States (11.9 percent) by 1990. Third, Japan overtook the United States as the largest foreign investor in the world. In 1989, Japan's investment reached US$43.08 billion, whereas U.S. investment was merely US$40.5 billion (Das 1993, 120).

Subsequently, there emerged a new "flying geese" ideology, which portrayed Japan as playing the lead position in the East Asian region owing to its most advanced level of technological sophistication. Ranked behind Japan in a spreading "V" were first the NIEs (Hong Kong, Singapore, South Korea, and Taiwan) and then the Southeast Asian states (Malaysia, Thailand, Indonesia, and the Philippines). The "geese" behind Japan would learn from the progress of those up ahead, move into their positions, and eventually close the technological gap. This model predicted that the NIEs would follow the Japanese pattern, while the Southeast Asian states would follow the NIEs. As a result, every player could supposedly improve its position in following Japanese leadership in East Asia (Hill and Fujita 1995). The ideological implication was that Japan replaced the United States as the new model of development for many peripheral states. In response, the United States imposed protectionist measures against Japanese imports and sought to open up the Japanese market for U.S. imports, while the U.S. mass media started "Japan-bashing."

Another event with implications for the centrality of the East Asian region was the Chinese national reunification project. From an economic viewpoint, the national reunification of the three Chinese states could be seen as mutually beneficial and greatly enhancing the competitiveness of their respective economies. On the one hand, mainland China helped solve the developmental problems of Taiwan and Hong Kong by providing them with cheap labor, resources, and investment opportunities. On the other hand, Taiwan and Hong Kong contributed to mainland China's development by providing employment opportunities, market stimuli to local enterprises, and vital information and contacts for reentering the world-market. In the late 1980s, mainland China's state managers successfully developed close business partnerships with the capitalists of Hong Kong and Taiwan, and this "unholy alliance" will likely continue into the next century (Hsiao and So 1993).

Nevertheless, the prospects for this national reunification project remain problematic. This project deepened the division between the right-wing Guomindang conservatives and the radical independence faction in Taiwan, while in Hong Kong the 1997 issue led to the emergence of a new Hong Kong ethnic identity and a massive emigration of the new middle class. The people of Taiwan and Hong Kong tend to distrust mainland China's promise of establishing highly autonomous special administrative zones after national reunification. On the other hand, mainland China still remains suspicious of the Guomindang for promoting Taiwan's independence and the Hong Kong government for laying the ground-work for the continuation of British rule after 1997 (So and Kwok 1995).

If we locate the Chinese triangle of mainland China-Taiwan-Hong Kong in the wider context of the CWE, it seems that national economic integration is the best strategy for the three Chinese states to overcome the "middle squeeze" from the core's protectionism and the periphery's intense competition. Should this national reunification project succeed, this Chinese triangle may evolve into a new regional economic power, challenging the dominance of Japan and the United States in East Asia.

From a theoretical angle, what are the merits of the above interpretation over the current theories on East Asian development?

THEORETICAL REPRISE

Using the large-scale, holistic, and long-time-span heuristic devices of world-systems analysis, this article highlights the crucial role played by geopolitics in East Asian development.

Toward a Geopolitical Explanation

In terms of large-scale analysis, this article studies interstate dynamics in the East Asian region. China, Japan, and Korea greatly influenced one another's development owing to the geopolitics of their strategic locations in East Asia. From the sixteenth century to the mid–nineteenth century, Japan escaped from colonization largely because it was far away from European trade routes, and because the core states were attracted instead to the huge Chinese hinderland. The rise of Japan in the early twentieth century, in turn, led to the colonization of South Korea and Taiwan as well as the acceleration of the communist revolution's triumph in China. Then the 1949 Chinese Communist Revolution caused the United States to reverse its Japanese occupation policy so as to build up its former enemy as a junior partner in the domination of East Asia. The transborder expansion of Japanese industries in the early 1970s accelerated the economic growth of Hong Kong, South Korea, and Taiwan, which in turn prompted China to reenter the CWE.

Through holistic analysis, this study shows that geopolitics is often intertwined with emerging cultural constructs and changing regional dynamics. For instance, when Japan started to develop a regionalization project to conquer East Asia in the early twentieth century, there arose the ideology of anti-imperialist pan-Asianism in

Japan proper. When China was forced to withdraw from the CWE because of the cold war in the mid-twentieth century, Maoism and revolutionary socialism were at their height, preaching mass mobilization and self-reliance. And when mainland China pushed for national reunification in the 1980s, new ethnic identities quickly arose among the residents of Taiwan and Hong Kong, revealing their political distrust of the national reunification project.

Through long-term analysis, this article shows that contemporary East Asia must be understood in terms of its pre-World War II geopolitical development. For instance, the rapid growth of Japanese heavy industry in the 1960s and the 1970s had historical origins in the wartime munitions industry in the 1930s and the 1940s. Links between big Japanese factories and small businesses developed in the munitions industry became the basis for the postwar subcontracting system. Wartime controls left a legacy of administrative guidance by government ministries—a fundamental characteristic of the postwar developmental state. As for South Korea, its robust *minjung* movement owes much to the strong nationalist resistance movement against Japanese colonial rule in the early twentieth century.

Contextualizing Current Theories

The above emphasis on geopolitical factors, needless to say, is not aimed at refuting entirely the current economic, cultural, statist, and dependency perspectives of East Asian development. Obviously, market and private enterprises, Confucianism and corporate management, the developmental state, and the "dark side" of dependent development are important factors of East Asian development. What this study hopes to contribute, however, is the often neglected geopolitical context necessary for reinterpreting these important factors.

First, while developmental states were instrumental in promoting industrialization in South Korea and Taiwan, they owed much to the repressive policies of the Japanese colonial government in the early decades of the twentieth century, to the purging of leftist forces in South Korea and of local Taiwanese elites in the late 1940s, to hypermilitarism during the cold war, as well as to the strong influence of U.S. advisors in the 1950s.

Second, while markets and private enterprises provided the dynamics for East Asian exports, their existence depended highly on special privileges granted to the East Asian states by the United States in gaining access to its market, on the "alliance for profit" between the South Korean state and the *chaebols*, on the monopolization of state sector by the GMD in Taiwan, and on the influx of refugee entrepreneurs in Hong Kong.

Third, while Confucianism was a common cultural trait in the East Asian region, it became fused with Japanese Shintoism in the late nineteenth century, severely criticized during the 1919 May Fourth Movement and the Chinese Cultural Revolution in the 1960s, evoked in Japan in the 1950s when the *zaibatsu* faced serious challenges from the labor movement, and existed side by side with sexism, anticommunism, the Three People's Principles, the ideology of economic growth, and refugee mentality in the NIEs.

Finally, while East Asian economic growth did indeed have a dark side, the NIEs overcame their dependency and moved into semiperipheries largely because of the upward phase of the CWE and generous U.S. and advice in the 1950s, the influx of foreign earnings into the NIEs during the Vietnam War, the NIEs' privileged access to the U.S. market in the 1960s, and the cheap credit offered by U.S. banks as well as the transborder expansion of the Japanese subcontracting system in the 1970s.

In sum, world-systems analysis highlights the salience of geopolitical factors in shaping the historical development of East Asia since its incorporation into the CWE in the early nineteenth century.

NOTE

This article presents some of the key arguments of our book *East Asia and the World-Economy* (So and Chiu 1995). We want to thank Thomas Hall and two anonymous referees for their helpful comments and criticisms on an earlier draft.

REFERENCES

Amsden, Alice. 1989. *Asia's Next Giant: South Korea and Late Industrialization.* New York: Oxford University Press.

Arrighi, Giovanni. 1994. *The Long Twentieth Century.* London: Verso.

Arrighi, Giovanni, Satoshi Ikeda, and Alex Irwan. 1993. "The Rise of East Asia: One Miracle or Many?" Pp. 41–65 in *Pacific-Asia and the Future of the World-System,* edited by Ravi Palat. Westport, CT: Greenwood.

Balassa, Bela. 1988. "The Lessons of East Asian Development: An Overview." *Economic Development and Cultural Change* 36(3, supplement):S273–S290.

Bello, Walden, and Stephanie Rosenfeld. 1990. "Dragons in Distress: The Crisis of the NICs." *World Policy Journal* 7:431–468.

Bianco, Lucien. 1971. *Origins of Chinese Revolution, 1915–1949.* Stanford: Stanford University Press.

Borthwick, Mark. 1992. *Pacific Century: The Emergence of Modern Pacific Asia.* Boulder, CO: Westview.

Chiu, Stephen. 1994. "The Changing World Order and the East Asian Newly Industrialized Countries." Pp. 75–114 in *Old Nations, New World: The Evolution of a New World Order,* edited by David Jacobson. Boulder, CO: Westview.

Cumings, Bruce. 1987. "The Origins and Development of the Northeast Asian Political Economy." Pp. 44–83 in *The Political Economy of the New Asian Industrialism,* edited by Frederic Deyo. Ithaca, NY: Cornell University Press.

Das, Phillip, 1993. *The Yen Appreciation and the International Economy.* London: Macmillan.

Dixon, Chris. 1991. *South East Asia in the World-Economy.* Cambridge: Cambridge University Press.

Duus, Peter. 1988. "Introduction." Pp. 1–54 in *The Cambridge History of Japan.* Vol. 6, *The Twentieth Century*, edited by Peter Duus. Cambridge: Cambridge University Press.

Fitzpatrick, John. 1992. "The Middle Kingdom, the Middle Sea, and the Geographical Pivot of History." *Review* 15:477–521.

Gereffi, Gary. 1992. "New Realities of Industrial Development in East Asia and Latin America: Global, Regional, and National Trends." Pp. 85–112 in *State and Development in the Asian-Pacific Rim*, edited by Richard Appelbaum and Jeffrey Henderson. Newbury Park, CA: Sage.

Gibney, Frank. 1992. "Introduction: Arrival of the Black Ships." Pp. 119–127 in *Pacific Century*, edited by M. Borthwick. Boulder, CO: Westview.

Hall, Thomas D. 1986. "Incorporation in the World-system: Toward a Critique." *American Sociological Review* 51:390–402.

Hill, Richard Child, and Kuniko Fujita. 1995. "Product Cycles and International Divisions of Labor: Contrasts Between the United States and Japan." Pp. 91–108 in *A New World Order? Global Transformation in the Late Twentieth Century*, edited by David A. Smith and Jozsef Borocz. Westport, CT: Praeger.

Hsiao, Hsin-Huang Michael, and Alvin Y. So. 1993. "Ascent Through National Integration: The Chinese Triangle of Mainland-Taiwan-Hong Kong." Pp. 133–150 in *Pacific-Asia and the Future of the World-Economy*, edited by Ravi Palat. Westport, CT: Greenwood.

Johnson, Chalmers. 1962. *Peasant Nationalism and Communist Power.* Stanford: Stanford University Press.

Koo, Hagen. 1993. "Strong State and Contentious Society." Pp. 231–249 in *State and Society in Contemporary Korea*, edited by Hagen Koo. Ithaca, NY: Cornell University Press.

McMichael, Phillip. 1987."Foundations of U.S./Japanese World-Economic Rivalry in the Pacific Rim." *Journal of Developing Societies* 3:62–77.

Moulder, Frances. 1977. *Japan, China, and the Modern World-Economy.* Cambridge: Cambridge University Press.

Norman, E. H. 1975. *Origins of the Modern Japanese State.* New York: Pantheon.

Nakamura, Takafusa. 1988. "Depression, Recovery, and War, 1920–1945." Pp. 451–493 in *The Cambridge History of Japan*. Vol. 6, *The Twentieth Century*, edited by Peter Duus. Cambridge: Cambridge University Press.

Naughton, Barry. 1991. "Industrial Policy During the Cultural Revolution: Military Preparation, Decentralization, and Leaps Forward." Pp. 153–182 in *New Perspectives on the Cultural Revolution*, edited by William A. Joseph, Christine Wong, and David Zweig. Cambridge: Council on East Asian Studies, Harvard University.

Onis Ziya. 1991. "Review Article: The Logic of Developmental State." *Comparative Politics* 24:109–126.

Palat, Ravi Arvind. 1993. "Introduction: The Making and Unmaking of Pacific-Asia." Pp. 3–20 in *Pacific-Asia and the Future of the World-System*, edited by Ravi A. Palat. Westport, CT: Greenwood.

Peattie, Mark. 1988. "The Japanese Colonial Empire, 1895–1945." Pp. 217–270 in *The Cambridge History of Japan*. Vol. 6, *The Twentieth Century*, edited by Peter Duus. Cambridge: Cambridge University Press.

Rozman, Gilbert. 1992. "The Confucian Faces of Capitalism." Pp. 310–318 in *Pacific Century*, edited by Mark Borthwick. Boulder, CO: Westview.

Sanderson, Stephen K. 1994. "The Transition from Feudalism and Capitalism: The Theoretical Significance of the Japanese Case." *Review* 17:15–55.

Selden Mark. 1992. *The Political Economy of Chinese Development*. Armonk: M. E. Sharpe.

So, Alvin Y. 1992. "The Dilemma of Socialist Development in the People's Republic of China." *Humboldt Journal of Social Relations* 18:163–194.

———. 1990. *Social Change and Development*. Newbury Park, CA: Sage.

So, Alvin Y., and Stephen Chiu. 1995. *East Asia and the World-Economy*. Newbury Park, CA: Sage.

So, Alvin Y., and Reginald Kwok. 1995. "Socio-economic Core, Political Periphery: Hong Kong's Uncertain Transition Toward 1997." Pp. 251–258 in *Hong Kong-Guangdong Link: Partnership in Flux,* edited by Reginald Kwok and Alvin Y. So. Armonk: M. E. Sharpe.

Thomas, Stephen C. 1984. *Foreign Intervention and China's Industrial Development, 1870–1911*. Boulder, CO: Westview.

Wallerstein, Immanuel. 1991. *Unthinking Social Science: The Limits of Nineteenth-Century Paradigms*. Cambridge: Polity.

———. 1979. *The Capitalist World-Economy*. Cambridge: Cambridge University Press.

Part V

Future Visions

15

From State Socialism to Global Democracy: The Transnational Politics of the Modern World-System

Terry Boswell and Christopher Chase-Dunn

> "Socialism is the path from capitalism to capitalism."

> "Capitalism is the exploitation of humans by humans; socialism is exactly the opposite."

> "Lenin defined socialism as all power to the Soviets and a program of electrification—the Soviets had all the power, but the people were still waiting for the damn electricity."

These definitions of socialism were popular jokes in Eastern Europe prior to the revolutions of 1989.[1] Given the widespread antipathy toward Communist Party rule, along with the deterioration of economic conditions in Eastern Europe, the derision of the "actually existing" state socialism as experienced in the East is not unexpected.

The initial, perhaps less expected, response was an embrace in Eastern Europe of unfettered market capitalism, including fondness for the likes of Margaret Thatcher, Milton Friedman, and even Augusto Pinochet in some circles. As the market has since worked its miracle of creative destruction and capital concentration, revulsion against growing poverty, crime, and inequality has tempered the amour. Former communists are now winning the elections that they formerly opposed. Nevertheless, most of the former communists coming back into power are nationalists first, closer to Pat Buchanan than to Karl Marx.

The collapse of the Soviets' self-proclaimed "second world" of state socialism has left Marxists and other leftists numbed at the difficulty, if not impossibility, of changing the capitalist world-system. Stalin soon betrayed the socialist principles of the Russian Revolution. But starting with Trotsky there was always the hope that genuine democratization could restore the second world to a viable alternative. Democratic socialists supported political revolution to overthrow the Stalinist dictatorship

289

but opposed the destruction of the Soviet Bloc itself. It had to be defended because the USSR proved that socialism was possible, if only it could be made democratic. Gorbachev, to his credit, tried to steer a path to a more democratic socialism. He failed, and despite his lionization in the West, most Russians think he only made things worse. Other socialists and social democrats wanted the Soviet bogeyman out of the way, supporting much of the cold war to focus on issues of class and poverty in domestic politics without being painted pink. The second world was for them a setback and an impediment to social progress. Regardless of whether democratic socialists or social democrats were right in the past, both proved wrong about what would happen next.

"Socialism is dead," shouts Ralf Dahrendorf.[2] The fall of the regimes in Eastern Europe has ushered in a great deal of shouting about the triumph of liberalism, as well as a heavy load of talk about the "end of ideology" (Fukuyama 1992). Nor is the antipathy confined to the former Soviet bloc. Eurocommunist parties have changed their names and endorsed social democracy, or disappeared. Communists in the periphery are also disappearing or where in power, such as Cuba, China, and Vietnam, becoming the managers of state capitalism. Nor did social democrats reap their long hoped-for windfall from the disappearance of their erstwhile competitor on the left. Without the communist alternative, social democrats have lost much of their hold on moderates seeking a party of reasonable compromise. Instead, all socialist rhetoric has become suspect. The "third way," which included varieties of democratic socialism, evolutionary socialism, social democracy, and welfare states, endures derision and despair by association, despite the long-ago break with the Stalinist path and critique of Soviet tyranny.

Is socialism dead? Is any fundamental social change possible, or are we better off seeking individual self-fulfillment and leaving the trajectory of human history to the invisible hands of uncontrollable circumstances? Our answer is no. *World revolutions* have repeatedly challenged and eventually changed the political rules that govern capitalist relations over the five hundred years of its existence. Abolishing slavery, liberating colonies, and winning democracy have been the three most progressive changes in the world order. This answer requires seeing how socialist and other progressive movements have changed the system in the past and what the possibilities are for the future. As such, we premise the answer on the worldview offered by a *world-systems perspective*.

As discussed in all the chapters in this book, the world-systems perspective (i.e., Wallerstein 1984; Arrighi 1994; Chase-Dunn and Hall 1997; Chase-Dunn 1998) starts with viewing the world-economy as the unit of analysis. World-systemic trends and cycles are discernible only over the long term and from comparing different zones of the world economy—core, semiperiphery, and periphery. The world's major inequality is between a high-technology industrial core and an underdeveloped periphery of former colonies, with an industrializing semiperiphery falling in between, dominated by its core neighbors and returning the favor to the surrounding periphery. World trade and other international interactions form a linked "hub and spoke" pattern (analogous to airline routes) that is visible only at the global level, with core states having multiple sources of exchange and peripheral states dependent on their core links.

By lifting the unit of analysis from societies to the world-system, we place the short life history of state socialism in a systemic context. Eastern European specialists will discuss, debate, and describe in great detail the unique sources of revolt in each country for decades to come. Our mission is to focus on a different level of reality and a different scale of time. A world-system conceptualization of socialism and progressive social change differs from the societal notions of socialism that have heretofore been understood as the transformation of nations. None of the efforts to construct socialism at the level of national societies were successful in building a self-sustaining socialist mode of production. Given the strength of larger forces in the capitalist world-economy, this was never feasible in practice. Revolutionaries of all stripes have long faced the conundrum of seeking to overthrow a particular state when the political-economic system as a whole is global. Attempting to transform the system through revolution seems pointless, as revolutions only change one state at a time, and since no state or bloc of states has been able to change or exit the system, transformation of the world-system has been thwarted before it could be reached. In attempting to build a "second world," the Soviet bloc found it could not separate itself from military competition or economic influence from the West, and where it did find autarky, it was cut-off from scientific and technical development that feeds on open and wide exchange.

Soviet-style communism, revolving around using the state to force economic development, largely at the expense of the peasantry, was a product of the semiperiphery. In the periphery, however, national liberation movements based on mobilizing the peasantry dominated, while in the core a social democratic compromise emerged between capital and labor (Arrighi, Hopkins, and Wallerstein 1989a, 1992; Korpi 1982). Each set of left-wing social movements emerged in the confluence of development over time and of uneven development between zones of the world economy. Historically, the origins of each can be traced to the uneven progress of past world revolutions, all of which were failures at socialism even where they captured state power, but which made huge global changes. The three modal revolutions that initiated socialist movements were the following:

- 1848, European revolutions, initiated social democratic movements in the core;
- 1917, Russian Revolution, initiated state socialist movements in the semiperiphery;
- 1949, Chinese Revolution, initiated national liberation movements in the periphery.

World revolutions, in which new global institutional arrangements are initiated and past institutional changes are consolidated, result when the effects of social revolts are widespread or widely emulated throughout the core of the world-system. World revolutions tend to be progressive, in the sense that they shift labor market competition away from lowest wages and toward highest productivity—such as the world revolution of 1848 that ushered in the end of slavery and the beginnings of labor unions. Since 1848, world revolutions have revolved around the issues of socialism.

Over the long term, the expansion of capitalism interacts with the efforts of people to resist domination and to control exploitation. Globally, expansion and resistance produce a *long-run spiraling interaction* between expanding capitalism and socialist organizational forms. It is this larger world-system that develops, not national societies. What has been called national development is, in world-systems theory, upward mobility in the core/periphery hierarchy. The history and developmental trajectory of the socialist states can be explained as counterhegemonic movements in the semiperiphery that attempted to transform the basic logic of capitalism but that ended up using socialist ideology to mobilize industrialization for the purpose of catching up with core capitalism.

The spiraling interaction between capitalist development and socialist movements can be seen in the history of labor movements, socialist parties, and socialist states over the last two hundred years. This long-run comparative perspective enables us to see the events in the late 1990s in China, the Soviet Union, and Eastern Europe in a framework that has important implications for the future of socialism. The metaphor of the spiral means this: both capitalism and socialism affect one another's growth and organizational forms. Capitalist exploitation and domination spur socialist responses, and socialist organizations spur capitalism to expand market integration, to revolutionize technology, and to reform politically. Class conflict oscillates along the spiral among open struggle, managed compromise, and repressive domination. While class conflict of all types is omnipresent, open struggles proliferate in the semiperiphery, managed compromise is a hallmark of the core, and repressive class domination plagues the periphery.

The capitalism referred to here is the phenomenon not only of capitalist firms producing commodities but also of capitalist states and the modern interstate system, which is the political backdrop for capitalist accumulation. The world-systems perspective has produced an understanding of capitalism in which geopolitics and interstate conflict are normal processes of capitalist political competition. Socialist movements are, defined broadly, those antisystemic political and organizational efforts in which people try to protect themselves from and gain control over market forces, capitalist exploitation, and state domination. The series of industrial revolutions in which capitalism restructured production and the control of labor have stimulated a succession of political organizations and institutions created by workers and communities to protect their livelihoods. This happened differently under particular political and economic conditions in different parts of the world-system, but viewing them on a global scale, we can see the commonalities and temporal relationships of all these movements. Skilled workers created guilds and craft unions. Less skilled workers created industrial unions. In many countries, these coalesced into labor parties that played central roles in the development of welfare states. In other regions workers were less politically successful but managed at least to protect access to rural areas or subsistence plots for a fallback or hedge against the insecurities of employment in capitalist enterprises.[3]

The varying success of workers' organizations has had an impact back on the further development of capitalism. In some areas, workers or communities were successful at raising the wage bill or protecting the environment in ways that raised the costs of production for capital. When this happened, capitalists either employed more productive technology or migrated to where fewer constraints allowed cheaper production. Where skilled labor and advanced technologies increase productivity, living standards rise, setting a new and higher floor of social standards. The alternative of capital flight depends on whether the technologies and skills are transferable to low-wage areas. If so, the profit gains are even greater. The process of capital flight is not a new feature of the world-system. It has been an important force behind the uneven development of capitalism and the spreading scale of market integration for centuries.

But like all facets of contemporary capitalism, the pace has quickened, particularly in the last thirty years. During the long post–World War II expansion, labor unions and socialist parties were able to obtain some power in certain states. Fordism, the employment of large numbers of assembly-line workers in centralized production locations, had facilitated industrial union organizing. The response by capital was to become yet more international. Firm size increased. International markets became more and more important to successful capitalist competition. By the 1980s "flexible accumulation" was supplanting Fordism. Under flexible accumulation, new technologies such as computer-controlled systems allow for short production lines, quick style changes, mobile capital, and global sourcing. Flexible accumulation reduces the viability of traditional labor organization strategies based on stable assembly lines and huge investments in fixed capital within national states. Just as the local craft unions in the 1890s were unable to organize the then new assembly-line industries, national industrial unions in the 1990s are faltering in the face of global capitalism. In the 1890s, the meaning and goals of socialism changed in tandem, from the "cooperative commonwealth," to the socialist state. In the 1990s, the meaning and goals of socialism are changing again.

The most recent world revolution occurred in 1989. Applying a world-systems perspective to the revolutions of 1989 compares these events to prior social transformations in the governance structure of the world economy. For the first forty years, state socialism was a successful strategy for promoting rapid industrial development. To third world countries that had been shackled by Western colonialism, the "socialist path" once appeared to promise development, equality and independence. Yet, the socialist states faltered badly since the late 1960s, finally resulting in an economic crisis so deep as to make them vulnerable in 1989 to the spread of popular revolution. What happened in the last twenty years to undermine and eventually topple state socialism? Why was their industrial rise temporary? What can we learn from this failed experiment, both in why it failed and where it managed some success?

STATE SOCIALISM IN THE WORLD-SYSTEM

Socialists were able to take state power in certain semiperipheral and peripheral states, and they used this power to create political mechanisms of protection against competition with core capital. This was not a wholly new phenomenon. As discussed later, capitalist semiperipheral states had done, and were doing, similar things. But the socialist states claimed a fundamentally oppositionist ideology in which socialism was a superior system that would eventually replace capitalism. Ideological opposition is a phenomenon that the capitalist world-economy has also seen before. The geopolitical and economic battles of the Thirty Years War were fought in the name of Protestantism against Catholicism. The content of the ideology makes important differences for the internal organization of states and parties, but every contender must be able to legitimate itself in the eyes and hearts of its cadre. The claim to represent a qualitatively different and superior socioeconomic system on the part of the socialist states is not evidence that they were indeed structurally autonomous from, or qualitatively different from, world capitalism.

This constraint on the mobility of capital was an important force behind the post–World War II wave of the increasing scale of market integration and a new revolution of technology. In certain areas capitalism was driven to further revolutionize technology or to improve living conditions for workers and peasants because of the demonstration effect of propinquity to a socialist state. U.S. support for state-led industrialization of South Korea (in contrast to U.S. policy in Latin America) is only understandable as a geopolitical response to the Chinese revolution (Cumings 1987). The existence of "two superpowers"—one capitalist and one communist—in the period since World War II provided a fertile context for the success of international liberalism within the "capitalist" bloc (Wallerstein 1995). This was the political/military basis of the rapid growth of transnational corporations and the latest revolutionary "time-space compression" (Harvey 1989). This technological revolution has once again restructured the international division of labor and created a new regime of labor regulation called "flexible accumulation." The processes by which the socialist states have become reintegrated into the capitalist world-system have been long, as described later. But, the final phase of reintegration since 1989 was provoked by the inability to be competitive with the new forms of capitalist regulation (Boswell and Peters 1990; Chirot 1991). Thus, capitalism spurs socialism, which spurs capitalism, which spurs socialism again in a wheel that turns and turns while getting larger—the spiral.

One might say that we perform a "biopsy" on state socialism, rather than the autopsy that most specialists are performing, because a world-systems view shows that the communist states never left the system. Communist states never constituted an alternative "second world," but rather were always states pursuing a specific political strategy for development within the system.

Socialists have long envisioned that each national revolution would give inspiration and support to the next until every state was socialist—a progressive domino theory. A truly new world order would then be created a piece at a time. The problem with the state socialist countries, from our perspective, is that *they went down the path backward.* Rather than socialist states cumulating until they produced world socialism, the institutions and relations at the global level must be changed to generate equality and end exploitation in every state.

Marx always conceived of socialism in global terms. In his last writings he condemned the program of the German Social Democratic Party for inadequate internationalism that failed to place the state "economically 'within the framework' of the world market" and "politically 'within the framework' of the system of states" (1875, 544). And in some of his first scientific work he was prophetic:

> Without this [world socialism] (1) Communism could only exist as a local event; (2) the forces of intercourse themselves could not have developed as universal, hence intolerable powers: they would have remained home-bred superstitious conditions; and (3) each extension of intercourse would abolish local communism. Empirically, communism is only possible as the act of the dominant peoples "all at once" or simultaneously, which presupposes the universal development of productive forces and the world-intercourse bound up with them. . . . The proletariat can thus only exist *world-historically,* just as communism, its movement, can only have a "world-historical existence" (Marx 1846–47,178–179)

The unanswered question is, How is socialism created "all at once"? Our answer and fundamental starting point is one of *global democracy.* Global democracy has a dual meaning—democracy at the global level, with democratic institutions governing the ever-more integrated world economy, and local democracy, with all economic and social as well as political administration and management, open to democratic participation. Democracy includes civil and individual human rights, without which democratic institutions are meaningless. It encompasses political, social, and economic realms, rather than posing an artificial separation among them.[4]

As a political and theoretical concept, the term *global democracy* is a global analog for the societal term *social democracy* as it was understood prior to the rise of the communist states. Social democracy has been a theory and a practice of progress toward the goals of steadily raising the living standards and insuring the basic needs of the working class, expanding the public sphere and community life, and eliminating all forms of oppression and exploitation. Prior to World War I and the Russian Revolution, *socialism* and *social democracy* were interchangeable. Since then, progressive movements took several different paths, with the main distinction being between reform and revolution. We cannot discuss the historical twists and turns of socialist politics in this essay. What is important here is to note that, from a world-systems point of view, the split between the Second and the Third Internationals that accompanied the Russian Revolution should have been a tactical difference, not a strategic one. Building socialism in the core can proceed legally because core politics are usually democratic. Building socialism in the semiperiphery usually requires the revolutionary taking of state power because semiperipheral states have rarely been democratic (although this is changing). This difference of means evolved into a difference of ends in which the revolutions in the semiperiphery that gave us communist states never achieved democracy, either political or social. They instituted a centralized command economy, based on a military model. It was justified, if at all, by geopolitical necessity and the desire to catch up with the capitalist core states, but bore only a rhetorical resemblance to socialism. We are not so foolish as to ignore the tragedies of the communist states or to advocate any sort of command economy, no matter how "new and improved."

In the core, socialists exercised power through the combination of electoral politics and union bargaining in a "democratic class struggle" (Korpi 1982). The best examples are found in Sweden and Norway, to a lesser extent in Germany and Austria, and in some odd ways, France. Although far from constructing complete socialism, of all the countries in the world, these have come closest to attaining socialist goals listed earlier. In this sense, social democrats have come closer to achieving socialist goals than did any of the countries in which communist parties took power. Even in classic Marxist terms of "surplus value," rates of class exploitation were higher in the former socialist states than among the social democracies (Boswell and Dixon 1993).

Redistributive and protective welfare state programs include poor relief, old age pensions, child labor prohibitions, health insurance, occupational safety codes, environmental protection, and educational expansion. These now constitute a greater portion of core state budgets than does the military/judicial side that defines states as the institution with a monopoly of legitimate violence. By setting a national floor on living standards, the welfare state shifted labor market competition away from lower wages and toward greater productivity. This was a structural shift among core states of tremendous value to average workers. Following World War II, rapidly increasing technological and human capital productivity in the core paid for both high wages for industrial workers and a slowly rising national floor. Unions led the creation of the welfare state, much more so than in increasing individual liberation, although success in one aided the other.

While far short of socialist or even liberal expectations, the combination of individual liberation and egalitarian welfare nevertheless dramatically improved the lives of working people in the core over the last hundred years. The extent of improvement is clear when compared not to a utopian ideal of human relations but to the rapacious "market despotism" that characterized the nineteenth century. Through state enforcement of welfare, the labor movement raised living standards for all workers, instead of just pursuing the select interests of union members. In fact, success has been sufficient to forestall revolutionary socialism in favor of reform wherever strong labor movements can make the state guarantee a class compromise (Przeworski 1987; Boswell and Dixon 1993). The problem for the twenty-first century is that market despotism is making a rather impressive comeback.

Just as the communist states failed to build socialism one state at a time, and probably could not have succeeded, social democratic states have reached clear limits on what they can achieve within the current parameters of the world-system. Rather than continuing to inch down poverty and exploitation, social democratic states in particular and welfare states in general have been in retreat. Welfare policies that once cut poverty now drive up unemployment and protective policies make companies uncompetitive. The decline of state efficacy, as we well know by now, has its source in the remarkable increase in the pace of world integration, "globalization," that has occurred over the last two decades. Increased economic and cultural interpenetrating across state boundaries is obvious to most observers. The surprise is short-lived when we find such ironies as that both sides in the 1992 Gulf War followed the battles on CNN, that Chinese students raised a "Statue of Liberty" during the 1989 protest in Tiananmen Square, or that in 1994 the U.S. dollar became legal tender in Cuba. The latest phase of globalization has reduced the ability of any state to manage its share of the world economy. Most evident recently in East Asia, but true everywhere, is the tender vulnerability of formerly robust states to the vicissitudes of global markets.

During the last quarter of the twentieth century, the welfare side of the progressive agenda has come under increasing attack and steady reversal. With globalization has come a neoliberal world order in which economic growth and state welfare are increasingly at odds.[5] Most explanations of stagnant wages, declining unions, increased child labor, corporate restructuring, government deficits, and shrinking welfare provisions point in varying degrees to increased world integration. Hardest hit in the core have been the less skilled workers, especially low-skill assembly line or piece-rate workers, who had over the past half-century used their electoral power to garner state protection from market despotism. With the decline in state efficacy has come a decline in the legitimacy of states and of political action. Older generations lament noticeable losses of state power and benefits. Many among the younger generation have spurned youthful idealism for public cynicism and private nihilism. Globalization of cultural idioms further undermines the protection that national traditions and cultural diversity provide domestic markets from international competition. As such, decline of the state is seen as a crisis in national *values*.

States have always faced the trade-off in which raising domestic living standards through welfare programs or protective legislation reduces the ability of domestic firms to compete with those in less generous or less developed societies. What differs now is that world economic integration has risen to such a level, and continues to escalate so rapidly, that the trade-off has shifted markedly toward competition for ever lower labor costs. This is a shift in the global division of labor between core and periphery as well as a reorganization of industrial relations in the core. A much greater portion of core assembly line and service workers are now in direct competition with each other and with semiperipheral labor, a situation that is similar to the period prior to the "second" industrial revolution at the end of the nineteenth century. Highly skilled and professional workers still fare well, as their skills are even more valuable in a rapidly changing market and technological environment. These valued employees increasingly differ from their more easily replaceable brethren and income inequality increases (Piore and Sabel 1984). As a result, the welfare state's floor of national living standards has cracked and is crumbling.

WHAT IS POSSIBLE?

An enduring distinction within social democracy is whether it is possible to achieve socialism through progressive reform of a society or whether reformed capitalism is the best that can be achieved. The latter position has been in the ascendance since World War II (with an added boost since 1989). This has led to the use of the term *democratic socialists* by those who hold to the possibility of a truly socialist system. From a global perspective, we agree that reformed capitalism is the best one can hope for within a single society in the overall capitalist world-system. However, democratic socialism is a real possibility for the world-system as a whole. In a halting and still far from entirely complete fashion, slaves, ethnicities, races, and women became legally equal citizens in core and most other states. Their liberation is now entrenched in the world order, such that egregious states are global pariahs. At times, these juridical and national forms of liberation have also meant loss of the dependent security of paternalism. Of greater lasting importance, however, is that each liberation strengthens the foundation for removing politically coercive sources of wage determination in the labor market and "superexploitation" of minorities and women in the workplace, raising the living standards of all workers.

These global victories may now seem distant. As we discussed at the beginning of this essay, the collapse of the former state socialist countries in Europe, along with the rejection of Marxist-Leninist parties throughout most of the rest of the developed world, has left the impression that defeating capitalism and eliminating exploitation is utopian. The current hegemony of neoclassical economics prescribes only one bromide for developing countries: accept "austerity" budgets and keep labor costs low to play ball with the big firms and compete for the global markets. The world-systems perspective, as we interpret it, provides a different view. For socialism of any type to be a reasonable topic, one must demonstrate not only that capitalism is exploitative and socialism would be a better alternative but also that achieving world socialism is *possible*.

Goran Therborn's (1980) classic study of ideology explains that any worldview is defined by the answers to the following three questions:

- What exists?
- What is good?
- What is possible?

Determining "what is possible" is the ultimate defense of the status quo. One can empirically demonstrate that exploitation exists and that it is unfair (even by capitalist standards of justice), but the goal of eliminating exploitation is irrelevant if that is not possible. The "end of ideology" does not occur, as Fukuyama (1992) suggests, because the evils of capitalism have been muted (quite the contrary) or because the class struggle is no longer a central facet of capitalism (even more the contrary). Rather, socialism has become a suspect ideology, no matter how carefully one constructs a societal model of "real socialism," because past theories no longer seem to have a viable path toward its future realization. The lack of a "utopian" goal against which to organize criticism and, more importantly, to direct progress has led erstwhile progressives and leftist intellectuals into the nihilism and endless relativism of post-modernism (Jameson 1990). Getting past this impasse requires a theory of a realistic alternative at the global level, which we find in the idea of global democracy.

Academics are often reluctant to speculate or predict what is possible. Our primary job is to explain and educate. Simple extrapolations from what happened before often produce a mechanical notion of history that belies the importance of new departures or cumulative developments. Earlier efforts to scientifically understand the prospects for socialism have been flawed by assumptions of teleology or inevitability. We cannot predict outcomes exactly or entirely, but structural theories can predict when social conflicts or contingent actions are more likely to influence the course of history.

A good example is the revolutions of 1989, which world-systems theory could have predicted, and some did hint at, but all failed to proclaim.[6] Frank (1980) and Chase-Dunn (1982) explain that the communist states never escaped the powerful military and economic forces of the larger capitalist world-system despite their strong effort to create an alternative and separate socialist world-system. Instead, state communism had become a political program for catching up with the core of the capitalist world-system. Frank and Chase-Dunn further predicted that the communist states were being increasingly reintegrated into the world market and would become more like other capitalist countries over time, but they did not offer a time frame or predict the revolutions of 1989. Arrighi, Hopkins, and Wallerstein (1989b) in a collection of papers written during the 1980s explicitly predicted that the decline of U.S. hegemony would have a mirror effect on the East, with the implication the Soviet bloc would break up. They too failed to offer a time frame or to predict revolutions. Boswell and Peters (1990), writing in the summer of 1989 about the revolts in Poland and China, predicted that they would spread to other state socialist countries (a prediction that became an outcome by the time of publication; see their footnote 1). They also hinted that ethnic nationalism would break up Yugoslavia and the Soviet Union. While they predicted revolts, their proximity to the events makes this less surprising, and they did not predict that the other states would fall so quickly.

The failure of world-systems theorists to predict adequately the fall of Soviet communism is one of our motivations for applying world-systems theory to the prospects for building global democracy. This failure was partly a reticence to recognize the importance of making predictions—as much a lack of will power as an absence of insight. Explanation of the limits on future actions and predictions about probable outcomes given differing conditions has immense value not just for policy makers but for each of us seeking greater self-determination and a more progressive direction to world history. We are properly cautious regarding statements about the future. And we eschew inevitability and teleology. We also are careful with the notion of "progress," though we do not throw it away as have so many others.

The association of the idea of progress with legitimating capitalism and imperialism has led many contemporary thinkers to discard it altogether. We recognize that progress is a normative concept, not a scientific one. We also recognize that the ideology of progress has been used to justify exploitation and domination. Yet, it is nonsense to think that no progress has occurred under capitalism or that no progress is possible. Sanderson (1995, 337–356) provides a sensible discussion of the notion of progress in connection with long-term social evolution. He contends reasonably that considerable consensus is possible among humans regarding the basic elements of a desirable life (health, longevity, autonomy) and that movement toward attaining the conditions that make these elements possible can be called progress. Sanderson also contends that the evolution from hunter-gatherer societies to agrarian civilizations produced little in the way of progress in this sense for the masses of mankind. However, we agree with Sanderson, and with Marx, that capitalism and the resistance to exploitation that it has engendered, have produced changes that ought to be called progressive.

Capitalist accumulation requires economic expansion yet produces recurring cycles of recession and stagnation. Economic development and social progress are neither linear, as some modernization theorists might have it, nor cyclical, as some Marxists might imply, but rather both. The combination of linear and cyclic processes produces an upward *spiral.* The spiraling growth in productive forces, which however unevenly, always offers periods with the possibility of improving people's lives. How much they improve depends on how much of the increase in productivity people can claim, which is the twin result of market competition and class conflict.

Hegemonic and economic cycles repeat the opportunity for social change as existing organizational structures break down. The conditions of cyclic breakdown and of the opportunities they produce are always different, depending on the current state of the ongoing development of the system. The future is never predetermined, and it often gets worse instead of better, but the opportunities for transformation are greater at each critical turn in the spiral. While workers in general, and core workers in particular, have improved their lives, it does not follow that capitalism is the best possible system or even that continued progress can be assumed. The working classes had to fight for progress, winning it where their struggles produced democratic states. It is through democratizing global institutions that an even more progressive civilization can be built.

While some may contest that capitalism was global in its origins, few dispute the transnational nature of contemporary capitalism. The degree of capital mobility has increased dramatically in the last twenty years, producing a qualitative change in class relations and production systems comparable to the development of assembly line mass production a century ago. State socialism, which originated as a response to the mass society of assembly line industrialization, has followed its progenitor into the grave. The left has been left in a postindustrial, postmodern haze without a vision of what is ahead. Our theoretical perspective tells us that the solution to current woes lies in *transnational politics*. This focus on world politics means more than just a shift in emphasis. We are also suggesting that politics at the world-system level are more important than was true previously.

Changes in world capitalist production, combined with the decline in American hegemony and end of the cold war, reveal the impotence of any state, including any communist state, to fully control its domestic portion of the world-economy. In the past, mobility offered international capital an escape from political control. This was the defining difference between world capitalism and world empires. The upsurge in capital mobility and interdependence, while lessening the importance of national states, is increasing the importance of, and benefits from, interstate authority, choking off capital's own escape route from political authority. The result has been a concomitant intensification of global institutions with intrastate authority. We have seen a meteoric rise of international governmental and nongovernmental organizations since World War II (Boli and Thomas 1997). The world capitalist system is now constituted as a single world-economy with an emerging global state. With the increasing development of global intrastate institutions comes a rapid rise in the transnational politics of world governance. Most important are transnational labor movements, as labor has always been the prime mover in major social transformations.

The combination of structural constants, cycles, and trends produces a model of world-system structure that is reproducing its basic features while growing and intensifying (Chase-Dunn 1998). For socialism to transform capitalism, it too must be a global system. World-systems theories imply greater systemic volatility and increased opportunities for structural change during periods of economic transition and when no core state is hegemonic over the world economy. Having entered such a period, the value of making predictions about "what is possible" increases dramatically for social and global movements. Even though we remain scientifically skeptical of predictions, especially of events, the world-systems perspective gives us a powerful tool for visualizing future structural alternatives. Global institutions, more than any time since the origins of capitalism, are directly governing the modern world-system. This has ironically become almost a truism, and yet the muddle of current events and discussions of globalization has produced little clarity of vision regarding possible structural outcomes and sensible strategies.

World socialism is the only possible form of socialism because only a global democracy can govern transnational relations in a progressive direction. As a global phenomenon, global democracy is inherently limited to very broad parameters of directing capital investment and economic development within a market framework.

While a command economy has proven to be a societal failure, globally it would be absurd. World market socialism would leave great variation among states in forms and functions but would attack uneven development, unequal social standards, and undercutting wage competition where they reproduce at the global level. Both the means and the goals of socialism are important. Basic needs, sustainable development, social justice, and peace are the goals. Global democracy is both a means and goal.

CONCLUSIONS: TEN POINTS

Our overall theoretical and political stance can be summarized by the following ten points:

1. Capitalism is a global system with a single world-economy but multiple competing states. The system is marked by long-term cycles and trends and is most subject to change by war and revolution during world divides, when the cycles are in transition. The combination of winning ears and pacifying rebellious populations has produced a series of *world orders* in which the rules and standards of international relations are enforced by hegemonic power and/or core consensus.

2. World revolutions have altered the world order. From the elimination of slavery to the end of colonialism, a rough and tumbling spiral between socialist progress and capitalist expansion has resulted in higher living standards and greater freedom for working peoples. Fundamental change in the system happens only at the global level. For socialism to replace capitalism, it too must be a global system that embraces a democratic world polity.

3. Building socialism one state at a time is doomed by the costs of transition. State socialism was primarily a strategy for economic development within the system, not an alternative to it. Communist parties came to power in semiperipheral regions and employed strategies that were similar to those of other semiperipheral states that were attempting upward mobility (e.g., trade protectionism, import substitution).

4. State socialism successfully produced economic development until the long stagnation that began around 1968. Stagnation forced industries to restructure to restore growth. In communist countries, where the state owned the means of production, this also required restructuring the state itself.

5. The communist states of Eastern Europe housed three internal contradictions that imploded under the pressures of a stagnant world economy and changing social structure of production. Those contradictions were:

- a polity that failed to represent popular interests yet bore sole responsibility for all social grievances,
- a development strategy whose initial gains were based on investing in a mass production model but continued application of the same model produced economic stagnation and social dissent, and

- a distribution system that guaranteed jobs and welfare but generated consumer shortages and technological backwardness.

6. In the core industrial countries of Western Europe, North America, and Japan, the long stagnation of the 1970s to early 1990s drove manufacturing to increase dramatically the pace of transnational integration. The growing integration of the world-economy has a contradictory effect on states, encouraging them to both shrink and expand at the same time. One result is a growing number of small nations that have their own state. The other is the development of multistate organizations, such as the European Union, that have the potential to become the next hegemon. Growing integration and international organization are also producing an increasingly powerful world polity of nongovernmental organization. A strengthening world polity increases the possibility of transnational organizing.

7. Increased transnational capital mobility produces new waves of short-term immiserization and long-term environmental degradation. Labor, environmental, and women's movements are increasingly transnational in response. At the global level, these progressive movements share key interests and goals that make them inherently intertwined, with democratization at the center of the cord. Compared across world zones, the struggle for labor rights in transnational corporations is a feminist struggle as women workers face the highest exploitation rates, and it is an environmental struggle as pollution is highest where democracy is curtailed, and democracy follows unionization—and so on. Global democracy is not the only strategy open to contest transnational capital. The first response has typically been nationalist, which has often devolved into racist and xenophobic politics. Global social movements cannot deny national identities as this will only inflame ethnic resistance and stall the growth of a true world polity.

8. Global democracy is the equivalent of social democracy at the world level. It is both the means and the first goal of world socialism. All other benefits for working people, from environmental protection to banning child labor, flow from their ability to contest the standards and rule of the world order. Most important is the development of a true World Bank that would direct investment and adjust interest rates to support environmentally safe production for human needs and work to balance development worldwide. Global democracy also means the democratization of all national states, which along with the development of a united peacekeeping force, could stop the cycle of world wars. No Soviet-style world state could "command" the world-economy without wrecking it or succeed at denying national identity without fostering nationalist revolts. Global democracy is the best protection against any such attempt at global repression, whether by religious zealots, transnational capital, or resurgent Stalinists. To make progress toward the goal of global democracy, we need to show that this goal is possible as well as desirable.

9. Peripheral and semiperipheral workers in industrializing countries are the most motivated agents of global democratic relations; core labor has been the strongest potential force for building more egalitarian institutions worldwide. A strong core/

periphery alliance of popular forces is a necessary requisite for global democracy. Missing are the institutions that could organize labor transnationally. This is quite different from the labor internationalism of old in which leaders occasionally have meetings. Rank and file understanding and interaction across national, cultural, racial and religious borders can facilitate cooperative action. This kind of direct linking is increasingly feasible with the dramatic decline in communication costs, but it will not happen automatically. One start would be to focus on issues of wide concern, such as child labor and environmental destruction, forming alliances with transnational women's and environmental organizations.

10. The contemporary transnational drive toward heightened exploitation can only be checked by transnational politics. Global labor standards, environmental regulations, and women's rights are a single intertwined starting point. Through global democracy, transnational labor and allied movements can direct market competition away from cheaper wages and toward increasing human productivity. In world historical terms, this is the essence of the term *progressive*.

NOTES

A fuller and more historical treatment of the arguments put forth here can be found in Boswell and Chase-Dunn (1999). This chapter offers a synopsis of our world-systemic perspective on progressive politics, drawing mainly on chapter 1 of the book. An earlier version was presented at the 1998 meetings of the American Sociological Association.

1. These jokes were repeated by a first-time panel of Eastern European scholars convened at the 1990 Convention of the International Studies Association in Washington, D.C.

2. Stated in 1990, as quoted in Kumar (1992, 309).

3. To some extent, the burgeoning contemporary "informal sector" provides such a fallback.

4. We agree with Robinson's (1996) critique of "polyarchy," the sanitized and constrained definition of democracy as limited to political rights of participation in the election of leaders. This definition of democracy, though undoubtedly preferable to authoritarian regimes, functions as a legitimization for the continuation of huge inequalities in societies. We support a popular democracy that includes social and economic as well as political rights.

5. For an excellent overview of this particular issue, see Cox (1987, especially chap. 8).

6. Randall Collins, utilizing a geopolitical model, did predict the demise of the Soviet Union within fifty years but offered little insight into its social origins (Collins and Waller 1992).

REFERENCES

Arrighi, Giovanni. 1994. *The Long Twentieth Century: Money, Power and the Origins of Our Times*. New York: Verso.

Arrighi, Giovanni, Terence K. Hopkins, and Immanuel Wallerstein. 1989a. "1968: The Great Rehearsal." Pp. 19–33 in *Revolution in the World-System*, edited by Terry Boswell. New York: Greenwood.

―――. 1989b. *Antisystemic Movements*. New York: Verso.

―――. 1992. "1989, the Continuation of 1968." *Review* 15:2(Spring):221–242.

Boli, John, and George M. Thomas. 1997. "World Culture in the World Polity." *American Sociological Review* 62:2(April):171–190.

Boswell, Terry, and Christopher Chase-Dunn. 1999. *The Spiral of Capitalism and Socialism: The Decline of State Socialism and the Future of the World-System*. Boulder, Colo.: Reinner.

Boswell, Terry, and William Dixon. 1993. "Marx's Theory of Rebellion: A Cross-National Analysis of Class Exploitation, Economic Development, and Violent Revolt." *American Sociological Review* 58:5(Oct.):681–702.

Boswell, Terry, and Ralph Peters. 1990. "State Socialism and the Industrial Divide in the World-Economy: A Comparative Essay on the Rebellions in Poland and China." *Critical Sociology* 17:1(Spring):3–35.

Chase-Dunn, Christopher, ed. 1982. *Socialist States in the World-System*. Beverly Hills: Sage.

―――. 1998. *Global Formation: Structures of the World-Economy*. 2d ed. Lanham, Md.: Rowman & Littlefield. (Originally published 1989, Cambridge, Mass.: Blackwell.)

Chase-Dunn, Christopher, and Thomas D. Hall. 1997. *Rise and Demise: Comparing World-Systems*. Boulder, Colo.: Westview.

Chirot, Daniel, ed. 1991. *The Crisis of Leninism and the Decline of the Left*. Seattle: University of Washington Press.

Collins, Randall, and David Waller. 1992. "What Theories Predicted the State Breakdown and Revolution of the Soviet Bloc?" Pp. 31–47 in *Research in Social Movements, Conflict and Change*, vol. 14, edited by Louis Kreisberg and David Segal. Greenwich, Conn.: JAI.

Cox, Robert. 1987. *Production, Power and World Order: Social Forces in the Making of History*. New York: Columbia University Press.

Cumings, Bruce. 1987. "The Origins and Development of the Northeast Asian Political Economy: Industrial Sectors, Product Cycles and Political Consequences." Pp. 44–83 in *The Political Economy of New Asian Industrialism*, edited by Fred Deyo. Ithaca, N.Y.: Cornell University Press.

Frank, Andre Gunder. 1980. "Long Live Transideological Enterprise: The Socialist Economies in the Capitalist International Division of Labor and West-East-South Political Economic Relations." Chapter 4 of *Crisis: In the World Economy*, edited by Andre Gunder Frank. New York: Holmes & Meier.

Fukuyama, Francis. 1992. *The End of History and the Last Man*. New York: Free Press.

Harvey, David. 1989. *The Condition of Postmodernity*. Cambridge, Mass.: Blackwell.

Jameson, Frederic. 1990. *Postmodernism or the Cultural Logic of Late Capitalism*. Durham, N.C.: Duke University Press.

Korpi, Walter. 1978. *The Democratic Class Struggle*. London: Routledge.

Kumar, Krishan. 1992. "The Revolutions of 1989: Socialism, Capitalism, and Democracy." *Theory and Society* 21:3(June):309–356.

Marx, Karl. 1846–47. [1983]. "The German Ideology." Pp. 162–197 in *The Portable Karl Marx*, edited by Eugene Kamenka. New York: Penguin.

―――. 1875. [1983]. "Critique of the Gotha Programme." Pp. 533–566 in *The Portable Karl Marx*, edited by Eugene Kamenka. New York: Penguin.

Piore, Michael J., and Charles F. Sabel. 1984. *The Second Industrial Divide: Possibilities for Prosperity*. New York: Basic Books.

Przeworski, Adam. 1985. *Capitalism and Social Democracy*. Cambridge: Cambridge University Press.

Robinson, William I. 1996. *Promoting Polyarchy: Globalization, U.S. Intervention, and Hegemony.* New York: Cambridge University Press.

Sanderson, Stephen J. 1995. *Social Transformations: A General Theory of Historical Development.* New York: Blackwell. (Reprinted 1999, Lanham, Md.: Rowman & Littlefield.)

Therborn, Goran. 1980. *The Power of Ideology and the Ideology of Power.* London: Verso.

Wallerstein, Immanuel. 1984. *The Politics of the World-Economy.* Cambridge: Cambridge University Press.

———. 1995. *After Liberalism.* New York: New Press.

16

World-System and Ecosystem

Albert J. Bergesen and Tim Bartley

"Think global, act local." That's an old adage in the environmental movement. Usually it means think of planet Earth, but thinking globally should also mean thinking of the world-system whose operation might have detrimental effects on planet Earth. Clearly the dependent variable, as the object to be studied is called in social science, is global—planet Earth. But the independent variable should also be thought of as global—the functioning of the modern world-system. The great French sociologist Emile Durkheim had a maxim something to the effect that in explaining large-scale things the explanation should also be large scale—social facts should have social explanations. The analog here is that when studying the effects of economic systems on the environment, which of course is worldwide, one should look at the effect had by something worldwide, like the modern world-system. And that, of course, means one should employ some version of world-systems analysis.

Studying the environment and world-system, then, go hand in hand. The environment does not recognize national boundaries, and neither does the world-system. Both are global phenomena. World-systems theorists remind us that the human economy is global in scope; politics have a strong international aspect; and culture is increasingly globalized. Human social organization, then, is increasingly global in nature. Actually, it has been that way for quite a while, but we are certainly aware of the increasing interconnectedness of our economic, political, and cultural lives. And the environment, as we all know, is global. We begin by reviewing what is presently known about the effects of the world-system on the environment.

ENVIRONMENTAL DEGRADATION

Deforestation

Several quantitative studies have shown the semiperiphery to be the site of the most intense deforestation (Burns, Kick, Murray, and Murray 1994; Kick, Burns, Davis, Murray, and Murray 1996). First, there is a long history of exploitation of peripheral and semiperipheral forests by core countries, and as Chew (1996) notes, the association between colonialism and deforestation in Southeast Asia has a long history. Spain and Portugal, Holland, Britain, and the United States have all exploited Asian forests during their periods of dominance in the world-system. Increased timber consumption seems to accompany ascendancy to a hegemonic world-system position, as may be occurring for Japan, which has dramatically increased its timber consumption recently (Chew 1996).

Population growth leads to deforestation in all sectors of the world-system, but its effects are exacerbated in the semiperiphery (Kick et al. 1996). Yet Burns et al. (1994) and Kick et al. (1996) find that, for semiperipheral countries, *rural* population growth is a better predictor of deforestation than is total population growth, arguing that urban concentration in the semiperiphery causes landless people to migrate out of the city into forested areas—what is called the process of rural encroachment. Since these migrants possess little knowledge of agricultural practices, they end up contributing to deforestation (Burns et al. 1994, 225). Although the process of rural encroachment occurs *within* a society, the urbanization that leads to outmigration is a consequence of rapid uneven development of semiperipheral countries in the world system.

In addition, semiperipheral countries deforest more than others because of their position of potential upward mobility in the world system, which leads them to place more weight on industrialization than on environmental protection.[1] Newly industrializing countries (NICs) tend to have lax environmental regulations (Smith 1994) and because of their *potential* for economic development, they are more eager to reap the economic benefits of forest exploitation than are developed countries. Furthermore, semiperipheral countries have a greater technological capability to deforest than do peripheral countries (Burns et al. 1994; Kick et al. 1996).

Such semiperipheral states have historically allowed or even encouraged deforestation in attempting to economically develop. Chew (1996) provides an example in his analysis of postcolonial Southeast Asia. He argues that attempts to build export-led economies and Western-style states have secured the cooperation of political elites and transnational corporations in exploiting forests. Nazmi (1991), though not espousing a world-systems perspective, offers a similar example for the case of Brazil. He notes that government incentives for cattle ranching have increased deforestation; badly defined property rights have encouraged small-scale, destructive agriculture; and an emphasis on pig iron production has necessitated deforestation to allow the planting of eucalyptus trees used in iron production. If Southeast Asian and

Brazilian examples of state-facilitated deforestation are generalizable to other semiperipheral countries, then the study of deforestation illustrates a more general process. States are important units of analysis, but since they act in the context of the world system, they cannot be treated as self-contained entities.

States are tied to the world-system through international trade, and trade in forest products is another factor related to deforestation. One would expect that major exporters of forest products would experience high levels of deforestation. Conversely, one would expect that when a country imports forest products, it should not need to deplete its own forests. Yet Kick et al. (1996) find that these expectations hold up only for core countries, not for semiperipheral ones. Core countries are able to export forest products without high rates of deforestation because they often use reforestation practices. (While reforestation results in old-growth forests being replaced by young trees, these are nonetheless counted as forests.) Interestingly, in semiperipheral countries both the export and import of forest products leads to deforestation—the former because of a lack of reforestation programs, the latter because imports are an indication of infrastructure building, which contributes to deforestation itself. In sum, whether core countries import or export forest products, they experience less deforestation than semiperipheral countries, which deforest regardless of their patterns of import and export.[2]

Global Warming and the Curvilinear Hypothesis

A second environmental problem that has been studied from a world-systems perspective is global warming. The two "greenhouse gases" are emitted through different processes—carbon dioxide directly through fossil fuel use and indirectly through deforestation, methane through wet rice agriculture, livestock, uncontrolled coal mine emissions, and petroleum and natural gas leakages (Burns, Davis, and Kick 1997).

In their study of emissions of these two gases, Burns et al. (1997) find that carbon dioxide is produced mostly in highly developed and methane in less developed countries. They create the category "semicore" to distinguish between stronger (semicore) and weaker (semiperipheral) states that have previously been lumped together as semiperipheral. "This strength can be seen primarily in terms of global network ties, but typically is reflected domestically as well" (Burns et al. 1997, 10). The semicore includes the weak, noncore countries of Eastern and Western Europe, as well as China, Israel, Australia, and Brazil. The semiperiphery thus consists of the remaining countries that have normally been called semiperipheral.

The social dynamics by which greenhouse gases are differentially emitted vary with countries' world-system position (Burns et al. 1997). Core countries produce the most carbon dioxide, followed by the semicore, semiperiphery, and periphery. The high level of energy consumption in the core, which results from a high standard of living, explains the primacy of the core in producing carbon dioxide and the decreasing levels of such production for each less affluent sector of the world system. The semicore produces the most methane, followed by the core, semiperiphery, and periphery. Burns et al. (1997) suggest that the movement of commercial cattle ranching from core to semicore countries and the association of agriculture with methane production explain the high levels of this greenhouse gas produced there.

These findings suggest that the relation between world-system position and greenhouse gas emission does not mirror the relation between world-system position and economic development. Economically, there is a clear hierarchy from core to periphery, but the core is not the exclusive emitter of greenhouse gases, nor is it able to transfer all production of these gases to less developed areas. "[W]orld-system dynamics (e.g., concentration of wealth and power in the core) are likely to manifest themselves in a number of ways, which may include environmental outcomes that are most severe somewhere other than the core" (Burns et al. 1997, 32). Economic relations structure the world-system, but they cannot be simplistically projected onto areas such as environmental degradation. For instance, while the relationship between economic and sociopolitical development is essentially linear, when we look at toxic emissions we find that the wealthiest countries are not always the most intense emitters.

Interestingly, it is only in the last twenty years that the relationship between development and carbon dioxide emissions has become increasingly curvilinear in the shape of an inverted U (Roberts and Grimes 1997; Grimes and Roberts 1995; Grimes, Roberts, and Manale 1994). That is, currently, emission of carbon dioxide per unit GDP increases with increases in GDP per capita up to about $8,000 to $10,000 GDP per capita, at which point the relationship begins to curve downward (Roberts and Grimes 1997; Grimes and Roberts 1995; Dietz and Rosa 1997; Hettige, Lucas, and Wheeler 1992; Lucas, Wheeler, and Hettige 1992). In other words, emissions are most intense in moderately developed (semiperipheral) countries and less intense in less developed (peripheral) and highly developed (core) countries.[3] Roberts and Grimes (1997) thus liken this pattern to the observed Kuznets curvilinear relation between income inequality and national development that indicates that as a country develops, its inequality increases up to a level at which it begins to decrease. Yet the "environmental Kuznets curve" is not exactly analogous to the original Kuznets curve because the environmental curve is found by looking at different countries with different levels of development, not at a particular country's process of development. "The emergence of an inverted U-curve . . . is the result not of individual countries passing through stages of development, but of a relatively small number of wealthy ones becoming more efficient since 1970 while the rest of the world worsens" (Roberts and Grimes 1997, 196). While more affluent countries still contribute the most to overall carbon dioxide emissions, what the environmental Kuznets curve findings suggest is that they pollute less intensely or exhibit more "societal efficiency" than less developed countries (Grimes and Roberts 1995). Bergesen, Parisi, and Downey (1996) identify two possible explanations for the decreased intensity of emissions in core countries: (1) developed countries possess newer and cleaner technologies than do core countries; (2) they have moved from manufacturing toward service-oriented economies. Lucas et al. (1992) and Hettige et al. (1992) support the latter possibility by showing that, when pollution intensity is measured per unit of industrial output rather than unit of GDP, the inverted U-shaped relation nearly disappears. "[Intensity of] manufacturing output rises steadily with income, at most tapering off somewhat at very high incomes" (Hettige et al. 1992, 479). In other words, new technologies do not seem to be cleaning up industrial production.

Natural Resources and Development

So far we have looked at the environmental impact of development without admitting the impact of a degraded environment on the course of further development. Bunker (1984, 1985) shows the impact of both ecological degradation and natural resources on political and economic development/underdevelopment. He argues that Brazil's underdevelopment, environmental degradation, and state bureaucratic procedures are tightly linked, each impacting the other since Brazil's incorporation into the world-system in the sixteenth century. Imperialist processes of natural resource extraction decimated the indigenous population and their ecologically sound technologies. Huge numbers of slaves died on expeditions for sugar and spices that had to expand outward as local resources were depleted. In addition, European trade in animal oils reduced the indigenous population's food supply and disturbed fragile ecosystems. This depletion of the rural population meant that during the Brazilian "rubber boom" of the mid to late nineteenth century, capitalists' costs were increased by the necessity of recruiting workers from urban areas. These high costs eventually led to the downfall of the Brazilian economy, as rubber came to be produced more efficiently on English plantations in Asia. In the wake of the rubber boom, the Brazilian environment was further depleted and degraded as those laborers remaining from the rubber boom overfarmed fragile land and traded in animal skins, which further disturbed the ecosystem. Environmental conditions here are seen as both consequence and cause of Amazonian underdevelopment:

> The depopulation, environmental disruption, and demographic and economic dislocations brought about by the previous modes of extraction created the conditions for both large-scale capitalist enterprise and government economic planners to treat the Amazon as an empty frontier from which profits could be rapidly and wastefully extracted with little regard for, or sustained economic participation by, existing socioeconomic or environmental systems. (Bunker 1985, 77)

Bunker's theoretical notions are also closely tied to processes in nature, as his "ecological model of uneven development" (1985, 49) is inspired by the second law of thermodynamics. This principle of entropy states that while there can be no net loss of energy, the transformation of energy from one form to another will result in it becoming increasingly disorganized, or degraded (Erlich, Erlich, and Holdren 1993). Underdevelopment in economies based on natural resource extraction is a function of the core's ability to obtain *useful* forms of energy from the periphery and semi-periphery, degrading the latter. "If energy and matter necessarily flow from extractive to productive economies, it follows that social and economic processes will be intensified and accelerated in the productive economy and will become more diffuse and eventually decelerate in the extractive economy" (Bunker 1985, 47). In addition, Bunker (1985) argues that economies based on the extraction of natural resources necessitate a theory of development based on the mode of extraction rather than one based on the mode of production since forms of energy cannot be reproduced and sustained in the same sense as labor and capital.

Access to natural resources in the rise to hegemonic status is explored by Bunker and Ciccantell (1997), who argue that ascendancy requires the implementation of new strategies for entering raw material markets and accommodating to the natural environment—conquest of resource-rich peripheries, technological innovations that change the "relations between economy and environment," and the shifting of capital costs and financial risks to peripheries (Bunker and Ciccantell 1997, 10). Technological innovations in Holland, Britain, and the United States served to increase the scale at which natural resource extraction and transport occurred. Japan's recent ascent has also hinged largely on its ability to convince semiperipheral countries, which occupy a weak bargaining position, to assume the costs of tailoring transport infrastructures to Japanese specifications as well as the costs of environmental protection (Bunker and Ciccantell 1997, 1995; Bunker 1994).

Environmental Constraints and Social Change

Chase-Dunn and Hall (1997a, 1997b) bring the environment into an abstract theory explaining changes in world-systems in the twelve thousand years since the establishment of sedentary societies. They propose an ecological and evolutionary theory of the formation and expansion of hierarchical world systems that recognizes the ways in which environmental constraints direct the formation of world-systems and more energy-intensive economic practices. Their "iterative" model of long-term social change identifies recurring processes linking population pressure, environmental degradation, hierarchy formation, and economic intensification, essentially arguing that social change occurs as a consequence of populations expanding beyond their ecological base.[4]

The iterative model accounts for the formation of hierarchical world-systems as follows: Population growth leads to ecological degradation, which limits the natural resources available to the population at the existing level of effort. When such a condition of population pressure exists, people tend to emigrate to new regions. If emigration is inhibited by geographical features or other populations—environmental or social circumscription—people may develop new technologies of production to sustain themselves in the face of scarcity, or they may come into conflict over the existing resources. If conflict takes the form of war, enough people may be killed that population pressure is reduced. Alternatively, conflict may lead to the formation of hierarchical polities, which foster technological intensification. Technological intensification contributes directly to further population growth and environmental degradation. These processes occur repeatedly, and thus there is an increase in both the scale of world-systems and the scale of environmental degradation.

Chase-Dunn and Hall's model applies even to complex societies characterized by capitalism and large markets. When societies become more complex, several new paths of change become possible. Institutional structures allow population pressure to lead directly to new hierarchies and technologies, bypassing the basic path of the model through circumscription and conflict. Unlike other theories of long-term change, here the development of a capitalist world-system does not radically alter the logic of the theory. In periods of system expansion, "the superstructure of geopolitical accumulation overrides the underlying demographic and resource scarcity constraints of the iteration model" (Chase-Dunn and Hall 1997b, 112). Yet in periods of contraction, the superstructural dynamics fade and these constraints re-emerge.

Even the contemporary global world-system is subject to demographic and ecological constraints, and Chase-Dunn and Hall suggest that such constraints could turn periods of contraction into crises allowing for a transformation to a socialist world system. The crucial point here, from both a theoretical and environmental standpoint, is that current environmental factors do not lose their importance just because local constraints are increasingly replaced by global constraints in the contemporary world-system.

Another approach is taken by O'Connor (1994), who theorizes the issue of an ecological crisis for capitalism from a perspective different from Chase-Dunn and Hall. He brings the environment into a Marxist analysis of global capitalism, arguing that environmental degradation produced through capitalist enterprise constitutes a "second contradiction" of capitalism. In a basic sense, Marx's first contradiction asserted that capitalism's exploitation of the proletariat would lead to the overthrow of the system. O'Connor's eco-Marxism now shifts the focus to capitalism's exploitation of the natural environment. As capital degrades the environment, it increases the costs of future expansion and thus leads to its own demise. O'Connor says:

Cost-side crises originate in two ways. The first is when individual capitals defend or restore profits by strategies that degrade or fail to maintain over time the material conditions of their own production, for example, by neglecting work conditions (hence raising the health bill), degrading soils (hence lowering the productivity of land), or turning their backs on decaying urban infrastructures (hence increasing congestion costs). The second is when social movements demand that capital better provides for the maintenance and restoration for these conditions of life. (1994, 162)

Regimes, Movements, and World Polities

Environmental degradation is as much a political problem as an ecological one. In fact, while political action directed toward environmental regulation is obviously related to the increasing scale and intensity of degradation in the last century, increased degradation does not fully explain the rise of international environmental regimes or social movements. Frank (1994) argues that increasing participation in environmental treaties is due *not* simply to increased degradation but to a reconstitution of the concept of nature into an ecosystem paradigm. He shows that international discourse has gone from viewing nature as "a realm of chaos and savagery, and away from conceptions of nature as a "cornucopia of resources" to recognizing "planet wide interdependencies" (Frank 1994, 2). In addition, he shows that the increasing coherence of a world polity has led to more treaties, while the consolidation and routinization of intranational mechanisms for dealing with environmental problems tend to decrease the number of international treaties. So as the politics of the world-system have become more global, international environmental treaties have become more prevalent. Frank (1995) argues for the primacy of international social ties over economic, political, or even ecological characteristics, since social ties to world society are related more consistently and strongly to the ratification of environmental treaties than are measures of natural degradation, economic linkages to the world-system, or economic, scientific, or political infrastructure. As an ecosystem cultural frame has gained strength global associations, states and organizations have used world-system social networks to pressure others to participate, and environmental regulation has become more closely associated with state legitimacy.

While Frank (1994, 1995) focuses on world-system cultural constructions and network ties, Roberts (1996) explains environmental treaty participation strictly on the basis of political and economic factors. He finds that indicators of world-system position and internal political climate are both determinants of environmental treaty participation. The likelihood of participation decreases with national wealth, indebtedness, dependence on few trading partners, and the existence of a repressive state government. Several explanations underlie these findings. For one, peripheral and semiperipheral countries are simply less able to participate in treaties. Indebtedness also discourages a country from risking its ability to produce raw materials by participating in treaties, and countries that are dependent on few trading partners are more economically vulnerable and therefore less likely to alienate their trading partners by signing a treaty. Repressive regimes are also not sensitive to popular demands for regulation and therefore not likely to sign environmental treaties. While the finding that repressive states are unlikely to participate argues for the importance of democracy, Roberts finds that world-system position is consistently the best predictor of participation.

Although peripheral and semiperipheral countries seem unlikely to participate in environmental treaties, they may spawn environmental social movements. Smith (1994) has made preliminary arguments for connections between levels of development/world-system position and movement emergence. He points out the contradictory situation of NICs: their upward mobility leads to environmental degradation yet creates conditions of urbanization and education that are favorable toward the emergence of social movements acting in opposition to polluting industries. Smith provides only introductory remarks on environmental antisystemic movements, but it is possible to hypothesize that the combination of severe environmental degradation and urbanization in semiperipheral countries would make them the most fertile ground for the growth of environmental movements. Schaeffer (1997) makes a similar hypothesis arguing that the environmental movement finally began to grow in peripheral countries in the 1980s because of democratization and increased perceptions of resource scarcity caused by the debt crisis and austerity programs. During the 1970s environmental movements did not take hold in the periphery where population control efforts, for example, were often associated with U.S. foreign policy. Peripheral economies were also being boosted by the same wave of inflation that helped create the perception of resource scarcity in core countries.

TOWARD ECOSOCIOLOGY

Having examined what is known about the effects of the world-system on the environment, we turn to some implications of environmental ethics for world-system theory. Ethical beliefs may seem a far cry from the material political/economic structures that constitute the world-system, but environmental philosophy raises the fundamental question of whether the world-system is the widest possible web of extant social relations. If there is no fundamental ontological divide between human and nonhuman living things, then according to tenets of deep ecology environmental philosophy, all living things should be accorded the same moral worth (Devall and Sessions 1985; Drengson and Inoue 1995; Naess 1989; Sessions 1995). From this assumption comes the following sociological possibility: relations between morally equal beings constitute a moral order at a minimum, and an ecosocial system at the maximum. This conclusion impacts world-system theory directly, for it means the world-system is no longer the widest and most encompassing web of social relations, and hence no longer the most determinate social structural frame for human behavior.

The heart of world-system theory was its macro-ness, and its operational logic was wider than the principles of traditional development sociology with its focus upon national societies. States, modes of production, and national cultures, the world-economy and international state system all lie within the boundaries of the world-system. The world-system, then, was the society of societies, the system of systems, the most macro of macroanalysis, and as such no other societally predicated social science paradigm could subsume the world-system's operational logic. But the deep ecology assumption of bioequality poses a new challenge to formulating a theory of social systems by suggesting the inclusion of nonhumans. If only humans remain the focus of social theory and other living things remain unequal, then world-systems analysis remains the most macro of all analyses. But if the deep ecology assumption stands, then the theory of the world-system has to be abandoned and replaced by an ecosociological framework that would demote the world-system to being only a system of intraspecies, rather than interspecies, relationships (Bergesen 1995a, 1995b). This was not a problem as long as there was a taken for granted assumption that humanity was not only different but morally superior to other living things. That assumption is now falling. The field of ethics, from the rights of animals to whether trees have legal standing, is being rethought, and here we are rethinking what should constitute the most determinate social order. Theorizing the internal structure of one class (working class, upper class, peasant class, etc.) in a society is helpful, and important, but it is only one part of a larger societal whole, and we believe this class structure is determined by the larger mode of production in which it is embedded. Such a theory of social embeddedness is a principle of present day sociology.

Theorizing the internal structure of one zone of the world-system (core, periphery, semiperiphery) is also helpful, and important, but it too is only one part of a larger whole. We argue the structure of this zone to be determined by its location in the larger world-system core-periphery division of labor. Globalization and global embeddedness are principles of present day world-system theory. Finally, theorizing the internal structure of humanity as a world-systemic totality had been thought to be the final macrounit, but world-system structure is determined by its location in an ecosociological position within the larger ecosystem. Understanding the full nature of such ecoembeddedness will be the central principle of a future ecosociology.

This idea of widening our conception of social order beyond the bounds of the world-system is but another paradigm shift, where what was once understood as the collective totality is now demoted to being a functioning part within an ever larger complex. In the seventeenth and eighteenth centuries, the theory of society began with assumptions about individuals whose patterns of interaction were aggregated to produce social institutions and society. Then, in the late nineteenth century, this paradigm was challenged by the new field of sociology, which argued just the opposite. Social formations, classes, institutions, modes of production, and society as a whole came first, as an inherited collective environment, whose constraint, power, norms, rules, classes, cultures, groups, and institutions made individuals behave the way they did. Sociology turned utilitarian rational choice theory on its head. Utilitarianism started with the individual and derived society; sociology begins with society and derives the behavior of individuals. This theoretical inversion is captured in a phrase attributed to Marx—that it is not humans buying and selling that makes capitalism, but capitalism that makes humans buy and sell. By the middle of the twentieth century, though, this sociological paradigm was itself turned on its head by world-systems theory, which assumed an even larger functioning system, thereby demoting societies, modes of production, nations, and all other societal entities, to functioning parts within the global order. From the world-systems point of view, it is not some countries exporting raw materials and others industrial manufactures that makes the world-economy, but the world-economy that makes some countries export raw materials and others export manufactures.

What we want to suggest here is that the impact of environmental thinking will result in yet another paradigm revolution. Eco-Marxists and social ecologists theorize that humans, transforming themselves (e.g., rebuilding the capitalist economy), will alter human relations with nature; ecofeminists (King 1993; Merchant 1992, 1993) theorize that humans deciding to change their gender relations and destroying patriarchy will alter human relations with nature; and deep ecologists theorize that humans, deciding to change their self-identity to be one with nature, will alter human relations with the environment. While the specific human relations and identities differ—capitalism, patriarchy, self-identity—these theoretical orientations all share the common assumption that human self-transformation must precede changing the encompassing framework of power relations between humans and nature.

This can be characterized as the species self-transformation assumption, where from the point of view of the larger ecosystem of all living things, it is assumed that one species out of thousands is somehow capable of self-transformation, deciding to alter itself from capitalism, patriarchy, and anthropocentric identity to something else, without regard to its structural position in the larger system of interspecies relations. This assumption is found in almost all environmental thinking: we (humans) must do something (on our own) to change our relationship with nature. But compare this argument with an earlier one about self-transformation of individuals within a class-structured society, found in rational choice theory. This argument about freely deciding individuals was challenged by sociology that persuasively noted the necessity of placing individuals within their social structural context. The same point needs to be made today about various environmental theories: humanity, as an aggregate rational choice actor, needs to be placed in its ecostructural context to fully understand the constraints on human action. The notion that we can, without regard to our ecoposition, engage in successful self-transformation as a species (i.e., overthrow capitalism, overthrow patriarchy, and change to an ecocentric identity) seems just as naive as earlier rational choice arguments about the possibility of individual self-transformation without understanding the social position within which this supposedly autonomous person is freely deciding. Today it is the idea of a supposedly autonomous humanity that can purportedly self-transform without first transforming the ecosocial relations it shares with other species.

All of this is speculation. But what we do know is that the moral status of non-human living things is being raised and human distinctiveness is disappearing as a legitimate basis for social and political theory. This ethical equality among living things also has a direct implication for sociology and its widest extension, world-systems theory. A universe of equals that interact in a patterned and systematic fashion comprises a social order. If this is so, the present definition of a social system as relations among just humans is too narrow, and the study of sociology can no longer concern itself with just the structure of human life.

It is also incumbent on us to recognize some very obvious reservations over what has been proposed here. Given the varying cognitive endowments of different species, how will they be able to interact, exchange, and communicate, let alone vote and politically participate? Social orders may very well be limited to within-species structural relations, given the limits species specific communication place upon constructing transspecies institutions. We know order and hierarchy exist in communities of ants and chicken pecking orders, but a larger order in which ants and chickens are class occupants may not exist in the traditional sociological sense. If, though, transspecies order is considered from a biological point of view, then ecological order between species does exist, as between species predator-prey relations and niche survival rates for different species have been identified. Our interest, though, is not to naturalize social relations, absorbing sociology/world-systems theory into biology, but to open the intellectual door to thinking about socializing ecological relations. The key is the ecoequality assumption, for once it is recognized that all living things have the same moral status, then it is impossible to leave nonhumans out of a conception of a moral community and some kind of social order.

This discussion has been not only speculative, but it is also just a beginning. Environmental ethics is just starting to change the way ethical issues in philosophy and law are approached, and these ideas are being extended here to the question of what constitutes ecosocial order and whether or not the world-system is the widest of all social systems. It would seem there is no going back, once the notion of ecoequality is accepted, as a community of equals will require a theory of how that community is structured. At present the communicative limits intraspecies cognitive machinery impose has kept the notion of social system at the level of single species: packs of dogs, schools of fish, flocks of birds, societies of humans, and so forth. If the cognitive constraint cannot be lifted, then transspecies social systems may not be possible. If, though, there are other ways in which political intent and material interest can be expressed allowing politics to be organized in a transspecies manner, then a new ecosociological theory to accompany our present biologically based ecology theory may indeed be possible.

NOTES

1. See Leonard (1988), Walter (1982), and Low and Yeats (1992) for tests of the "pollution-haven" and "industrial flight" hypotheses that polluting industries have moved to the countries with the fewest environmental regulations.

2. Rudel (1989) finds that for countries with large amounts of forested land, industrialization sustains high rates of deforestation, more as a result of rural encroachment than of capital investment. He uses these findings to cast doubt on a world-systems analysis that would not take the amount of forested land to be as important as economic variables. Yet

Kick et al. (1996) do take into account a country's amount of forest and still find that economic factors such as import and export dynamics have an effect on deforestation.

3. Grimes, Roberts, and Manale (1994) show that carbon dioxide produced through deforestation accounts for much of the variation around the inverted U-curve. Consequently, they identify several variables that help explain the differing levels of deforestation-emissions both across and within world-system positions. They find the greatest intensity in "those countries in the lower periphery having large forest areas, and where pressure for land in the countryside is also high" (1994, 33). The findings seem to contradict Burns et al. (1994) and Kick et al. (1996), who locate the most intense deforestation in *semiperipheral* countries. Grimes et al. use a measure of carbon dioxide from deforestation while the others use deforestation itself, but it is still not clear why the results should be so different.

4. Chase-Dunn and Hall combine and extend the role accorded to ecological factors in circumscription (1970, 1981, 1987, 1988), resource stress (Harris 1977, 1979), and population pressure (Cohen 1977) theories. These theories account for the transition to agriculture and the development of hierarchical societies but not for the development of a world-system (see also Sanderson 1995).

REFERENCES

Bergesen, Albert. 1995a. "Deep Ecology and Moral Community." Pp. 193–213 in *Rethinking Materialism: Perspectives on the Spiritual Dimension of Economic Behavior,* edited by Robert Wuthnow. Grand Rapids, Mich.: Erdmanns.

————. 1995b. "Eco-Alienation." *Humboldt Journal of Social Relations* 21:1:1–14.

Bergesen, Albert, and Laura Parisi. 1999. "Ecosociology and Toxic Emissions." Pp. 43–57 in *Ecology and the World-System,* edited by Walter L. Goldfrank, David Goodman, and Andrew Szasz. Greenwich, Conn.: Greenwood.

Bergesen, Albert, Laura Parisi, and Liam Downey. 1996. "Development and Pollution: Is the Relationship Curvilinear?" Paper presented at the annual meeting of the International Studies Association, San Diego, Calif., April.

Bunker, Stephen G. 1984. "Modes of Extraction, Unequal Exchange, and the Progressive Underdevelopment of an Extreme Periphery: The Brazilian Amazon." *American Journal of Sociology* 89:5(Sept.):1017–1064.

————. 1985. *Underdeveloping the Amazon: Extraction, Unequal Exchange, and the Failure of the Modern State.* Urbana: University of Illinois Press.

————. 1994. "Flimsy Joint Ventures in Fragile Environments." Pp. 261–296 in *States, Firms, and Raw Materials: The World Economy and Ecology of Aluminum,* edited by Bradford Barham, Stephen Bunker, and Denis O'Hearn. Madison: University of Wisconsin Press.

Bunker, Stephen, and Paul S. Ciccantell. 1995. "Restructuring Space, Time and Competitive Advantage in the Capitalist World-Economy: Japan and Raw Materials Transport after World War II." Pp. 109–129 in *A New World Order? Global Transformation in the Late Twentieth Century,* edited by David A. Smith and József Böröcz. Westport, Conn.: Greenwood.

————. 1997. "Economic Ascent and the Global Environment: World Systems Theory and the New Historical Materialism." Paper presented at the Political Economy of the World System XXI Conference, Santa Cruz, Calif., April.

Burns, Thomas J., Byron L. Davis, and Edward L. Kick. 1997. "Position in the World-System and National Emissions of Greenhouse Gases." Revision of paper presented at the National Third-World Studies Conference, Omaha, Neb., October.

Burns, Thomas J., Edward L. Kick, David A. Murray, and Dixie A. Murray. 1994. "Demography, Development and Deforestation in a World-System Perspective." *International Journal of Comparative Sociology* 35:3–4(Sept.):221–239.

Carneiro, Robert L. 1970. "A Theory of the Origin of the State." *Science* 169 (August):733–738.

———. 1981. "The Chiefdom: Precursor of the State." Pp. 37–79 in *The Transition to Statehood in the New World,* edited by Grant D. Jones and Robert R. Kautz. New York: Cambridge University Press.

———. 1987. "Further Reflections on Resource Concentration and its Role in the Rise of the State." Pp. 245–260 in *Studies in the Neolithic and Urban Revolutions,* edited by Linda Manzanilla. BAR International Series 349. Oxford: BAR.

———. 1988. "The Circumscription Theory: Challenge and Response." *American Behavioral Scientist* 31:4(March):497–511.

Chase-Dunn, Christopher, and Thomas D. Hall. 1997a. "Ecological Degradation and the Evolution of World-Systems." *Journal of World-Systems Research* 3(Fall):403–431 (electronic journal: http://csf.colorado.edu/wsystems/jwsr.html).

———. 1997b. *Rise and Demise: Comparing World-Systems.* Boulder, Colo.: Westview.

Chew, Sing C. 1996. "Wood, Environmental Imperatives and Developmental Strategies: Challenges for Southeast Asia." Pp. 206–226 in *Asia—Who Pays for Growth? Women, Environment, and Popular Movements,* edited by J. Lele and W. Tettey. Brookfield, Vt.: Darmouth.

Cohen, Mark. 1977. *The Food Crisis in Prehistory.* New Haven, Conn.: Yale University Press.

Devall, Bill, and George Sessions. 1985. *Deep Ecology.* Salt Lake City, Utah: Gibbs Smith.

Dietz, Thomas, and Eugene A. Rosa. 1997. "Effects of Population and Affluence on CO_2 Emissions." *Proceedings of the National Academy of Sciences of the USA* 94:1(Jan. 7):175–179.

Drengson, Alan, and Yuichi Inoue, eds. 1995. *The Deep Ecology Movement: An Introductory Anthology.* Berkeley, Calif.: North Atlantic Books.

Ehrlich, Paul R., Anne H. Ehrlich, and John P. Holdren. 1993. "Availability, Entropy, and the Laws of Thermodynamics." Pp. 69–73 in *Valuing the Earth: Economics, Ecology, Ethics,* edited by Herman Daly and Kenneth Townsend. Cambridge, Mass.: MIT Press.

Frank, David John. 1994. "Science, Nature, and the Globalization of the Environment, 1870–1990." Paper presented at the annual meeting of the American Sociological Association, Los Angeles, August.

———. 1995. "The Nation-State and the Natural Environment, 1900–1995." Paper presented at the annual meeting of the American Sociological Association, August, Washington, D.C.

Grimes, Peter E., and J. Timmons Roberts. 1995. "Carbon Dioxide Emissions Efficiency and Economic Development." Paper presented at the annual meeting of the American Sociological Association, Washington, D.C., August

Grimes, Peter E., J. Timmons Roberts, and Jodie L. Manale. 1994. "Social Roots of Environmental Damage: A World Systems Analysis of Global Warming and Deforestation." Revision of paper presented at the annual meeting of the American Sociological Association, Miami Beach, August.

Harris, Marvin. 1977. *Cannibals and Kings: The Origins of Cultures.* New York: Random House.

_____. 1979. *Cultural Materialism: The Struggle for a Science of Culture.* New York: Random House.

Hettige, Hemamala, Robert E. B. Lucas, and David Wheeler. 1992. "The Toxic Intensity of Industrial Production: Global Patterns, Trends, and Trade Policy." *American Economics Review* 82:2(May):478–481.

Kick, Edward L., Thomas J. Burns, Byron Davis, David A. Murray, and Dixie A. Murray. 1996. "Impacts of Domestic Population Dynamics and Foreign Wood Trade on Deforestation: A World-System Perspective." *Journal of Developing Societies* 12:1(June):68–87.

King, Ynestra. 1993. "Toward an Ecological Feminism and a Feminist Ecology." Pp. 70–80 in *Radical Environmentalism,* edited by Peter C. List. Belmont, Calif.: Wadsworth.

Leonard, H. Jeffrey. 1988. *Pollution and the Struggle for the World Product: Multinational Corporations, Environment, and International Comparative Advantage.* Cambridge: Cambridge University Press.

Low, Patrick, and Alexander Yeats. 1992. "Do 'Dirty' Industries Migrate?" Pp. 89–104 in *International Trade and the Environment,* edited by Patrick Low. World Bank Discussion Papers, no. 159. Washington, D.C.: International Bank for Reconstruction and Development/World Bank.

Lucas, Robert E. B., David Wheeler, and Hemamala Hettige. 1992. "Economic Development, Environmental Regulation and the International Migration of Toxic Industrial Pollution: 1960–88." Pp. 67–86 in *International Trade and the Environment,* edited by Patrick Low. World Bank Discussion Papers, no. 159. Washington, D.C.: International Bank for Reconstruction and Development/World Bank.

Merchant, Carolyn. 1992. *Radical Ecology: The Search for a Livable World.* New York: Routledge.

_____. 1993. "Ecofeminism and Feminist Theory." Pp. 49–55 in *Radical Environmentalism: Philosophy and Tactics,* edited by Peter C. List. Belmont, Calif.: Wadsworth.

Naess, Arne. 1989. *Ecology, Community and Lifestyle.* Translated and edited by David Rothenberg. Cambridge: Cambridge University Press.

Nazmi, Nader. 1991. "Deforestation and Economic Growth in Brazil: Lessons from Conventional Economics." *Centennial Review* 35:2(Spring):315–322.

O'Connor, James. 1994. "Is Sustainable Capitalism Possible?" Pp.152–175 in Is *Capitalism Sustainable? Political Economy and the Politics of Ecology,* edited by Martin O'Connor. New York: Guilford.

Roberts, J. Timmons. 1996. "Predicting Participation in Environmental Treaties: A World-system Analysis." *Sociological Inquiry* 66:1(Winter):38–57.

Roberts, J. Timmons, and Peter E. Grimes. 1997. "Carbon Intensity and Economic Development 1962–91: A Brief Exploration of the Environmental Kuznets Curve." *World Development* 25:2(Feb.):191–198.

Rudel, Thomas K. 1989. "Population, Development, and Tropical Deforestation: A Cross-National Study." *Rural Sociology* 543:(Fall):327–338.

Sanderson, Stephen K. 1995. *Social Transformations: A General Theory of Historical Development.* Cambridge, Mass.: Blackwell. (Reprinted 1999, Lanham, Md.: Rowman & Littlefield.)

Schaeffer, Robert K. 1997. "Success and Impasse: The Environmental Movement in the United States and Around the World." Paper presented at the Political Economy of the World System XXI Conference, Santa Cruz, Calif., April.

Sessions, George, ed. 1995. *Deep Ecology for the 21st Century: Readings on the Philosophy and Practice of the New Environmentalism.* Boston: Shambhala.

Smith, David A. 1994. "Uneven Development and the Environment: Toward a World-System Perspective." *Humboldt Journal of Social Relations* 20:1:151–175.

Walter, Ingo. 1982. "Environmentally Induced Industrial Relocation to Developing Countries." In *Environment and Trade: The Relation of International Trade and Environmental Policy,* edited by Seymour Rubin and Thomas R. Graham. Totowa, N.J.: Littlefield, Adams.

Appendix 1

Political-Economy of the World-System Annuals

Kaplan, Barbara Hockey, ed. 1978. *Social Change in the Capitalist World Economy.* Immanuel Wallerstein, series editor, Political-Economy of the World-System Annuals, 01. Beverly Hills: Sage.

Goldfrank, Walter L., ed. 1979. *The World-System of Capitalism: Past and Present.* Immanuel Wallerstein, series editor, Political-Economy of the World-System Annuals, 02. Beverly Hills: Sage.

Hopkins, Terence K., and Immanuel Wallerstein, eds. 1980. *Processes of the World-System.* Immanuel Wallerstein, series editor, Political-Economy of the World-System Annuals, 03. Beverly Hills: Sage.

Rubinson, Richard, ed. 1981. *Dynamics of World Development.* Immanuel Wallerstein, series editor, Political-Economy of the World-System Annuals, 04. Beverly Hills: Sage.

Friedman, Edward, ed. 1982. *Ascent and Decline in the World-System.* Immanuel Wallerstein, series editor, Political-Economy of the World-System Annuals, 05. Beverly Hills: Sage.

Bergesen, Albert, ed. 1983. *Crises in the World-System.* Immanuel Wallerstein, series editor, Political-Economy of the World-System Annuals, 06. Beverly Hills: Sage.

Bergquist, Charles, ed. 1984. *Labor in the Capitalist World-Economy.* Immanuel Wallerstein, series editor, Political-Economy of the World-System Annuals, 07. Beverly Hills: Sage.

Evans, Peter, Dietrich Rueschemeyer, and Evelyne Huber Stephens, eds. 1985. *States versus Markets in the World-System.* Immanuel Wallerstein, series editor, Political-Economy of the World-System Annuals, 08. Beverly Hills: Sage.

Tardanico, Richard, ed. 1987. *Crises in the Caribbean Basin.* Immanuel Wallerstein, series editor, Political-Economy of the World-System Annuals, 09. Newbury Park, Calif.: Sage.

Ramirez, Francisco O., ed. 1988. *Rethinking the Nineteenth Century: Contradictions and Movements.* Immanuel Wallerstein, advisory editor, Studies in the Political-Economy of the World-System, 10. New York: Greenwood.

Smith, Joan, Jane Collins, Terence K. Hopkins, and Akbar Muhammad, eds. 1988. *Racism, Sexism, and the World-System.* Immanuel Wallerstein, advisory editor, Studies in the Political-Economy of the World-System, 11. New York: Greenwood.

323

Boswell, Terry, ed. 1989. *Revolution in the World-System.* Immanuel Wallerstein, advisory editor, Studies in the Political-Economy of the World-System, 12a. New York: Greenwood.

Schaeffer, Robert K., ed. 1989. *War in the World-System.* Immanuel Wallerstein, advisory editor, Studies in the Political-Economy of the World-System, 12b. New York: Greenwood.

Martin, William G., ed. 1990. *Semiperipheral States in the World-Economy.* Immanuel Wallerstein, advisory editor, Studies in the Political-Economy of the World-System, 13. New York: Greenwood.

Kasaba, Resat, ed. 1991. *Cities in the World-System.* Immanuel Wallerstein, advisory editor, Studies in the Political-Economy of the World-System, 14. New York: Greenwood.

Palat, Ravi Arvind, ed. 1993. *Pacific-Asia and the Future of the World-System.* Immanuel Wallerstein, series adviser, Studies in the Political-Economy of the World-System, 15. Westport, Conn.: Greenwood.

Gereffi, Gary, and Miguel Korzeniewicz, eds. 1994. *Commodity Chains and Global Capitalism.* Immanuel Wallerstein, series adviser, Studies in the Political-Economy of the World-System, 16. Westport, Conn.: Greenwood.

McMichael, Philip, ed. 1995. *Food and Agrarian Orders in the World-Economy.* Immanuel Wallerstein, series adviser, Studies in the Political-Economy of the World-System, 17. Westport, Conn.: Greenwood.

Smith, David A., and József Böröcz, eds. 1995. *A New World Order? Global Transformations in the Late Twentieth Century.* Immanuel Wallerstein, series adviser, Studies in the Political-Economy of the World-System, 18. Westport, Conn.: Greenwood.

Korzeniewicz, Roberto Patricio, and William C. Smith, eds. 1996. *Latin America in the World-Economy.* Immanuel Wallerstein, series adviser, Studies in the Political-Economy of the World-System, 19. Westport, Conn.: Greenwood.

Ciccantell, Paul S., and Stephen G. Bunker, eds. 1998. *Space and Transport in the World-System.* Immanuel Wallerstein, series adviser, Studies in the Political-Economy of the World-System, 20. Westport, Conn.: Greenwood.

Goldfrank, Walter, David Goodman, and Andrew Szasz, eds. 1999. *Ecology and the World-System.* Immanuel Wallerstein, series adviser, Studies in the Political-Economy of the World-System, 21. Greenwich, Conn.: Greenwood.

Derlugian, Georgi, and Scott L. Greer, eds. 2000. *Dilemmas of Globalization: Political Projects in a Changing World-System.* Immanuel Wallerstein, series adviser, Studies in the Political-Economy of the World-System, 22. Greenwich, Conn.: Greenwood (in press).

Appendix 2

Political-Economy of the World-System, Section of the American Sociological Association Winner of Annual Book Prizes

1989 Bunker, Stephen. 1987. *Peasants against the State.* Urbana: University of Illinois Press. (Reprinted in paper, 1991, Chicago: University of Chicago Press.)

1990 Abu-Lughod, Janet. 1989. *Before European Hegemony: The World System A.D. 1250–1350.* New York: Oxford University Press.

1991 Tomich, Dale W. 1990. *Slavery in the Circuit of Sugar: Martinique and the World Economy, 1830–1848.* Baltimore, Md.: Johns Hopkins University Press.

1992 Chase-Dunn, Christopher. 1989. *Global Formation: Structures of the World-Economy.* London: Blackwell. (Reprinted 1998, Lanham, Md.: Rowman & Littlefield.)

1993 Suter, Christian. 1992. *Debt Cycles in the World-Economy: Foreign Loans, Financial Crises, and Debt Settlements, 1820–1990.* Boulder, Colo.: Westview.

1994. Foran, John. 1993. *Fragile Resistance: Social Transformation in Iran from 1500 to the Revolution.* Boulder, Colo.: Westview.

1995 Arrighi, Giovanni. 1994. *The Long Twentieth Century: Money, Power and the Origins of Our Times.* London: Verso.

1996 Evans, Peter B. 1995. *Embedded Autonomy: States and Industrial Transformation.* Princeton, N.J.: Princeton University Press.

1997 Robinson, William I. 1996. *Promoting Polyarchy: Globalization, U.S. Intervention, and Hegemony.* New York: Cambridge University Press.

1998 Paige, Jeffery. 1997. *Coffee and Power: Revolution and the Rise of Democracy in Central America.* Cambridge, Mass.: Harvard University Press.

1999 Stark, David, and Laszlo Bruszt. *Postcolonial Pathways: Transforming Politics and Property in East Central Europe.* Cambridge: Cambridge University Press.

Index

About the Contributors

Tim Bartley is a graduate student in sociology at the University of Arizona. He is currently studying relations between the American lumber industry and the state in the late nineteenth and early twentieth centuries. Other interests include the sociology of culture and contemporary changes in forms of work.

Albert J. Bergesen, professor of sociology at the University of Arizona, took his Ph.D. from Stanford in 1973. He has long-standing research interests in cultural, world-systems, and environmental sociology. He has published numerous articles on global cultural and political cycles and edited books, such as *Studies of the Modern World-System* (1980), *Crises in the World-System* (1983), and *America's Changing Role in the World-System* (with T. Boswell) (1987). He is presently preparing a book with the tentative title *The World-System of Art.*

Terry Boswell is professor and chair of sociology at Emory University. His current research is on the pattern of core revolutions over the last five hundred years. He attempts to explain how the structure and dynamics of the world-system are affected from "below" by world revolutions and the social movements they spawn. He co-authored a manuscript related to this topic with Christopher Chase-Dunn, from which their chapter in this volume is drawn.

Christopher Chase-Dunn is professor of sociology at Johns Hopkins University and founding editor of the *Journal of World-Systems Research.* He received his Ph.D. in 1975 from Stanford University. He is the author of *Global Formation* (Blackwell, 1979; 2d ed., Rowman & Littlefield, 1998) and (with Thomas D. Hall) *Rise and Demise: Comparing World-Systems* (Westview, 1997). His current research is on the rise and fall of empires, globalization, the future of warfare, and popular responses to neoliberalism.

Stephen W. K. Chiu received his Ph.D. from Princeton University. He is an assistant professor at the Chinese University of Hong Kong. His research interests are in social movements, industrial sociology, and East Asia. His books include *East Asia the World Economy* (with Alvin Y. So, Sage, 1995) and *City States in the Global Economy: Industrial Restructuring in Hong Kong and Singapore* (with K. C. Ho and Tai-lok Lui, Westview, 1997).

Wilma A. Dunaway experienced childhood close to the Cherokee culture. Her grandmother was only two generations away from the forced removal of the Cherokees from their Appalachian homelands. She completed her Ph.D. at the University of Tennessee in 1994 and received the American Sociological Association Dissertation Award in 1995. She is also the recipient of a Woodrow Wilson Fellowship, the dissertation award from the ASA Political Economy of the World-System section, and the Szymanski Award from the ASA Marxist section. Her book, *The First American Frontier: Transition to Capitalism in Southern Appalachia, 1700–1860* (University of North Carolina, 1996), has received critical acclaim. She is currently assistant professor of sociology at the University of Virginia, Blacksburg.

Colin Flint is assistant professor of geography at The Pennsylvania State University. His research interests range from world-systems theory through political geography to spatial statistics. He has recently finished coauthoring (with Peter Taylor) the fourth edition of *Political Geography* (Longman), a textbook that uses a world-systems framework to discuss geopolitics, imperialism, the state, nationalism, electoral geography, and localities. Other research includes the role of world-systems theory (and other social theories) to inform electoral geography and other spatial analyses. Future projects include a study of hate crimes in Appalachia that will integrate world-systems theory, place-specific context, and spatial statistics.

Peter Grimes received his doctorate in sociology from Johns Hopkins and has had a long-standing interest in linking the history of global capital accumulation with that of ecological degradation. In 1992 he received a grant from the National Science Foundation to apply a world-systems analysis to the issues of global warming. His recent publications include "World-Systems Analysis" (with Christopher Chase-Dunn) in the 1995 *Annual Review of Sociology;* "Carbon Intensity and Economic Development 1962–1991: A Brief Exploration of the Environmental Kuznets Curve"(with Timothy Roberts) in *World Development,* February 1997; and "The Horseman and the Killing Fields," a chapter in *Ecology and the World System* (1999). He is currently working on a book on social evolution and the environment.

Thomas D. Hall received his Ph.D. from the University of Washington in 1981 and is Lester M. Jones Professor of Sociology at DePauw University in Greencastle, Indiana. In 1999–2000 he was A. Lindsay O'Connor Professor of American Institutions in the Sociology and Anthropology Department at Colgate University in Hamilton, New York. He is coauthor with Christopher Chase-Dunn of *Rise and Demise: Comparing World-Systems* (Westview, 1997) and *Social Change in the South-*

west, 1350–1880 (University Press of Kansas, 1989). In addition to studies of Native Americans, he is studying social change over the last three millennia in Southeast Asia.

Leslie S. Laczko is associate professor of sociology at the University of Ottawa, where he is also associate dean academic and secretary of the Faculty of Social Sciences. He continues to teach and write about intergroup relations, Canadian society, research methods, and social change in the global system. Recent publications include *Pluralism and Inequality in Quebec* (St. Martin's Press and University of Toronto Press, 1995); "Feelings of Fraternity toward Old and New Canadians: The Interplay of Ethnic and Civic Factors," in Jean Laponce and William Safran (eds.), *Ethnicity and Citizenship: The Canadian Case* (1996); "Feelings of Fraternity in Canada: An Empirical Exploration of Regional Differences," in *Asian and Pacific Migration Journal* (1997); and "Inégalités et État-providence: le Québec, le Canada, et le monde," in *Recherches sociographiques* (1998). His current research is on changing religious boundaries and welfare state restructuring in comparative perspective, using international survey data, as well as on the methodology of interviewing in multilingual settings.

Joya Misra is an assistant professor of sociology at the University of Massachusetts, Amherst. Her research centers on the political economy of gender. She argues in much of her scholarship that gender ideologies shape both economic and political processes in important and significant ways. Recent research on the world-system focuses on gender and the state in East Asia, as well as an ongoing project on hegemony in the modern world-system with Terry Boswell. In addition to her work in world-systems theory, she explores gender and welfare state theorizing in recent papers in *Social Politics* and *Gender & Society*.

Peter N. Peregrine received his Ph.D. in anthropology from Purdue University in 1990 and is currently associate professor of anthropology at Lawrence University. His research focuses on the evolution of complex societies and the origins of political centralization through archaeological fieldwork in the midwestern United States and in north Syria. Recent publications include "Networks of Power: The Mississippian World-System," in Michael Nassaney and K. Sassaman, eds., *Native American Interactions* (University of Tennessee Press, 1995); *Archaeology of the Mississippian Culture* (Garland, 1966); and (with Gary M. Feinman, eds.) *Pre-Columbian World-Systems* (Prehistory Press, 1996).

Fred M. Shelley, Ph.D., is professor of geography and graduate program coordinator in the Department of Geography and Planning at Southwest Texas State University in San Marcos. He received his Ph.D. from the University of Iowa in 1981 and has taught at the University of Oklahoma, University of Southern California, University of Iowa, and Florida State University. His research interests include political and electoral geography, locational conflicts, popular culture, and the geography of the United States. He is the author of over one hundred books, refereed journal articles, book chapters, and other publications on these and related topics. He is review editor of *Social Science Quarterly and Political Geography,* a former president of the Southwest Division of the Association of American Geographers, and a former president of the Political Geography Specialty Group of the Association of American Geographers.

David A. Smith is an associate professor of sociology and urban planning at the University of California–Irvine. He is the author of *Third World Cities in Global Perspective* (Westview, 1996) and coeditor of *A New World Order?* (with József Böröcz, Greenwood, 1995) and *States and Sovereignty in the Global Economy* (Routledge, 1999), as well as articles in *Social Forces, American Sociological Review, Urban Affairs Quarterly, International Journal of Urban and Regional Research,* and others. His current research focuses on industrial upgrading in East Asia, global trade and commodity chains, and comparative urbanization.

Alvin Y. So received his Ph.D. from UCLA and is now head of the Social Science Division at the Hong Kong University of Science and Technology. He received the Regents' Medal for Excellence in Teaching at the University of Hawaii in 1989. His research interests are in development, class theory, and East Asia. His books include *Social Change and Development* (Sage, 1990), *East Asia and the World Economy* (with Stephen W. K. Chiu, Sage, 1995), *Hong Kong–Guangdon Link: Partnership in Flux* (Sharpe, 1997), *Asia's Environmental Movements: Comparative Perspective* (coedited with Yok-Shiu Lee, Sharpe, 1999), and *Hong Kong's Embattled Democracy: A Societal Analysis* (Johns Hopkins University Press, 1999).

Yodit Solomon is a research associate at the Women's Research Institute. She holds a bachelor of science degree in international relations and a master's in sociology from Brigham Young University. Her thesis examined sociocultural differences in the effects of education on contraceptive use among women in Ghana and Senegal. She is currently working on research assessing the impact of development projects in Mali, West Africa. She has worked closely with Dr. Carol Ward on projects dealing with the relationship of spirituality to healing from domestic violence and substance abuse, and dropout rates among Northern Cheyenne high school students.

Elon Stander is a Ph.D. graduate student in sociology at Brigham Young University. Currently ABD, her dissertation topic is recovery strategies of Native American women, exploring avenues of recovery from substance abuse, domestic abuse, sexual abuse, or other trauma in the lives of these women. Her degrees include a bachelor of science degree from Utah State University in education and psychology and a master's degree in psychology from Brigham Young University. Her pedagogical activities include sociology courses at Brigham Young University and Utah Valley State College, where she is currently an adjunct instructor. Active in community service, she has been involved with the Shelter for Women and Children in Crisis and the Utah County Rape Crisis Team. She also served as an instructor at the Native

American Alcohol Recovery Center. Her domestic life has consisted of mothering seven children.

Debra Straussfogel has a joint appointment as associate professor of geography and natural resources at the University of New Hampshire in Durham. Her research areas include the application of complex systems approaches to world-systems theory, and bioregional and systems approaches to sustainable development in resource-based economies, with particular concern for community and demographic factors. Her regional interests focus on New England, Quebec, and the Maritimes. Dr. Straussfogel teaches courses in economic geography, urban geography, the geography of population and development, the geography of the Western world, and the geography of the United States and Canada. She received her Ph.D. in geography from the Pennsylvania State University.

William R. Thompson is professor of political science at Indiana University. He is a former coeditor of *International Studies Quarterly* and the author of a number of books and articles primarily about conflict processes. His current interests are centered on evolutionary approaches to world politics, the development of rivalry-centric theories in international relations, and very long-term processes of conflict and economic development.

Carol Ward is an associate professor of sociology and coordinator of the Cross-Cultural and International Research Initiative at the Women's Research Institute. She received her Ph.D. in sociology from the University of Chicago in 1992. She has done extensive research on minority education issues, rural community development, gender and intergenerational role change, and the relationship of spirituality to healing from domestic violence and substance abuse. Her recent work addressing educational stratification and human capital development among American Indian high school students has been published in *Family Perspectives* (1993), *Rural Sociology* (1995, 1998), *Sociological Inquiry* (1998), and *American Indian Development,* coedited with C. Matthew Snipp (1996). Her monograph *The Role of Native Capital in American Indian High School Completion: Culture, Structure, and Community in Human Capital Development* is forthcoming.